Anthocyanins

Kevin Gould · Kevin Davies · Chris Winefield
Editors

Anthocyanins

Biosynthesis, Functions, and Applications

 Springer

Editors
Kevin Gould
School of Biological Sciences
Victoria University of Wellington
Wellington
New Zealand
kevin.gould@vuw.ac.nz

Kevin Davies
Crop and Food Research
Palmerston North
New Zealand
daviesk@crop.cri.nz

Chris Winefield
Lincoln University
Canterbury
New Zealand
winefiec@lincoln.ac.nz

ISBN: 978-0-387-77334-6   e-ISBN: 978-0-387-77335-3
DOI: 10.1007/978-0-387-77335-3

Library of Congress Control Number: 2008933450

Printed on acid-free paper

springer.com

Preface

Amongst the range of naturally occurring pigments, anthocyanins are arguably the best understood and most studied group. Research into their occurrence, inheritance and industrial use encompasses hundreds of years of human history, and many volumes are dedicated to describing the prevalence, type and biosynthesis of anthocyanins. Only recently have studies begun to explain the reasons for the accumulation of these red pigments in various tissues of plants. Indeed it has only been within the last 20 years, coinciding with the ability to genetically manipulate plants, that we have begun to tease out the multitude of roles that these compounds play within plants. Alongside these fundamental advances in understanding the functional attributes of anthocyanins *in-planta* we are now beginning to realise the potential of anthocyanins as compounds of industrial importance, both as pigments in their own right and also as pharmaceuticals.

With this backdrop, the 4th International Workshop on Anthocyanins was convened in Rotorua, New Zealand in January 2006. The programme was designed to bring together a wide range of researchers across an array of disciplines to highlight the increasing importance of these pigments as a research area but also as chemicals of wider importance to human activity. The chapters in this book represent a collection of recent and highly relevant reviews prepared by participants in this symposium who are internationally recognised experts in their respective fields.

The book is divided into 10 chapters that address a wide range of topics including the proposed roles of anthocyanins *in vivo*, methods for manipulating the biosynthesis of anthocyanins to both produce new pigments as well as pigments that exhibit greater stability, use of plant and microbial cell cultures for large scale production of anthocyanins for industrial uses and the effectiveness of anthocyanins as pharmaceutical compounds.

Chapter 1 reviews the current thinking about the role of anthocyanins in leaves, stems roots and other vegetative organs. The authors point out that current roles for anthocyanins such as participation in photoprotective, UV-B protective, and protection from oxidative stresses do not adequately explain the range of spatial and temporal distributions of anthocyanins observed in plants. They hypothesise that anthocyanins possess a more indirect role in signalling and developmental regulation in response to oxidative stress.

The discussion in Chapter 1 is expanded on in Chapter 2, whose authors proceed to examine the potential roles of anthocyanins in plant/animal interactions. In particular this chapter reviews current thought and recent experimental evidence on plants' use of anthocyanins as visual clues that provide information to animals about palatability of plant structures and the potential role of anthocyanins in camouflage, undermining insect crypsis and in mimicry of defensive structures.

The molecular basis for spatial and temporal regulation of anthocyanin biosynthesis forms the focus of Chapter 3. This review concentrates on the recent advances in our understanding of the biosynthesis and molecular regulation of this pathway and how this information has been used in conjunction with recombinant DNA technologies to manipulate anthocyanin production in plants for both scientific and commercial applications.

Picking up again the theme of *in vivo* roles for anthocyanins, Chapter 4 reviews the role of anthocyanin pigmentation in fruits and adaptive advantages accumulation of these pigments confers to plants. In particular the author concentrates on the accumulation of anthocyanin pigments in fruits in response to environmental factors, seed disperser visual systems and fruit quality parameters. Accumulation of pigments contributing differing hues to fruit is discussed with respect to the interaction with animal dispersers and as a measure of fruit ripeness and quality.

Chapter 5 provides an in-depth review of the use of plant cell cultures for the industrial production of anthocyanins for use as high quality food pigments. A wide range of plant species are reviewed as to their ability to produce cell cultures capable of production of anthocyanins in cell culture, the types of cultures obtained and the pros and cons of using these cell types as production systems for anthocyanins. Methodology to increase the production of anthocyanins from these cultures, and limitations of these cultures for anthocyanin production are discussed alongside potential methods for overcoming production barriers that currently prevent large scale anthocyanin production from plant derived cell cultures.

Anthocyanin stability and the colour imparted to plant tissues by anthocyanin accumulation are in part due to the extent and nature of secondary modifications to the anthocyanin aglycone. Glycosylation, acylation and methylation are especially important in altering the chemical characteristics of anthocyanins both *in vivo* and *in vitro*. In Chapter 6 the authors review the current advances in our understanding of the biochemical pathways that lead to these chemical modifications and how this information may be utilized to modify and stabilize anthocyanins both in plants and for industrial uses such as those described in Chapter 5 and later in Chapter 9.

While production of anthocyanins in plant cell cultures is well documented, production of these compounds in microbial cell culture systems is a relatively new concept. With the increasing knowledge of the biosynthesis of anthocyanins in plant systems it has become feasible to engineer microbial species to contain a functional anthocyanin pathway. Chapter 7 reviews current advances in this area and is split into two sections. The first deals with advances in the metabolic engineering of bacterial and yeast species for anthocyanin production while the second section reviews the endogenous biotransformations carried out by host species and provides a excellent counterpoint to natural modification schemes found in plants that is reviewed in Chapter 6.

Continuing the theme of industrial application of flavonoids and anthocyanins developed in earlier chapters, Chapter 8 turns our attention to the utility of this ubiquitous chemical group in agricultural systems. In particular the roles of anthocyanins, related flavonoids and their derivatives in forage and forage legume species is discussed in the context of the value of these compounds in these crops from an animal health and nutrition perspective. Potential methods for manipulating the levels of important compound classes are reviewed from both a genetic modification and traditional breeding standpoint.

Returning to anthocyanins as food colorants, Chapter 9 develops themes outlined in Chapter 5 with a review of how anthocyanins are currently utilized in the food industry from sources thorough isolation and analytical methodologies employed. While industrial utilization of anthocyanins is still to be realized on a wide scale, this review covers the potential and perspectives for anthocyanins and their derivatives in food products.

Finally Chapter 10 concludes the book with a discussion of anthocyanins and other flavonoids as phytochemicals that promote human health. This very relevant topic is reviewed with a particular emphasis on the interaction of these compounds with other components of diet to protect and enhance human heath. The chapter describes how plant cell cultures and research models are being used to increase our understanding of the complex and multi-faceted roles that interacting phytochemicals play in the human body, specifically in the context of developing novel insights into the competing mechanisms of action, bioavailability and distribution *in situ*.

The editors, Associate Professor Kevin Gould, Dr. Kevin Davies and I hope that this book will both provide a valuable reference resource and provide inspiration for new researchers in this exciting and rapidly expanding field.

Dr. Chris Winefield
January 2008

Contents

Contributors

Joseph A. Chemler
State University of New York at Buffalo, Chemical and Biological Engineering,
jchemler@buffalo.edu

Kevin M. Davies
New Zealand Institute for Crop & Food Research Ltd, Private Bag 11600,
Palmerston North, New Zealand,
daviesk@crop.cri.nz

Victor de Freitas
Department of Chemistry, University of Porto, CIQ, Rua do Campo Alegre, 687,
4169-007 Porto, Portugal,
vfreitas@fc.up.pt

Simon Deroles
New Zealand Institute for Crop & Food Research Ltd, Private Bag 11600,
Palmerston North, New Zealand,
deroless@crop.cri.nz

Kevin S. Gould
School of Biological Sciences, Victoria University of Wellington,
P.O. Box 600, Wellington, New Zealand,
kevin.gould@vuw.ac.nz

Jean-Hugues B. Hatier
School of Biological Sciences, University of Auckland, Auckland.
Current address: AgResearch Limited, Grasslands Research Centre, Tennent Drive,
Private Bag 11008, Palmerston North, New Zealand,
jimmy.hatier@agresearch.co.nz

Mattheos A.G. Koffas
State University of New York at Buffalo, Chemical and Biological Engineering,
mkoffas@buffalo.edu

Effendi Leonard
State University of New York at Buffalo, Chemical and Biological Engineering,
eleonard@buffalo.edu

Simcha Lev-Yadun
Department of Biology Education, Faculty of Science and Science Education,
University of Haifa – Oranim, Tivon 36006, Israel,
levyadun@research.haifa.ac.il

Mary Ann Lila
Department of Natural Resources & Environmental Sciences, University of Illinois,
imagemal@uiuc.edu

Nuno Mateus
Department of Chemistry, University of Porto, CIQ, Rua do Campo Alegre,
687, 4169-007 Porto, Portugal,
nbmateus@fc.up.pt

Toru Nakayama
Department of Biomolecular Engineering, Tohoku University,
nakayama@seika.che.tohoku.ac.jp

Susanne Rasmussen
AgResearch, Grasslands, Tennent Drive, Palmerston North, NZ,
Susanne.rasmussen@agresearch.co.nz

Kazuki Saito
Graduate School of Pharmaceutical Science, RIKEN Plant Science Center/Chiba
University, ksaito@faculty.chiba-u.jp

W.J. Steyn
Department of Horticultural Science, University of Stellenbosch,
wsteyn@sun.ac.za

Mami Yamazaki
Graduate School of Pharmaceutical Science, Chiba University,
mamiy@p.chiba-u.ac.jp

Keiko Yonekura-Sakakibara
RIKEN Plant Science Center,
keikoys@psc.riken.jp

1

Anthocyanin Function in Vegetative Organs

Jean-Hugues B. Hatier[1] and Kevin S. Gould[2]

[1] School of Biological Sciences, University of Auckland, Auckland. Current address: AgResearch Limited, Grasslands Research Centre, Tennent Drive, Private Bag 11008, Palmerston North, New Zealand, jimmy.hatier@agresearch.co.nz

[2] School of Biological Sciences, Victoria University of Wellington, P.O. Box 600, Wellington, New Zealand, kevin.gould@vuw.ac.nz

Abstract. Possible functions of anthocyanins in leaves, stems, roots and other vegetative organs have long attracted scientific debate. Key functional hypotheses include: (i) protection of chloroplasts from the adverse effects of excess light; (ii) attenuation of UV-B radiation; and (iii) antioxidant activity. However, recent data indicate that the degree to which each of these processes is affected by anthocyanins varies greatly across plant species. Indeed, none of the hypotheses adequately explains variation in spatial and temporal patterns of anthocyanin production. We suggest instead that anthocyanins may have a more indirect role, as modulators of reactive oxygen signalling cascades involved in plant growth and development, responses to stress, and gene expression.

1.1 Introduction

"Yet it is difficult to find a hypothesis which would fit all cases of anthocyanin distribution without reduction to absurdity. The pigment is produced, of necessity, in tissues where the conditions are such that the chemical reactions leading to anthocyanin formation are bound to take place. For the time being we may safely say that it has not been satisfactorily determined in any one case whether its development is either an advantage or a disadvantage to the plant".

From Muriel Wheldale's *The Anthocyanin Pigments of Plants*, 1916.

The possible physiological roles of anthocyanins in vegetative tissues have perplexed scientists for well over a century. Anthocyanins are to be found in the vacuoles of almost every cell type in the epidermal, ground, and vascular tissues of all vegetative organs. They occur in roots, both subterranean and aerial, and in hypocotyls, coleoptiles, stems, tubers, rhizomes, stolons, bulbs, corms, phylloclades, axillary buds, and leaves. There are red vegetative organs in plants from all terrestrial biomes, from the basal liverworts to the most advanced angiosperms. Plants also show tremendous diversity in anthocyanin expression. In leaves, for example,

K. Gould et al. (eds.), *Anthocyanins*, DOI: 10.1007/978-0-387-77335-3_1,

anthocyanins may colour the entire blade, or else be restricted to the margins, stripes, patches, or seemingly random spots on the upper, lower or both lamina surfaces. In some leaves, only the petiole and major veins are pigmented red, in others it is the interveinal lamina tissue, or the stipules, or domatia that are anthocyanic. Some leaves turn red shortly before they abscise, others are red only while they are growing, yet others remain red throughout their lives. In many species, anthocyanins are produced only when the plant is unhealthy or has been exposed to environmental stress, but there are some that develop the red pigments even under optimal growth environments. Given this enormous variation in location, timing, and inducibility of anthocyanins in vegetative tissues, it is not surprising that a unified explanation for the presence of these pigments has thus far eluded scientific investigation. Muriel Wheldale's statement that a unified explanation would require "reduction to absurdity" remains as valid today as it was when she wrote it in 1916.

Although the selective pressure that has driven the evolution of anthocyanins in such disparate vegetative structures remains far from obvious, plant physiologists have nevertheless made significant progress over the past decade in elaborating the *consequences* of cellular anthocyanins on plant function. Reflecting the resurgence of scientific interest in anthocyanin (and betalain) function in vegetative organs, several reviews have been written on this topic in recent years (Chalker-Scott 1999, 2002; Hoch et al. 2001; Gould et al. 2002b; Lee and Gould 2002a, 2002b; Steyn et al. 2002; Close and Beadle 2003; Gould 2004; Gould and Lister 2005; Stintzing and Carle 2005; Manetas 2006). Those reviews provide a comprehensive summary of contemporary knowledge, particularly in relation to leaf physiology, on which most research has been done. It is not our intention to duplicate that information in this chapter, although for completeness we do briefly summarise the three leading hypotheses for anthocyanin function in leaves. Rather, with the use of selected examples, we hope to demonstrate the extraordinary versatility in anthocyanin function. Thus, any two species might benefit from anthocyanins in very different ways and to different degrees, even though the chemical nature and histological location of the pigment are identical. Finally, in acknowledgement of the recent paradigm shift in relation to the role of reactive oxygen species (ROS) in plants (see Foyer and Noctor 2005), we develop an argument for a novel function of anthocyanins in leaves – that of a modulator of signal transduction cascades in physiological responses to stress.

1.2 Anthocyanins and Stress Responses

Foliar anthocyanins most commonly occur as vacuolar solutions in epidermal and/or mesophyll cells, although in certain bryophytes these red pigments bind to the epidermal cell wall (Post 1990; Gould and Quinn 1999; Gould et al. 2000; Lee and Collins 2001; Post and Vesk 1992; Kunz et al. 1994; Kunz and Becker 1995; Hooijmaijers and Gould 2007). Irrespective of their cellular location, however, anthocyanin biosynthesis in many leaves is generally upregulated in response to one or more environmental stressors. These include: strong light, UV-B radiation, temperature extremes, drought, ozone, nitrogen and phosphorus deficiencies, bacterial and fungal infections, wounding, herbivory, herbicides, and various pollutants (McClure 1975; Chalker-Scott 1999). Because of their association with such biotic and abiotic stressors, anthocyanins are usually considered to be a stress symptom and/or part of a mechanism to mitigate the effects of stress. Much of the physiological work undertaken in recent years has attempted to unravel the phytoprotective functions of anthocyanins that would enhance tolerance to these stress factors.

1.3 Photoprotection

Photoprotective roles of foliar anthocyanin have probably received more attention in recent years than any other functional hypothesis. It had long been suggested that anthocyanins might shield photosynthetic cells from adverse effects of strong light (see Wheldale 1916), yet the first experimental confirmation of this was not achieved until the 1990s, following the advent of the field-portable pulse amplitude modulated (PAM) chlorophyll fluorometer which permitted non-invasive comparisons of the quantum efficiencies of photosynthesis in red versus green leaves (Gould et al. 1995; Krol et al. 1995). Although photosynthesis is driven by light, quanta in excess of the requirements of the light reactions can adversely affect the photosynthetic system components (antenna pigments, reaction centres, accessory proteins, and electron transport carriers), and can lead to secondary destructive and repair processes in thylakoid membranes (Adir et al. 2003). Photoinhibition, the term given to the decline in quantum yield of photosynthesis attributable to excessive illumination, can be quantified directly using PAM chlorophyll fluorometers (Genty et al. 1989; Krause and Weis 1991). One of the most useful parameters for this is the ratio of variable to maximum chlorophyll fluorescence (Fv/Fm) for dark-adapted leaves, which correlates to the maximum quantum yield of photosystem II (Maxwell and Johnson 2000). Fv/Fm values are typically around 0.83 for pre-dawn, healthy plants, but they can be considerably lower in plants under stress. Measurements of Fv/Fm values for red and green leaves before and after exposing them to photoinhibitory light fluxes provide a convenient method to compare their relative tolerances to light stress.

   Anthocyanic leaves typically absorb more light in the green and yellow wavebands than do acyanic leaves (Neill and Gould 1999; Gitelson et al. 2001). The fate of these absorbed quanta is unknown, but it is very clear that their energy is not transferred to the chloroplasts. Indeed, the chlorenchyma of red leaves may receive considerably less green light than do those of structurally comparable green leaves (Gould et al. 2002c), and red leaves may develop the morphological and physiological attributes of shade leaves (Manetas et al. 2003). This light-filtering effect of anthocyanins has been shown many times both to reduce the severity of photoinhibition and to expedite photosynthetic recovery in red as compared to green leaves (see reviews by Steyn et al. 2002; Gould and Lister 2005). In point of fact, sufficient experimental evidence of a photoprotective function of anthocyanins has accrued to justify its elevation from hypothesis to theory.

Anthocyanins confer measurable photoprotection when present both in senescing foliage of deciduous trees (Feild et al. 2001; Hoch et al. 2003) and in the mature, overwintering foliage of evergreen plants (Hughes et al. 2005). Young, developing leaves can also benefit significantly from these pigments (Cai et al. 2005). Indeed, nascent chloroplasts in immature leaves are particularly vulnerable to the effects of light stress (Pettigrew and Vaughn 1998; Choinski et al. 2003). Strong support for a photoprotective role of anthocyanins in developing leaves was provided recently by Hughes et al. (2007), who followed the timing of anthocyanin production and degradation across three unrelated species: *Acer rubrum*, *Cercis canadensis*, and *Liquidambar styraciflua*. In all three species, anthocyanins were produced early in leaf development, and persisted until leaf tissues had fully differentiated. The subsequent decline in anthocyanin levels occurred only after leaves had synthesised approximately 50% of the total chlorophylls and carotenoids, and had attained close to their maximum photosynthetic assimilation rates. The authors suggested that the strong coupling between the timing of anthocyanin reassimilation and those of leaf developmental processes indicates that anthocyanins serve to protect tissues until other photoprotective mechanisms mature.

In view of the immutable property of the coloured anthocyanins to absorb light that might otherwise strike chloroplasts, it is perhaps surprising that the degree to which anthocyanins contribute to the photoprotection of leaves seems to vary substantially from species to species. In *Galax urceolata*, for example, Fv/Fm values for green leaves decreased 36% more than did those for red leaves following exposure to photoinhibitory conditions (Hughes and Smith 2007). Differences of a similar magnitude were noted between yellow and red senescent leaves of *Cornus stolonifera* (Feild et al. 2001), and between green, flushing leaves of *Litsea dilleniifolia* and the red flushing leaves of *Litsea pierrei* and *Anthocephalus chinensis* (Cai et al. 2005). However, much larger differences (ca. 75%) have been reported for the green adult and red juvenile leaves of *Rosa* sp., and in *Ricinus communis* the decline in Fv/Fm for green leaves was almost double that of red leaves (Manetas et al. 2002). In contrast, photosynthetic efficiencies of young red leaves of *Quercus coccifera* were only marginally greater than those of young green leaves under photoinhibitory light flux (Karageorgou and Manetas 2006), and red-leafed species of *Prunus* actually performed worse than green-leafed species under saturating light (Kyparissis et al. 2007).

The reasons for these large interspecific differences in photoprotection by anthocyanin are unknown. It is uncertain whether they reflect true physiological differences, or else are the result of disparities in the experimental conditions under which measurements were taken. Karageorgou and Manetas (2006) suggested that the photoprotective capacity of foliar anthocyanins might vary simply as a function of leaf thickness. They argued that because green light contributes to photosynthesis only in the lowermost tissues of the leaf lamina (Sun et al. 1998; Nishio 2000), then those leaves which are relatively thick and whose mesophyll contain large amounts of chlorophyll would benefit most from the abatement of green light by anthocyanins. Photosynthesis of thinner leaves that contain low amounts of chlorophyll would be driven almost exclusively by red and blue light, and therefore their propensity for photoinhibition would not be affected greatly by the presence of anthocyanins. Accordingly, the immature red leaves of *Quercus coccifera*, which are less than 200 μm thick and hold 11 μg m^{-2} chlorophyll, show little evidence of photoprotection by anthocyanin (Karageorgou and Manetas 2006). In contrast, the mature leaves of that species, which are twice as thick and hold four times as much chlorophyll, show a sizeable benefit from anthocyanins (Manetas et al. 2003). The "leaf-thickness hypothesis" warrants further testing, although the recorded benefits of anthocyanins to the photosynthesis of thin, immature leaves in certain other species (Hughes et al. 2007) would suggest that the hypothesis is not universally applicable.

At least some of the variation among reports of photoprotection by foliar anthocyanins is likely to be attributable to differences in the experimental protocol. Photoinhibition is often intensified when, in addition to excess photon flux, plants experience other types of abiotic stressor (Long et al. 1994). By limiting the rates of CO_2 fixation, environmental factors such as chilling and freezing temperatures, high temperatures, and nitrogen deficiency have been shown to exacerbate the photoinhibitory responses to strong light. It seems possible, therefore, that the photoprotective capacity of anthocyanins would assume greater importance in plants that face combinations of such stressors. Evidence for this was presented recently in a comparison of green- and red-leafed genotypes of maize; the beneficial effects of anthocyanins were apparent only after the plants had experienced a combination of strong light (2000 μmol m^{-2} s^{-1}) and a 5°C chilling treatment (Pietrini et al. 2002). Light quality is also important; reductions in *Fv/Fm* have been found to be greater for green than red leaves when irradiated with white or green light, yet they are similar in magnitude under red light (Hughes et al. 2005). Thus, the experimental conditions under which leaves are tested for photoinhibition can have a significant bearing on measurements of chlorophyll fluorescence.

The interpretation of chlorophyll fluorescence signals can itself be problematic for red leaves. In a recent report describing the common pitfalls of chlorophyll fluorescence analysis, Logan et al. (2007) explained that anthocyanins may absorb a proportion of the measuring light issued from the PAM fluorometer, and therefore reduce the intensity of the emitted chlorophyll fluorescence that is collected for detection. This can lead to low signal to noise ratios, and therefore compromise the accuracy of the data. Fluorescence output can be improved by increasing the intensity of the measuring light, yet this runs the risk of the measuring beam becoming actinic (i.e. driving photosynthesis), which would artifactually reduce *Fv/Fm* values. Some machines perform better than others for measuring chlorophyll fluorescence in red leaves; Pfündel et al. (2007) showed that because anthocyanins attenuate about half of the incident radiation at 470 nm, a fluorometer that issues pulses of blue measuring light can be inferior to one that emits red pulses. It is also noteworthy that the chlorophyll fluorescence signals can alter as a leaf ages, and can even vary from region to region across a leaf lamina (Šesták and Šiffel 1997). Thus, the comparison of young (red) and old (green) leaves, or else red and green parts of the same leaf blade, may yield differences in chlorophyll fluorescence that are unrelated, or only partially related to the presence of anthocyanins.

There are in addition to anthocyanins many other mechanisms by which plants can avoid or dissipate excess light energy. These include morphological features, such as hairs or a waxy cuticle that reflect and scatter incident radiation from the lamina surface, and physiological processes such as thermal dissipation by the xanthophyll cycle pigments and the triplet chlorophyll valve, and the transfer of excess electrons to alternative sinks (Niyogi 2000). The degree to which each of these mechanisms is utilised apparently varies from species to species, as well as with the intensity and duration of exposure to abiotic stress (Demmig-Adams and Adams III 2006). Accordingly, the requirement for supplementary photoprotection, such as that provided by anthocyanins, would also vary. Consistent with this, the young leaves of *Rosa* sp. and *Ricinus communis* contain only low levels of xanthophyll pigments, yet they are resistant to photoinhibitory damage possibly because of their high anthocyanin concentrations (Manetas et al. 2002). Similarly, the combined effects of pubescence and anthocyanins in certain cultivars of grapevine (*Vitis vinifera*) apparently compensate for their reduced xanthophyll contents relative to levels in green, glabrous cultivars (Liakopoulos et al. 2006). In their analysis of mutants of *Arabidopsis thaliana*, Havaux and Kloppstech (2001) concluded that the flavonoids might actually be *more* important than the xanthophylls in regard to long-term protection from photoinhibitory damage, although the anthocyanins were less effective as photoprotectants in that system than were the flavonols and dihydroflavonols. Interspecific differences in requirements for supplementary photoprotection probably best explain why reports of the capacity of foliar anthocyanins to protect leaves from photoinhibiton vary so greatly.

1.4 Protection Against Ultraviolet Radiation

In addition to their capacity to protect plant tissues from excess visible radiation, anthocyanins have also been implicated in the protection from ultraviolet (UV) radiation. UV radiation is often classified as UV-A (320–390 nm), UV-B (280–320 nm) and UV-C (<280 nm). Stratospheric ozone (O_3) absorbs most of the UV-C and part of the UV-B. UV-A radiation, however, is not filtered by stratospheric O_3. With an absorption maximum ($A_{\lambda max}$) at 260 nm, DNA is particularly vulnerable to the adverse effects of highly energetic UV rays (Hoque and Remus 1999).

To fortify themselves against the harmful effects of UV radiation, plants have developed multifarious mechanisms to diminish UV penetration into plant tissues, including the synthesis of UV-absorbing phenolic compounds (Ryan and Hunt 2005). The biosynthesis of anthocyanins and other flavonoids is known to be activated in many plant species by UV exposure (Takahashi et al. 1991; Mendez et al. 1999; Singh et al. 1999), although exceptions have been noted (Jordan et al. 1994; Buchholz et al. 1995; Solovchenko and Merzlyak 2003). Most anthocyanins, especially those that are acylated, can absorb biologically-active UV radiation (Markham 1982; Giusti et al. 1999), and it has been suggested that their function in vegetative organs may be to buffer tissues against UV damage by attenuating the excess energy (Takahashi et al. 1991; Li et al. 1993; Koostra 1994). Support for a protective role of anthocyanins was provided by Burger and Edwards (1996), who noted that following exposure to UV-B or UV-C radiation, the photosynthetic capacities of green-leafed varieties of *Coleus* were lower than those of red-leafed varieties.

However, UV filtering is unlikely to be the primary role of anthocyanins in leaves. Foliar anthocyanins tend not to be acylated, and are therefore less effective absorbers of UV radiation than are certain other flavonoids (Woodall and Stewart 1998). Moreover, to be an efficient screen, anthocyanins must intercept incident UV radiation before it reaches the chloroplasts (Caldwell et al. 1983). In the case of leaves, this means that the pigments should reside in the vacuoles and/or cell walls of epidermal or hypodermal tissues (Day et al. 1992; Ålenius et al. 1995; Gorton and Vogelmann 1996; Olsson et al. 1999). In some species, anthocyanins can indeed be found in these superficial foliar tissues. More commonly, however, anthocyanins occur in the vacuoles of the chlorenchyma cells themselves (Wheldale 1916; Gould and Quinn 1999; Gould et al. 2000; Lee and Collins 2001), a suboptimal location for UV filtering.

It was shown recently that the presence of anthocyanins might in the long term be detrimental rather than beneficial to plants that face high UV levels (Hada et al. 2003). The authors found that in purple-leafed rice, anthocyanins absorb a portion of the blue/UV-A radiation that would otherwise activate the DNA-repairing enzyme photolyase. Such inhibition of DNA repair would offset any short-term gain from UV absorption by anthocyanins.

1.5 Free Radical Scavenging

Environmental stressors such as saturating light flux or high levels of UV radiation can augment the production of free radicals in plant cells (Foyer et al. 1994; Gould 2003). It has been suggested that by absorbing a proportion of the incident quanta, and by scavenging the free radicals thus formed, foliar anthocyanins might serve to abate this oxidative insult.

A free radical is any chemical species capable of independent existence that contains one or more unpaired electron (Halliwell and Gutteridge 1999). The oxygen radicals have been most extensively studied in plants, although there is increasing awareness of the roles of nitrogen-centred radicals. The collective term "reactive oxygen species" (ROS) or "reactive oxygen intermediates" (ROI) is often used to include both the oxygen radicals and non-radical derivatives of oxygen which have similar chemical properties. These include the superoxide radical (O_2^-), hydroxyl radical (OH), peroxyl radical (ROO), and alkoxyl radical (RO^-), as well as the non-radical intermediates such as singlet oxygen ($^1O_2$), hydrogen peroxide (H_2O_2), and ozone (O_3).

In plant cells, chloroplasts and mitochondria are the principal sources of ROS, which are generated via the aerobic reactions involved in photosynthesis and respiration (Mittler 2002; Rhoads et al. 2006). ROS are also produced in the peroxisomes during photorespiration and fatty acid oxidation (Corpas et al. 2001). Enzymatic sources of ROS have been identified, including NADPH oxidase in the plasma membrane (Grant and Loake 2000), oxalate oxidase and amine oxidase in the apoplasm (Allan and Fluhr 1997; Dat et al. 2000), and peroxidases in the cell wall (Kawano 2003). Under optimal growth conditions the production of ROS from routine metabolic processes is low: 240 $\mu M\ s^{-1}$ O_2^-, and a steady-state level of 0.5 μM H_2O_2 in chloroplasts (Polle 2001). However, environmental stressors can increase levels of ROS three-fold (Polle 2001).

A superabundance of ROS potentially causes cellular damage to phospholipid membranes, proteins, and nucleic acids, and this has traditionally been considered detrimental to plant functioning (Alscher et al. 1997). Guarding against oxidative damage, plants have evolved elaborate antioxidant defence mechanisms in the different intracellular compartments. These serve to control concentrations of ROS, to improve the plant's resistance to stressors, to repair damage to proteins, particularly those in photosystem II, and to re-activate key enzymes (Halliwell and Gutteridge 1999). An antioxidant may be defined as any substance which, when present at low concentrations compared with those of an oxidisable substrate, significantly delays or prevents oxidation of that substrate. The major antioxidants are enzymes, and include superoxide dismutase (SOD), catalase (CAT), various peroxidases such as ascorbate peroxidase (APX), and glutathione reductase (GR) (Polle 1997). There are in addition a number of low molecular weight antioxidants (LMWAs) in plant cells: ascorbate (vitamin C), tocopherols (vitamin E), glutathione, β-carotene, and phenolic compounds such as the flavonoids.

Certain flavonoids, including the more common anthocyanin pigments, have ROS-scavenging capacities up to four times greater than those of vitamin E and C analogues (Rice-Evans et al. 1997; Wang et al. 1997). Their potency stems from a high reactivity as proton and electron donors, from their ability to stabilize and delocalize unpaired electrons, and from their capacity to chelate transition metal ions (Rice-Evans et al. 1996; van Acker et al. 1996; Brown et al. 1998). Flavonoids have been shown *in vitro* to neutralise most of the biologically important ROS and nitrogen-centred radicals. Recently, compelling evidence was presented for the scavenging of ROS by flavonoids *in vivo*. Agati et al. (2007) infused leaves of *Phillyrea latifolia* with DanePy, a fluorochrome whose fluorescence is quenched exclusively by $^1O_2$. Microscopic examinations of cross-sections through those leaves revealed that the scavenging of $^1O_2$, which had been generated by subjecting the leaves to strong light, was largely attributable to flavonols and flavones specifically associated with chloroplasts in the mesophyll cells.

Could antioxidant activity explain the presence of anthocyanins in vegetative tissues? Two mechanisms by which anthocyanins might reduce the oxidative load in leaves have been proposed. First, by reducing the numbers of high-energy quanta incident on the photosynthetic cells, anthocyanins might prevent or moderate the light-driven reactions that generate ROS. This is an old concept. Indeed, Wheldale (1916) herself described an experiment in which a solution of chlorophyll, when illuminated behind a glass vessel containing a red solution, remained green for longer than when illuminated behind a colourless solution. Chlorophyll bleaching is a classic symptom of oxidative damage (Kato and Shimizu 1985). More recently, Neill and Gould (2003) showed that chloroplasts suspended in a buffered solution produced fewer O_2^- radicals, and were bleached less, when irradiated with monochromatic red light than with white light of comparable intensity. However, the benefits of anthocyanin as an optical shield have yet to be demonstrated *in situ*.

Second, anthocyanins might directly scavenge ROS. Anthocyanins are usually colourless or light blue at the pH of the cytoplasm, but they turn red after being transported into the vacuole. Both the colourless and the red tautomers of cyanidin glycosides have been demonstrated to scavenge O_2^- produced by a suspension of chloroplasts under light stress (Neill and Gould 2003). Clearly, cytosolic anthocyanins would be better located than would vacuolar anthocyanins for scavenging ROS produced by organelles such as chloroplasts, mitochondria, and peroxisomes. However, the question of the common occurrence of anthocyanins in both cytosol and vacuole is one that requires further attention. It remains unclear how anthocyanins are transported to the vacuole from their site of synthesis at the endoplasmic reticulum. If they move to the vacuole by diffusion, then they would transiently pass through the cytosol. Alternatively, there is growing evidence of a route from the endoplasmic reticulum directly into vesicles, which then migrate to the vacuole, completely bypassing the cytosol (Poustka et al. 2007). Irrespective of their intracellular location, however, anthocyanin-containing leaf cells have been observed under the microscope to remove H_2O_2 more swiftly than acyanic cells (Gould et al. 2002a). It is possible, therefore, that antioxidant activity may be one of the major functions of anthocyanins in vegetative tissues.

The available data indicate that anthocyanins contribute to the total antioxidant pool more in some species than in others. For example, in *Elatostema rugosum*, a sprawling understorey herb from New Zealand, extracts from red leaves had a significantly greater LMWA activity than did those from green leaves, which could be attributed primarily to the presence of anthocyanins (Neill et al. 2002b). In contrast, in the canopy plant *Quintinia serrata*, extracts from red and green leaves showed similar ranges in antioxidant potential (Neill et al. 2002c). Similarly, in the sugar maple (*Acer saccharum*), antioxidant activity correlated strongly with anthocyanin content in extracts from juvenile leaves, but the correlation was only weak in extracts from senescing leaves (van den Berg and Perkins 2007). Reasons for these differences are not known, though it has been suggested that the location of anthocyanic cells within the leaf tissues may be important. Kytridis and Manetas (2006) compared the effects of methyl viologen, a herbicide, on various species for which foliar anthocyanins were located in different cell types. Methyl viologen inhibits photosynthetic electron transport, generating ROS that lead to the destruction of chloroplast membranes. The authors claimed that red leaves for which anthocyanins were located in the mesophyll were more resistant to methyl viologen treatment than were those that held anthocyanin in the epidermis. Although antioxidant activities were not measured in that study, the data are consistent with the hypothesis that when anthocyanins are located in the mesophyll, they can contribute to the LMWA pool.

In addition to the anthocyanins, concentrations of other LMWAs, as well as certain enzymatic antioxidants, can also be higher in red than in green leaves. For example, the red-leafed morphs of *Elatostema rugosum* had higher levels of caffeic acid derivatives and greater SOD and CAT activities than had the green-leafed morphs (Neill et al. 2002b). Similarly, leaves of maize cultivars that had been exposed to toxic copper concentrations upregulated the production of anthocyanin as well as the activities of superoxide dismutase, ascorbate peroxidase, and glutathione reductase (Tanyolaç et al. 2007). Anthocyanins might well supplement the antioxidant potential in such plants, but they clearly do not substitute the major LMWA and enzymatic antioxidants.

1.6 Paradigm Shift

Recent genetic studies have indicated that the potential for ROS to cause unrestricted damage to plant cell components is realised far less commonly than had been previously thought. On the contrary, there is a growing body of empirical evidence to suggest that ROS may serve many useful functions in plants. Indeed, ROS appear to be actively produced by plant cells for their use as signalling molecules in processes as diverse as growth and development, stomatal closure, pathogen defence, programmed cell death, and abiotic stress responses. This evidence led Foyer and Noctor (2005) to state, "The moment has come to re-evaluate the concept of oxidative stress." They proposed that the processes by which ROS are generated and scavenged might better be described as "oxidative signalling", and should be regarded as "an important and critical function associated with the mechanisms by which plant cells sense the environment and make appropriate adjustments to gene expression, metabolism and physiology."

The arguments in favour of a signalling role for ROS have been expertly summarised in several reviews (Dat et al. 2000; Mittler 2002; Vranová et al. 2002; Mahalingam and Fedoroff 2003; Apel and Hirt 2004; Laloi et al. 2004; Mittler et al. 2004; Foyer and Noctor 2005; Pitzschke et al. 2006). Among the more compelling lines of evidence for ROS signalling is the work by Wagner et al. (2004) on the *flu* mutant of *Arabidopsis thaliana*. The *flu* mutant, when transferred from darkness to the light, generates singlet oxygen in the plastids, ultimately leading to chlorophyll bleaching and death of seedlings, or to the arrest of growth in more mature plants. However, when a single gene, *EXECUTER1*, is inactivated in this species (as in the *flu/executer1* double mutant), both seedlings and mature plants grow normally despite the continued production of $^1O_2$. The authors concluded that in wild type plants, $^1O_2$ does not damage cellular components directly, but rather, activates a genetic switch that initiates a signalling cascade leading to programmed cell suicide.

ROS-induced programmed cell death may be useful for plants facing pathogenic attack since it potentially limits the spread of disease from the point of infection. However, cell death would probably not be beneficial under conditions of abiotic stress such as those imposed by strong light and elevated UV-B. Mittler (2002) argued that the steady state levels of ROS may be used by plants as a gauge of intracellular stress. When levels of ROS rise in response to abiotic stress, plants face the challenge of removing excess ROS to avoid programmed cell death, yet retaining sufficient low levels of the different types of ROS for signalling purposes. This would require the finely-tuned modulation of ROS production and scavenging mechanisms. Specificity in response may be achievable by the coordinated production of LMWAs such as ascorbate and glutathione (Foyer and Noctor 2005). The flavonoids, too, seem likely to play a role in this.

1.7 Modulation of Signalling Cascades: A New Hypothesis

That ROS can be at once the products of plant stress as well as mediators in plant stress responses presents the possibility for a new functional hypothesis for the presence of anthocyanins in vegetative tissues. We propose that the anthocyanins, along with some other flavonoids, provide multifarious mechanisms for the modulation of signalling cascades that mitigate the effects of abiotic and biotic stressors. As explained below, this role is achievable in three interrelated ways: (i) by protecting antioxidant enzymes; (ii) by scavenging ROS directly; and (iii) by interactions with other molecules in the signal transduction pathways.

Many of the putative roles of anthocyanins in plant physiology could equally be achieved by antioxidant enzymes. For example, like the anthocyanins, the ROS-scavenging enzymes of the so-called "water-water cycle" (SOD and APX) in the chloroplasts result in a reduced propensity for photoinhibition and photo-oxidation (Asada 1999, 2000; Rizhsky et al. 2003). These enzymes scavenge O_2^- and H_2O_2 with extreme efficiency, and are undoubtedly key players in the modulation of ROS signalling cascades. Under certain conditions, however, these enzymes may be inactivated. Strong light combined with chilling stress, for example, reduces the efficiency of APX, leading to the accumulation of H_2O_2 to levels can inactivate APX, SOD, and CAT (Jahnke et al. 1991; Wise 1995; Casano et al. 1997; Streb et al. 1997; Asada 1999). For the water-water cycle to function properly, its enzymatic antioxidants need to be protected from free-radical attack.

It is perhaps no coincidence that the very conditions that can lead to the inactivation of such enzymes may also stimulate the formation of anthocyanins in plant tissues. Anthocyanins may prevent the inactivation of antioxidant enzymes by restricting the amount of light within a photosynthetic cell (thereby reducing the production of ROS). The strong antioxidant capacities of anthocyanins mean that they could also scavenge supernumerary ROS and therefore spare the antioxidant enzymes from inactivation. Thus, for a one-time investment in the production of anthocyanin pigments, plants might achieve the long-term protection of these pivotal components of plant stress responses. Indeed, the capacity of plants to maintain or enhance their antioxidant enzyme activities is regarded as a key feature in the acclimation of plant tissues to environmental stress (Bowler et al. 1992; Anderson et al. 1995; Pinhero et al. 1997; Scebba et al. 1999; Kuk et al. 2003).

It is possible that anthocyanins interact with stress signal transduction cascades more directly. This has been demonstrated already in human tumour cells; two anthocyanin aglycones, cyanidin and delphinidin, were found to inhibit tumour cell growth by shutting off downstream signalling cascades that would otherwise lead to the production of growth factors (Meiers et al. 2001). Interactions between phenolic compounds and ROS signalling have also been documented for plants. For example, the softening of plant cell walls, which is necessary for cell expansion, results partly from OH radical attack on cell wall polysaccharides (Fry 1998), and is terminated by the cross-linking of phenolic compounds (Rodríguez et al. 2002). Because anthocyanins can scavenge a variety of free radicals and oxidants such as H_2O_2 (Takahama 2004), they have the potential directly to influence the balance between ROS production and ROS scavenging in stress responses. H_2O_2 is considered a particularly important molecule in plant signalling because of its relative stability, as well as its ability to diffuse rapidly across membranes and between different cell compartments (Dröge 2002; Neill et al. 2002a). H_2O_2 is a known activator of MAP kinase cascades, and has been shown to regulate the expression of certain genes

(Bowler and Fluhr 2000). Despite its efficient scavenging by enzymes in the chloroplasts, mitochondria, and peroxisomes, H_2O_2 may leak into the cytosol and possibly the vacuole during periods of severe stress (Yamasaki et al. 1997). Vacuoles typically occupy more than 70% of the mature plant cell volume, and as a consequence of their size, vacuoles are one of the closest neighbours of all the major sources of organelle-derived ROS. For this reason, vacuolar LMWAs such as the anthocyanins are likely to have a crucial role, especially in preventing the symplastic movement of ROS from one cell to another (Mittler et al. 2004).

Finally, anthocyanins may interact with secondary messengers downstream of the ROS signalling pathway, or else be involved in the crosstalk with other response pathways. An intriguing possibility is the interaction between anthocyanins and sucrose. The anthocyanin biosynthetic pathway is strongly upregulated by sucrose in plants as diverse as radish (*Raphanus sativus*), English ivy (*Hedera helix*), and *Arabidopsis thaliana* (Murray and Hackett 1991; Hara et al. 2003; Solfanelli et al. 2006). Soluble sugars, especially sucrose, glucose, and fructose, are now known to play central roles in the control of plant development, stress responses, and gene expression (Gibson 2005). Sugar accumulation has been associated with improved tolerance to diverse stressors including drought, salinity, high light, cold, anoxia and herbicides (Roitsch 1999; Couée et al. 2006). It has been suggested that these roles relate to the regulation of the pro-oxidant and antioxidant balance in plant cells; sucrose is known to be involved in both ROS-producing and ROS-scavenging metabolic pathways (Couée et al. 2006). However, the mechanism by which sucrose, anthocyanins, and ROS might contribute to plant function remains to be established.

A signalling role for anthocyanins is attractive because it potentially also explains why anthocyanins often accumulate in organs that do not photosynthesise, or else for which photosynthetic carbon assimilation is not the primary function, such as stems petioles, and adventitious roots. It can also explain why anthocyanins are in some species preferentially produced at certain developmental stages, such as seed dormancy, leaf initiation or leaf senescence, or in certain seasons such as autumn or spring. Establishing possible relationships between the cellular redox balance and anthocyanin function presents the promise of an exciting, new line of investigation into this intriguing class of plant pigments.

References

Adir, N., Zer, H., Scholat, S. and Ohad, I. (2003) Photoinhibition – a historical perspective. Photosynth. Res. 76, 343–370.

Agati, G., Matteini, P., Goti, A. and Tattini, M. (2007) Chloroplast-located flavonoids can scavenge singlet oxygen. New Phytol. 174, 77–89.

Ålenius, C.M., Vogelmann, T.C. and Bornman, J.F. (1995) A three-dimensional representation of the relationship between penetration of U.V.-B radiation and U.V.-screening pigments in leaves of *Brassica napus*. New Phytol. 131, 297–302.

Allan, A.C. and Fluhr, R. (1997) Two distinct sources of elicited reactive oxygen species in tobacco epidermal cells. Plant Cell 9, 1559–1572.

Alscher, R.G., Donahue, J.L. and Cramer, C.L. (1997) Reactive oxygen species and antioxidants: relationships in green cells. Physiol. Plant. 100, 224–233.

Anderson, M.D., Prasad, T.K. and Stewart, C.R. (1995) Changes in isozyme profiles of catalase, peroxidase, and glutathione reductase during acclimation to chilling in mesocotyls of maize seedlings. Plant Physiol. 109, 1247–1257.

Apel, K. and Hirt, H. (2004) Reactive oxygen species: metabolism, oxidative stress, and signal transduction. Annu. Rev. Plant Biol. 55, 373–399.

Asada, K. (1999) The water-water cycle in chloroplasts: scavenging of active oxygens and dissipation of excess photons. Annu. Rev. Plant Physiol. Plant Mol. Biol. 50, 601–639.

Asada, K. (2000) The water-water cycle as alternative photon and electron sinks. Phil. Trans. R. Soc. Lond. B 355, 1419–1431.

Bowler, C. and Fluhr, R. (2000) The role of calcium and activated oxygens as signals for controlling cross-tolerance. Trends Plant Sci. 5, 241–246.

Bowler, C., van Montagu, M. and Inzé, D. (1992) Superoxide dismutase and stress tolerance. Annu. Rev. Plant Physiol. Plant Mol. Biol. 43, 83–116.

Brown, J.E., Khodr, H., Hider, R.C. and Rice-Evans, C.A. (1998) Structural dependence of flavonoid interactions with Cu^{2+} ions: implications for their antioxidant properties. Biochem. J. 330, 1173–1178.

Buchholz, G., Ehmann, B. and Wellmann, E. (1995) Ultraviolet light inhibition of phytochrome-induced flavonoid biosynthesis and DNA photolyase formation in mustard cotyledons (Sinapis alba L.). Plant Physiol. 108, 227–234.

Burger, J. and Edwards, G. (1996) Photosynthetic efficiency, and photodamage by UV and visible radiation, in red versus green leaf Coleus varieties. Plant Cell Physiol. 37, 395–399.

Cai, Z.-Q., Slot, M. and Fan, Z.-X. (2005) Leaf development and photosynthetic properties of three tropical tree species with delayed greening. Photosynthetica 43, 91–98.

Caldwell, M.M., Robberecht, R. and Flint, S.D. (1983) Internal filters: prospects for UV-acclimation in higher plants. Physiol. Plant. 58, 445–450.

Casano, L.M., Gómez, L.D., Lascano, H.R., González, C.A. and Trippi, V.S. (1997) Inactivation and degradation of CuZn-SOD by active oxygen species in wheat chloroplasts exposed to photooxidative stress. Plant Cell Physiol. 38, 433–440.

Chalker-Scott, L. (1999) Environmental significance of anthocyanins in plant stress responses. Photochem. Photobiol. 70, 1–9.

Chalker-Scott, L. (2002) Do anthocyanins function as osmoregulators in leaf tissues? Adv. Bot. Res. 37, 103–127.

Choinski, J.S. Jr., Ralph, P. and Eamus, D. (2003) Changes in photosynthesis during leaf expansion in Corymbia gummifera. Aust. J. Bot. 51, 111–118.

Close, D.C. and Beadle, C.L. (2003) The ecophysiology of foliar anthocyanin. Bot. Rev. 69, 149–161.

Corpas, F.J., Barroso, J.B. and del Río, L.A. (2001) Peroxisomes as a source of reactive oxygen species and nitric oxide signal molecules in plant cells. Trends Plant Sci. 6, 145–150.

Couée, I., Sulmon, C., Gouesbet, G. and El Amrani, A. (2006) Involvement of soluble sugars in reactive oxygen species balance and responses to oxidative stress in plants. J. Exp. Bot. 57, 449–459.

Dat, J., Vandenabeele, S., Vranová, E., Van Montagu, M., Inzé, D. and Van Breusegem, F. (2000) Dual action of the active oxygen species during plant stress responses. Cell. Mol. Life Sci. 57, 779–795.

Day, T.A., Vogelmann, T.C. and DeLucia, E.H. (1992) Are some plant life forms more effective than others in screening out ultraviolet-B radiation? Oecologia 92, 513–519.

Demmig-Adams, B. and Adams III, W.W. (2006) Photoprotection in an ecological context: the remarkable complexity of thermal energy dissipation. New Phytol. 172, 11–21.

Dröge, W. (2002) Free radicals in the physiological control of cell function. Physiol. Rev. 82, 47–95.

Feild, T.S., Lee, D.W. and Holbrook, N.M. (2001) Why leaves turn red in Autumn. The role of anthocyanins in senescing leaves of red-osier dogwood. Plant Physiol. 127, 566–574.

Foyer, C.H. and Noctor, G. (2005) Oxidant and antioxidant signalling in plants: a re-evaluation of the concept of oxidative stress in a physiological context. Plant Cell Environ. 28, 1056–1071.

Foyer, C.H., Lelandais, M. and Kunert, K.J. (1994) Photooxidative stress in plants. Physiol. Plant. 92, 696–717.

Fry, S.C. (1998) Oxidative scission in plant cell wall polysaccharides by ascorbate-induced hydroxyl radicals. Biochem. J. 332, 507–515.

Genty, B., Briantais, J.-M. and Baker, N.R. (1989) The relationship between quantum yield of photosynthetic electron transport and quenching of chlorophyll fluorescence. Biochim. Biophys. Acta 990, 87–92.

Gibson, S. (2005) Control of plant development and gene expression by sugar signaling. Curr. Opin. Plant Biol. 8, 93–102.

Gitelson, A.A., Merzylak, M.N. and Chivkunova, O.B. (2001) Optical properties and non-destructive estimation of anthocyanin content in plant leaves. Photochem. Photobiol. 74, 38–45.

Giusti, M., Rodriguez-Saona, L. and Wrolstad, R. (1999) Molar absorptivity and color characteristics of acylated and non-acylated pelargonidin-based anthocyanins. J. Agric. Food Chem. 47, 4631–4637.

Gorton, H. and Vogelmann, T.C. (1996) Effects of epidermal cell shape and pigmentation on optical properties of *Antirrhinum* petals at visible and ultraviolet wavelengths. Plant Physiol. 112, 879–888.

Gould, K.S. (2003) Free radicals, oxidative stress and antioxidants. In: Thomas, B., Murphy, D.J. and Murray, B.G. (Eds), *Encyclopedia of Applied Plant Sciences*. Elsevier, Amsterdam, pp. 9–16.

Gould, K.S. (2004) Nature's Swiss army knife: the diverse protective roles of anthocyanins in leaves. J. Biomed. Biotechnol. 2004, 314–320.

Gould, K.S. and Lister, C. (2005) Flavonoid functions in plants. In: Andersen, Ø.M. and Markham, K.R. (Eds), *Flavonoids: Chemistry, Biochemistry, and Applications*. CRC Press, Boca Raton, pp. 397–441.

Gould, K.S. and Quinn, B.D. (1999) Do anthocyanins protect leaves of New Zealand native species from UV-B? New Zeal. J. Bot. 37, 175–178.

Gould, K.S., Kuhn, D.N., Lee, D.W. and Oberbauer, S.F. (1995) Why leaves are sometimes red. Nature 378, 241–242.

Gould, K.S., Markham, K.R., Smith, R.G. and Goris, J.J. (2000) Functional role of anthocyanins in the leaves of *Quintinia serrata* A Cunn. J. Exp. Bot. 51, 1107–1115.

Gould, K.S., McKelvie, J. and Markham, K.R. (2002a) Do anthocyanins function as antioxidants in leaves? Imaging of H_2O_2 in red and green leaves after mechanical injury. Plant Cell Environ. 25, 1261–1269.

Gould, K.S., Neill, S.O. and Vogelmann, T.C. (2002b) A unified explanation for anthocyanins in leaves? Adv. Bot. Res. 37, 167–192.

Gould, K.S., Vogelmann, T.C., Han, T. and Clearwater, M.J. (2002c) Profiles of photosynthesis of red and green leaves of *Quintinia serrata*. Physiol. Plant. 116, 127–133.

Grant, J.J. and Loake, G.J. (2000) Role of reactive oxygen intermediates and cognate redox signalling in disease resistance. Plant Physiol. 124, 21–29.

Hada, H., Hidema, J., Maekawa, M. and Kumagai, T. (2003) Higher amounts of anthocyanins and UV-absorbing compounds effectively lowered CPD photorepair in purple rice (*Oryza sativa* L.). Plant Cell Environ. 26, 1691–1701.

Halliwell, B. and Gutteridge, J.M.C. (1999) *Free Radicals in Biology and Medicine*. Oxford University Press, Oxford.

Hara, M., Oki, K., Hoshino, K. and Kuboi, T. (2003) Enhancement of anthocyanin biosynthesis by sugar in radish (*Raphanus sativus*) hypocotyl. Plant Sci. 164, 259–265.

Havaux, M. and Kloppstech, K. (2001) The protective functions of carotenoid and flavonoid pigments against excess visible radiation at chilling temperature investigated in *Arabidopsis npq* and *tt* mutants. Planta 213, 953–966.

Hoch, W.A., Zeldin, E.L. and McCown, B.H. (2001) Physiological significance of anthocyanins during autumnal leaf senescence. Tree Physiol. 21, 1–8.

Hoch, W.A., Singsaas, E.L. and McCown, B.H. (2003) Resorption protection. Anthocyanins facilitate nutrient recovery in Autumn by shielding leaves from potentially damaging light levels. Plant Physiol. 133, 1296–1305.

Hooijmaijers, C.A.M. and Gould, K.S. (2007) Photoprotective pigments in red and green gametophytes of two New Zealand liverworts. New Zeal. J. Bot. 45, 451–461.

Hoque, E. and Remus, G. (1999). Natural UV-screening mechanisms of Norway spruce (*Picea abies* [L.] Karst.) needles. Photochem. Photobiol. 69, 177–192.

Hughes, N.M. and Smith, W.K. (2007) Attenuation of incident light in *Galax urceolata* (Diapensiaceae): concerted influence of adaxial and abaxial anthocyanic layers on photoprotection. Am. J. Bot. 94, 784–790.

Hughes, N.M., Neufeld, H.S. and Burkey, K.O. (2005) Functional role of anthocyanins in high-light winter leaves of the evergreen herb *Galax urceolata*. New Phytol. 168, 575–587.

Hughes, N.M., Morley, C.B. and Smith, W.K. (2007) Coordination of anthocyanin decline and photosynthetic maturation in juvenile leaves of three deciduous tree species. New Phytol. 175, 675–685.

Jahnke, L.S., Hull, M.R. and Long, S.P. (1991) Chilling stress and oxygen metabolizing enzymes in *Zea mays* and *Zea diploperennis*. Plant Cell Environ. 14, 97–104.

Jordan, B.R., James, P., Strid, Å. and Anthony, R. (1994) The effect of ultraviolet-B radiation on gene expression and pigment composition in etiolated and green pea leaf tissue: UV-B-induced changes are gene-specific and dependent upon the developmental stage. Plant Cell Environ. 17, 45–54.

Karageorgou, P. and Manetas, Y. (2006) The importance of being red when young: anthocyanins and the protection of young leaves of *Quercus coccifera* from insect herbivory and excess light. Tree Physiol. 26, 613–621.

Kato, M. and Shimizu, S. (1985) Chlorophyll metabolism in higher plants VI. Involvement of peroxidase in chlorophyll degradation. Plant Cell Physiol. 26, 1291–1301.

Kawano, T. (2003) Roles of the reactive oxygen species-generating peroxidase reactions in plant defense and growth induction. Plant Cell Rep. 21, 829–837.

Koostra, A. (1994) Protection from UV-B induced DNA damage by flavonoids. Plant Mol. Biol. 26, 771–774.

Krause, G.H. and Weis, E. (1991) Chlorophyll fluorescence and photosynthesis: the basics. Annu. Rev. Plant Physiol. Plant Mol. Biol. 42, 313–349.

Krol, M., Gray, G.R., Hurry, V.M., Oquist, G., Malek, L. and Huner, N.P.A. (1995) Low-temperature stress and photoperiod affect an increased tolerance to photoinhibition in *Pinus banksiana* seedlings. Can. J. Bot. 73, 1119–1127.

Kuk, Y.I., Shin, J.S., Burgos, N.R., Hwang, T.E., Han, O., Cho, B.H., Jung, S. and Guh J.O. (2003) Antioxidative enzymes offer protection from chilling damage in rice plants. Crop Sci. 43, 2109–2117.

Kunz, S. and Becker, H. (1995) Cell wall pigment formation of *in vitro* cultures of the liverwort *Ricciocarpos natans*. Z. Naturforsch. 50, 235–240.

Kunz, S., Burkhardt, G. and Becker, H. (1994) Riccionidins A and B, anthocyanidins from the cell walls of the liverwort *Ricciocarpos natans*. Phytochemistry 35: 233–235.

Kyparissis, A., Grammatikopoulos, G. and Manetas, Y. (2007) Leaf morphological and physiological adjustments to the spectrally selective shade imposed by anthocyanins in *Prunus cerasifera*. Tree Physiol. 27, 849–857.

Kytridis, V.-P. and Manetas, Y. (2006) Mesophyll versus epidermal anthocyanins as potential *in vivo* antioxidants: evidence linking the putative antioxidant role to the proximity of oxy-radical source. J. Exp. Bot. 57, 2203–2210.

Laloi, C., Apel, K. and Danon A. (2004) Reactive oxygen signalling: the latest news. Curr. Opin. Plant Biol. 7, 323–328.

Lee, D.W. and Collins, T.M. (2001) Phylogenetic and ontogenetic influences on the distribution of anthocyanins and betacyanins in leaves of tropical plants. Int. J. Plant Sci. 162, 1141–1153.

Lee, D.W. and Gould, K.S. (2002a) Anthocyanins in leaves and other vegetative organs: an introduction. Adv. Bot. Res. 37, 2–16.

Lee, D.W. and Gould, K.S. (2002b) Why leaves turn red. Am. Sci. 90, 524–531.

Li, J., Ou-Lee, T.M., Raba, R., Amundson, R.G. and Last, R.L. (1993) Arabidopsis flavonoid mutants are hypersensitive to UV-B irradiation. Plant Cell 5, 171–179.

Liakopoulos, G., Nikolopoulos, D., Klouvatou, A., Vekkos, K.-A., Manetas, Y. and Karabourniotis, G. (2006) The photoprotective role of epidermal anthocyanins and surface pubescence in young leaves of grapevine (*Vitis vinifera*). Ann. Bot. 98, 257–265.

Logan, B.A., Adams III, W.W. and Demmig-Adams, B. (2007) Avoiding common pitfalls of chlorophyll fluorescence analysis under field conditions. Funct. Plant Biol. 3, 853–859.

Long, S.P., Humphries, S. and Falkowski, P.G. (1994) Photoinhibition of photosynthesis in nature. Annu. Rev. Plant Physiol. Plant Mol. Biol. 45, 633–662.

Mahalingam, R. and Fedoroff, N. (2003) Stress response, cell death and signalling: the many faces of reactive oxygen species. Physiol. Plant. 119, 56–68.

Manetas, Y. (2006) Why some leaves are anthocyanic, and why most anthocyanic leaves are red. Flora 201, 163–177.

Manetas, Y., Drinia, A. and Petropoulou, Y. (2002) High contents of anthocyanins in young leaves are correlated with low pools of xanthophyll cycle components and low risk of photoinhibition. Photosynthetica 40, 349–354.

Manetas, Y., Petropoulou, Y., Psaras, G.K. and Drinia, A. (2003). Exposed red (anthocyanic) leaves of *Quercus coccifera* display shade characteristics. Funct. Plant Biol. 30, 265–270.

Markham, K.R. (1982) *Techniques of Flavonoid Identification*. Academic Press, London.

Maxwell, K. and Johnson, G.N. (2000) Chlorophyll fluorescence – a practical guide. J. Exp. Bot. 51, 659–668.

McClure, J.W. (1975) Physiology and functions of flavonoids. In: Harborne, J.B., Mabry, T.J. and Mabry, H. (Eds.), *The Flavonoids*. Chapman and Hall, London, pp. 970–1055.

Meiers, S., Kemény, M., Weyand, U., Gastpar, R., Von Angerer E. and Marko, D. (2001) The anthocyanidins cyanidin and delphinidin are potent inhibitors of the epidermal growth-factor receptor. J. Agric. Food Chem. 49, 958–962.

Mendez, M., Jones, D.G. and Manetas, Y. (1999) Enhanced UV-B radiation under field conditions increases anthocyanin and reduces the risk of photoinhibition but does not affect growth in the carnivorous plant *Pinguicula vulgaris*. New Phytol. 144, 275–282.

Mittler, R. (2002) Oxidative stress, antioxidants and stress tolerance. Trends Plant Sci. 7, 405–410.

Mittler, R., Vanderauwera, S., Gollery, M. and van Breusegem, F. (2004) Reactive oxygen gene network of plants. Trends Plant Sci. 9, 490–498.

Murray, J.R. and Hackett, W.P. (1991) Dihydroflavonol reductase activity in relation to differential anthocyanin accumulation in juvenile and mature phase *Hedera helix* L. Plant Physiol. 97, 343–351.

Neill, S.O. and Gould, K.S. (1999) Optical properties of leaves in relation to anthocyanin concentration and distribution. Can. J. Bot. 77, 1777–1782.

Neill, S.O. and Gould, K.S. (2003) Anthocyanins in leaves: light attenuators or antioxidants? Funct. Plant Biol. 30, 865–873.

Neill, S., Desikan, R. and Hancock, J. (2002a) Hydrogen peroxide signalling. Curr. Opin, Plant Biol. 5, 388–395.

Neill, S.O., Gould, K.S., Kilmartin, P.A., Mitchell, K.A. and Markham, K.R. (2002b) Antioxidant activities of red versus green leaves of *Elatostema rugosum*. Plant Cell Environ. 25, 539–547.

Neill, S.O., Gould, K.S., Kilmartin, P.A., Mitchell, K.A. and Markham, K.R. (2002c) Antioxidant capacities of green and cyanic leaves in the sun species, *Quintinia serrata*. Funct. Plant Biol. 29, 1437–1443.

Nishio, J.N. (2000) Why are higher plants green? Evolution of the higher plant photosynthetic pigment complement. Plant Cell Environ. 23, 539–548.

Niyogi, K.K. (2000) Safety valves for photosynthesis. Curr. Opin. Plant Biol. 3, 455–460.

Olsson, L.C., Veit, M. and Bornman, J.F. (1999) Epidermal transmittance and phenolic composition in leaves of atrazine-tolerant and atrazine-sensitive cultivars of *Brassica napus* grown under enhanced UV-B radiation. Physiol. Plant. 107, 259–266.

Pettigrew, W.T. and Vaughn, K.C. (1998) Physiological, structural, and immunological characterization of leaf and chloroplast development in cotton. Protoplasma 202, 23–37.

Pfündel, E.E., Ghozlen, N.B., Meyer, S. and Cerovic, Z.G. (2007) Investigating UV screening in leaves by two different types of portable UV fluorimeters reveals in vivo screening by anthocyanins and carotenoids. Photosynth. Res. 93, 205–221.

Pietrini, F., Iannelli, M.A. and Massacci, A. (2002) Anthocyanin accumulation in the illuminated surface of maize leaves enhances protection from photo-inhibitory risks at low temperature, without further limitation to photosynthesis. Plant Cell Environ. 25, 1251–1259.

Pinhero, R.G., Rao, M.V., Paliyath, G., Murr, D.P. and Fletcher, R.A. (1997) Changes in activities of antioxidant enzymes and their relationship to genetic and paclobutrazol-induced chilling tolerance of maize seedlings. Plant Physiol. 114, 695–704.

Pitzschke, A., Forzani, C. and Hirt, H. (2006) Reactive oxygen species signaling in plants. Antioxid. Redox Signal. 8, 1757–1764.

Polle, A. (1997) Defense against photooxidative damage in plants. In: Scandalios, J.G. (Ed.), *Oxidative Stress and the Molecular Biology of Antioxidant Defenses*. New York, Cold Spring Harbor Laboratory Press, pp. 623–666.

Polle, A. (2001) Dissecting the superoxide dismutase-ascorbate peroxidase-glutathione pathway in chloroplasts by metabolic modelling. Computer simulations as a step towards flux analysis. Plant Physiol. 126, 445–462.

Post, A. (1990) Photoprotective pigment as an adaptive strategy in the Antarctic moss *Ceratodon purpureus*. Polar Biol. 10, 241–245.

Post, A. and Vesk, M. (1992) Photosynthesis, pigments, and chloroplast ultrastucture of an antarctic liverwort from sun-exposed and shaded sites. Can. J. Bot. 70, 2259–2264.

Poustka, F., Irani, N.G., Feller, A., Lu, Y., Pourcel, L., Frame, K. and Grotewold, E. (2007) Trafficking pathway for anthocyanins overlaps with the endoplasmic reticulum-to-vacuole protein-sorting route in Arabidopsis and contributes to the formation of vacuolar inclusions. Plant Physiol. 145, 1323–1335.

Rhoads, D., Umbach, A.L., Subbaiah C.C. and Siedow J.N. (2006) Mitochondrial reactive oxygen species. Contribution to oxidative stress and interorganellar signaling. Plant Physiol. 141, 357–366.

Rice-Evans, C.A., Miller, N. and Paganga, G. (1996) Structure-antioxidant activity relationships of flavonoids and phenolic acids. Free Radical Biol. Med. 20, 933–956.

Rice-Evans, C.A., Miller, N. and Paganga, G. (1997) Antioxidant properties of phenolic compounds. Trends Plant Sci. 2, 152–159.

Rizhsky, L., Liang, H.J. and Mittler, R. (2003) The water-water cycle is essential for chloroplast protection in the absence of stress. J. Biol. Chem. 278, 38921–38925.

Rodríguez, A.A., Grunberg, K.A. and Taleisnik, E.L. (2002). Reactive oxygen species in the elongation zone of maize leaves are necessary for leaf extension. Plant Physiol. 129, 1627–1632.

Roitsch, T. (1999) Source-sink regulation by sugar and stress. Curr. Opin. Plant Biol. 2, 198–206.

Ryan, K.G. and Hunt, J.E. (2005) The effects of UVB radiation on temperate southern hemisphere forests. Environ. Pollut. 137, 415–427.

Scebba, F., Sebustiani, L. and Vitagliano, C. (1999) Protective enzymes against activated oxygen species in wheat (*Triticum aestivum* L.) seedlings: responses to cold acclimation. J. Plant Physiol. 155, 762–768.

Šesták, Z. and Šiffel, P. (1997) Leaf-age related differences in chlorophyll fluorescence. Photosynthetica 33, 347–369.

Singh, A., Selvi, M. and Sharma, R. (1999) Sunlight-induced anthocyanin pigmentation in maize vegetative tissues. J. Exp. Bot. 50, 1619–1625.

Solfanelli, C., Poggi, A., Loreii, E., Alpi, A. and Perata, P. (2006) Sucrose-specific induction of the anthocyanin biosynthetic pathway in *Arabidopsis*. Plant Physiol. 140, 637–646.

Solovchenko, A. and Merzlyak, M. (2003) Optical properties and contribution of cuticle to UV protection in plants: experiments with apple fruit. Photochem. Photobiol. Sci. 2, 861–866.

Steyn, W.J., Wand, S.J.E., Holcroft, D.M. and Jacobs, G. (2002) Anthocyanins in vegetative tissues: a proposed unified function in photoprotection. New Phytol. 155, 349–361.

Stintzing, F.C. and Carle, F. (2005) Functional properties of anthocyanins and betalains in plants, food and human nutrition. Trends Food Sci. Technol. 15, 19–38.

Streb, P.F., Feierabend, J. and Bligney, R. (1997) Resistance to photoinhibition of photosystem II and catalase and antioxidative protection in high mountain plants. Plant Cell Environ. 20, 1030–1040.

Sun, J., Nishio, J.N. and Vogelmann, T.C. (1998) Green light drives CO_2 fixation deep within leaves. Plant Cell Environ. 39, 1020–1026.

Takahama, U. (2004) Oxidation of vacuolar and apoplastic phenolic substrates by peroxidases: physiological significance of the oxidation reactions. Phytochem. Rev. 3, 207–219.

Takahashi, A., Takeda, K. and Ohnishi, T. (1991) Light-induced anthocyanin reduces the extent of damge to DNA in UV-irradiated *Centaurea cyanus* cells in culture. Plant Cell Physiol. 32, 541–547.

Tanyolaç, D., Ekmekçi, Y. and Ünalan, Ş. (2007) Changes in photochemical and antioxidant enzyme activities in maize (*Zea mays* L.) leaves exposed to excess copper. Chemosphere 67, 89–98.

van Acker, S.A.B.E., van den Berg, D.-J., Tromp, M.N.J.L., Griffioen, D.H., van Bennekom, W.P., van Der Vijgh, W.J.F. and Bast, A. (1996) Structural aspects of antioxidant activity of flavonoids. Free Radical Biol. Med. 20, 331–342.

van den Berg, A. and Perkins, T.D. (2007) Contribution of anthocyanins to the antioxidant capacity of juvenile and senescing sugar maple (*Acer saccharum*) leaves. Funct. Plant Biol. 34, 714–719.

Vranová, E., Inzé, D. and van Breusegem, F. (2002) Signal transduction during oxidative stress. J. Exp. Bot. 53, 1227–1236.

Wagner, D., Przybyla, D., op den Camp, R., Kim, C., Landgraf, F., Pyo Lee, K., Würsch, M., Laloi, C., Nater, M., Hideg, E. and Apel, K. (2004) The genetic basis of singlet oxygen-induced stress responses of *Arabidopsis thaliana*. Science 306, 1183–1185.

Wang, H., Cao, G. and Prior, R.L. (1997) Oxygen radical absorbing capacity of anthocyanins. J. Agric. Food Chem. 45, 304–309.

Wheldale, M. (1916) *The Anthocyanin Pigments of Plants*. Cambridge University Press, Cambridge.

Wise, R.R. (1995) Chilling-enhanced photooxidation: the production, action and study of reactive oxygen species produced during chilling in the light. Photosynth. Res. 45, 79–97.

Woodall, G.S. and Stewart, G.R. (1998) Do anthocyanins play a role in UV protection of the red juvenile leaves of *Syzygium*? J. Exp. Bot. 49, 1447–1450.

Yamasaki, H., Sakihama, Y. and Ikehara, N. (1997) Flavonoid-peroxidase reaction as a detoxification mechanism of plant cells against H_2O_2. Plant Physiol. 115, 1405–1412.

2

Role of Anthocyanins in Plant Defence

Simcha Lev-Yadun[1] and Kevin S. Gould[2]

[1] Department of Biology Education, Faculty of Science and Science Education, University of Haifa, Oranim, Tivon 36006, Israel, levyadun@research.haifa.ac.il
[2] School of Biological Sciences, Victoria University of Wellington, P.O. Box 600, Wellington, New Zealand, kevin.gould@vuw.ac.nz

Abstract. In addition to their well-documented beneficial effects on plant physiological processes, anthocyanins have also been proposed to function in a diverse array of plant/animal interactions. These include the attraction of pollinators and frugivores, as well as the repellence of herbivores and parasites. The optical properties of anthocyanins may serve as visual signals to potential herbivores, indicating a strong metabolic investment in toxic or unpalatable chemicals. Anthocyanins have also been implicated in the camouflage of plant parts against their backgrounds, in the undermining of insect crypsis, and in the mimicry of defensive structures. These hypotheses have in recent years attracted strong theoretical support and increasing experimental evidence. We emphasize that both the defensive and the physiological functions of anthocyanins may operate in plants simultaneously.

2.1 Introduction

In their natural environments, plants run the risk of multiple attacks by many different species of herbivores and pathogens. Phytophagous species feature in all major groups of vertebrates; mammals and birds are undoubtedly the most injurious to plants, but reptiles, amphibians, and fish also inflict damage (Schulze et al. 2002). Invertebrates, too, have the potential to devastate plant communities. In a collection of German woodlands, for example, Reichelt and Wilmanns (1973) identified numerous species of leaf chewers, excavators, leaf miners, bark borers, wood borers, sap suckers, bud and shoot eaters, root eaters, and seedling eaters. Some herbivores feed preferentially at the canopy, others focus on lower branches or seedlings; some parasitize the sugars and nitrogen flowing through leaf veins, others will eat the lamina tissues surrounding the veins; some cause the plants to develop galls in which the animals live and feed, yet others employ chemical signals to redirect the flow of plant nutrients in their direction (Karban and Baldwin 1997). Little wonder, therefore, that plants have evolved elaborate strategies of avoidance and/or a sophisticated armoury of morphological devices to counteract herbivore attacks. Chemical weaponry, too, is known to play a significant role in plant defence. A vast

K. Gould et al. (eds.), *Anthocyanins*, DOI: 10.1007/978-0-387-77335-3_2,
© Springer Science+Business Media, LLC 2009

assortment of secondary metabolites has been demonstrated to act as antifeedants, toxins, warning signals, or precursors to physical defence systems (Bennett and Wallsgrove 1994; Harborne 1997). Among them are the phenolics, a large group of structurally diverse compounds that includes terpenoids, cinnamic acids, catechols, coumarins, tannins, as well as certain flavonoids such as the anthocyanins, the subject of this chapter.

There are several different ways anthocyanins might assist plants in their defence against other organisms. These include both direct roles as chemical repellents and more indirect roles as visual signals. In common with other flavonoids, certain anthocyanins have demonstrable antiviral, antibacterial, and fungicidal activities (Konczak and Zhang 2004; Wrolstad 2004, and references therein). They have the potential, therefore, to protect plants from infections by pathogenic microorganisms. In general, however, the antimicrobial activities of anthocyanins are appreciably less effective than those of other phenolic compounds, such as key flavonols and hydroxycinnamic acids that are also likely to be present in the shoot (Padmavati et al. 1997; Werlein et al. 2005). Moreover, anthocyanins have not been found to be toxic to any higher animal species (Lee et al. 1987). Aphid survival rates, for example, are unaffected by anthocyanins in their diet (Costa-Arbulú et al. 2001). Thus, direct chemical defence is unlikely to be a major function of these pigments in plants. There is, in contrast, strong theoretical support and growing empirical evidence for a role of anthocyanins in the defence from "visually oriented" herbivores. This discussion focuses largely on the defensive roles of anthocyanins as visual cues, though some of these mechanisms also involve associated chemical or mechanical components such as poisons and thorns.

Although it is generally accepted that the colours of flowers and fruits enhance reproductive success by facilitating communication between plants, their pollinators, and seed-dispersers (Ridley 1930; Faegri and van der Pijl 1979; Willson and Whelan 1990; Weiss 1995; Schaefer et al. 2004), there is no *a priori* reason to assume that flower, fruit and leaf colours cannot also serve in defence from herbivory. This is achievable if the colours (i) undermine an herbivorous invertebrate's camouflage, (ii) are aposematic, (iii) mimic an unpalatable plant or animal, or (iv) serve in plant camouflage (see Hinton 1973; Givnish 1990; Cole and Cole 2005; Lev-Yadun 2006). Of course, anthocyanins are not the only class of pigments that might contribute to defence in these ways. In several species, leaf variegation caused by pigments unrelated to anthocyanins has been shown to correlate to reduced herbivory (Cahn and Harper 1976; Smith 1986). Such examples may, however, help to understand the principles that operate when anthocyanins serve in defence. Moreover, it has long been recognised that the non-photosynthetic plant pigments have the potential to serve more than one function concurrently (Gould et al. 2002; Lev-Yadun et al. 2002, 2004; Schaefer and Wilkinson 2004; Lev-Yadun 2006). The UV-absorbing dearomatized isoprenylated phloroglucinols, for example, serve a defensive role in the stamens and ovaries of *Hypericum calycinum*, but an attractive role in the petals of the same species (Gronquist et al. 2001). Thus, the various functional hypotheses concerning pigmentation in leaves and other plant parts need not contrast or exclude any other functional explanation for specific types of plant colouration, and those traits, such as colouration, that might have more than one type of benefit, may be selected for by several agents. Indeed, Armbruster (2002) suggested that plants for which anthocyanins are deployed in the defence of vegetative organs would be the *more likely* also to use anthocyanins to attract pollinators and seed dispersers. Combinations of traits that simultaneously enhance both pollination and defence would likely confer a disproportionate fitness advantage (Herrera et al. 2002). Such synergistic gains may act for a quicker and more common evolution of the red plant organ colour trait.

2.2 Hypotheses

Hypotheses for an anti-herbivory function of anthocyanic plant organs include:

(i) aposematism (conspicuous colouration serving to parry predators) in poisonous fruits and seeds (Cook et al. 1971; Harborne 1982; Williamson 1982), flowers (Hinton 1973), and thorns (Lev-Yadun 2001, 2003a, 2003b, 2006);

(ii) mimicry of dead or senescing foliage (Stone 1979; Juniper 1994), of thorns and spines (Lev-Yadun 2003a), and of ants, aphids, and poisonous caterpillars (Lev-Yadun and Inbar 2002);

(iii) camouflage of seeds against the background of the soil substrate (Saracino et al. 1997, 2004), and of variegated foliage in forest understory herbs (Givnish 1990);

(iv) the undermining of herbivorous insect crypsis by leaf variegation (Lev-Yadun et al. 2004; Lev-Yadun 2006);

(v) attraction of herbivores to young, colourful leaves, diverting them from the more costly older leaves (Lüttge 1997); and

(vi) signalling to insects by red autumn leaves that the trees are well defended (Archetti 2000; Hamilton and Brown 2001; Schaefer and Rolshausen 2006).

Each of these hypotheses is discussed in detail below.

2.3 Reluctance to Accept Hypotheses on Defensive Colouration

Prior to the year 2000, much of the published information on defensive plant colouration, including anthocyanin-based ones, had been largely anecdotal. As Harper (1977) commented in his seminal book about the possibility of defensive colouration operating in plants, botanists have been surprisingly reluctant to accept ideas that are commonplace for zoologists. The relative scarcity of papers on defensive colouration in botany as compared to those in zoology was highlighted in the annotated bibliography on mimicry and aposematism by Komárek (1998). It should be appreciated, however, that that it has taken zoologists more than a century to understand the defensive role and the genetic mechanisms of pigmentation in animals (Hoekstra 2006); the effort needed to achieve the same progress in botany would surely not be smaller. Thus, our explanations for the role of pigments in plant defences remain imperfect. Notwithstanding the difficulties involved in providing concrete evidence for plant defensive colouration, it has therefore been extremely encouraging to note a recent wave of interest in this area, particularly in relation to foliar anthocyanins.

2.4 Colour Vision in Animals

A frequent criticism of the anti-herbivory hypotheses for foliar anthocyanins is that herbivorous insects may lack ocular receptors for red light. Insects have up to five kinds of photoreceptors sensitive to different regions of the visible and UV spectrum (Kelber 2001; Kelber et al. 2003). Butterflies of the genera *Papilio* and *Pieris* have arguably the most sophisticated colour vision system of the insects studied so far, including a red receptor maximally sensitive around 610 nm (Arikawa et al. 1987; Shimohigashi and Tominaga 1991). However, most of the insects that have been examined to date – including the phytophagous aphid *Myzus persicae* – possess only three types of photoreceptors, maximally sensitive to green, blue, and ultraviolet light, respectively (Briscoe and Chittka 2001; Kirchner et al. 2005). In the absence of a red light receptor, it could be argued that insects would be unable to perceive the visual cue presented by anthocyanins.

There is nevertheless good evidence that red is recognised by insects. Döring and Chittka (2007) recently summarised the results from 38 studies in which the behavioural responses to red or green stimuli were compared in aphid species. In 28 of those studies, the aphids had been observed to move preferentially towards the green stimulus, and in only one of the reports had aphids not demonstrated a colour preference. In the remaining studies, the experimenters had varied the shade of the green stimulus, and found that insects moved preferentially towards whichever stimulus was the brighter, red or green. It is likely, therefore, that both chromatic (wavelength related) and achromatic (intensity related) information is involved, as is known also to be used by frugivorous birds (Schaefer et al. 2006). There are insufficient data to state with confidence whether aphids tend to avoid red light or, instead, are simply more attracted to green than to red light. Only one publication has addressed this issue, albeit indirectly: in a "no-choice" experiment, Nottingham et al. (1991) recorded positive phototaxis towards red targets by the bird cherry aphid *Rhopalosiphum padi*. Although response rates were very low, that experiment suggested that the insects were not innately repelled by red objects.

Detail of the mechanism by which insects lacking a red photoreceptor perceive red colours remains to be resolved. A colour opponency mechanism has been proposed, which may explain colour discrimination in certain aphid species (Döring and Chittka 2007). This requires negative excitation in the blue and UV, and positive excitation in the green waveband. It is unclear, however, how such a mechanism might facilitate perception of anthocyanic leaves which, compared to green leaves, typically reflect smaller quantities of both green light (Neill and Gould 1999) and UV radiation (Lee and Lowry 1980). It may simply be that red leaves are less attractive to insect herbivores because the excitation of their green receptor is lower than when excited by green leaves (Thomas Döring, personal communication). Contrasts in colour and/or brightness between red leaves and the visual background are likely also to be important (Dafni et al. 1997).

2.5 Anthocyanins and Other Red Pigments

Anthocyanins usually appear red in leaf cells, but depending on their chemical nature and concentration, the vacuolar pH, and interactions with other pigments, they can result in pink, purple, blue, orange, brown, and even black leaf colours (Schwinn and Davies 2004; Andersen and Jordheim 2006; Hatier and Gould 2007). Many of the published articles on plant defensive colouration have assumed red foliage to be the outcome of the production of anthocyanins, this despite the fact that other pigments – carotenoids, apocarotenoids, betalains, condensed tannins, quinones and phytomelanins – can also contribute to plant vermilion (Davies 2004). There is, moreover, a dearth of systematic information on the full complement of pigments in all plant organs at all developmental stages. This lack of data precludes detailed taxon-wide comparisons of the involvement of anthocyanin, or indeed any pigment, in plant defence. Clearly, if only visible cues (hue, lightness, and colour saturation) are involved in defence, the chemical nature of a pigment would be unimportant to a herbivore; red warnings would be similarly effective irrespective of whether they were generated by anthocyanins, carotenoids, or betalains. If, on the other hand, the efficacy of the warning relied on a combination of attributes, for example the reflection of red light *plus* the presence of a toxic or olfactory phenolic derived from an offshoot in the anthocyanin biosynthetic pathway, then the pigment type could be critical.

2.6 Olfactory Signals

An important, if not critical issue is whether or not olfactory signals are involved along with the visual ones. From studies of deceptive pollination, wherein insects are lured to flowers but receive no sugar reward, we know that signalling to animals can involve a combination of both visual and olfactory components (Dafni 1984; Ayasse et al. 2000; Schiestl et al. 2000); there are good reasons to think that the same may be true in the defence of vegetative organs (Pichersky and Gershenzon 2002). The identification of olfactory volatiles is achievable using modern laboratory equipment (e.g., Jürgens 2004; Jürgens et al. 2002, 2003; Pichersky and Dudareva 2007), but such procedures are difficult to accomplish in the field. It is, in addition, very difficult to identify the specific molecules that deter specific herbivores from among the many volatile molecules that plants omit, and there is the possibility that deterrence operates only if several molecules are sensed simultaneously. The fact that not all animals respond similarly to any chemical signal or cue, should also be considered.

2.7 Aposematic Colouration

Aposematic colouration, a well-known phenomenon in animals, has until recently been given little attention in plants. Often, a brightly-coloured animal (red, orange, yellow, white with black markings, or combinations of these colours) is dangerous or unpalatable to predators – a trait that confers a selective advantage because predators learn to associate the colouration with unpleasant qualities (Cott 1940; Edmunds 1974; Gittleman and Harvey 1980; Harvey and Paxton 1981; Wiklund and Järvi 1982; Ruxton et al. 2004). Although several authors had noted a similar association between conspicuous colouration and toxicity in plants (Cook et al. 1971; Hinton 1973; Harper 1977; Wiens 1978; Rothschild 1980; Harborne 1982; Williamson 1982; Knight and Siegfried 1983; Smith 1986; Lee et al. 1987; Coley and Aide 1989; Givnish 1990; Tuomi and Augner 1993), only in the past decade have the scope and significance of this phenomenon been appreciated. Indeed, the possibility of aposematic colouration was discounted in some of these earlier studies (Knight and Siegfried 1983; Smith 1986; Lee et al. 1987; Coley and Aide 1989). A related phenomenon, olfactory aposematism in poisonous plants, has also been proposed (e.g., Eisner and Grant 1981; Harborne 1982; Launchbaugh and Provenza 1993; Provenza et al. 2000) although this has received scant attention.

2.7.1 Poisonous Plants

The first detailed hypothesis for a possible defence from herbivory attributable to red colouration (and other colours) was published by Hinton (1973), who proposed that colourful poisonous flowers should be considered aposematic, and that they probably have mimics. His review about deception in nature was published in a book about illusion; this was not a biological book, but rather dealt with art. His hypothesis was briefly referred to by Rothschild (1980) in her discussion on the roles of carotenoids, but otherwise did not stir botanists or ecologists to pursue this issue. Indeed, Harper (1977), who had written the comment about botanists being reluctant to accept things that were commonplace for zoologists, omitted to explain why zoologists who dealt with animal aposematism, and who were also involved in research on plant-animal interactions, had not recognized how common are these phenomena in plants. Harborne (1982) proposed that the brightly coloured, purple-black berries of the deadly *Atropa belladonna* warn grazing mammals of the danger to consume them. Williamson (1982) also proposed that brightly coloured (red, or red and black) seeds lacking an arillate or fleshy reward (e.g. *Erythrina, Ormosia,* and *Abrus*) might be aposematically coloured to warn seed eaters of their toxicity. These hypotheses were written only as short paragraphs within long reviews, however, and there has been no further effort to study the function of their colouration.

2.7.2 Thorny Plants

In English there are three terms for pointed plant organs: spines (modified leaves), thorns (modified branches), and prickles (comprising cortical tissues, e.g. in roses). For the purposes of this discussion, we refer to plants as "thorny" if they produce any of the three types of sharp appendages. Thorny appendages provide mechanical protection against herbivory (Janzen and Martin 1982; Janzen 1986; Myers and Bazely 1991; Grubb 1992; Rebollo et al. 2002) because they can wound mouths, digestive systems (Janzen and Martin 1982; Janzen 1986), and other body parts of herbivores. They might also inject pathogenic bacteria into herbivores (Halpern et al. 2007). Thus, once herbivores learn to identify thorns – and their bright colours and associated markings should help in their recognition – they can avoid harmful plants displaying them.

The flora of countries such as Israel, which has a millennia-long history of large-scale grazing, clearly and "sharply" indicates the ecological benefit of being thorny when grazing pressure is high. A continuous blanket of spiny shrubs such as *Sarcopoterium spinosum*, as well as many types of thistles, covers large tracts of the land. The thorns effectively impede the rate at which an herbivore feeds within the canopy of the individual plant, and this presents an overall considerable advantage to such plants over non-defended ones. Spiny plants, such as *Echinops* sp. (Asteraceae), which normally grow as individuals or in small groups, sometimes become the most common perennial plant over many acres in heavily grazed lands. The same is true for many other taxa.

Thousands of thorny species have colourful or otherwise conspicuous markings (e.g. Fig. 2.1A), many of which can be considered to be aposematic (Lev-Yadun 2001). We will not discuss this common phenomenon in the thorniest taxon – the Cactaceae – since they lack anthocyanins, and use betalains instead (Stafford 1994). Lev-Yadun (2001) categorised two types of thorn ornamentation, which are typical of many thorny plant species: (i) colourful thorns, and (ii) white spots, or white and colourful stripes, in leaves and stems associated with the thorns. Both types have been recorded for approximately 2,000 species originating from several continents in both the Old and New World (Lev-Yadun 2001, 2003a, 2003b, 2006; Rubino and McCarthy 2004; Halpern et al. 2007; Lev-Yadun unpublished). It has been proposed

Fig. 2.1 Thorns and their mimics. (A) Anthocyanic thorns on a rose stem. (B) Red mucron at the apex of a leaf of *Limonium angustifolium*. (C) Red fruit of *Erodium laciniatum*. See Plate 1 for colour version of these photographs

that the pigmentation in thorns and associated organs (in many cases resulting from anthocyanins) are cases of vegetal aposematic colouration, analogous to such colouration of poisonous animals, that serves to communicate between plants and herbivores (Lev-Yadun 2001, 2003a, 2003b, 2006; Lev-Yadun and Ne`eman 2004; Rubino and McCarthy 2004; Ruxton et al. 2004; Speed and Ruxton 2005; Halpern et al. 2007). Interestingly, spiny animals also show the same phenomenon (Inbar and Lev-Yadun 2005; Speed and Ruxton 2005).

Colourful thorns are especially common in the genera *Agave*, *Aloe* and *Euphorbia*. *Agave* species can have two types of thorns in their leaves: spines at the leaf apex, and/or teeth along the leaf margins (Lev-Yadun 2001). In addition, many *Agave* species also have colourful stripes running along the margins that enhance spine and tooth visibility. The spines and the teeth along the margins of the leaves are brown, reddish, gray, black, white, or yellow; these colours are known to serve in aposematism. The same phenomenon is true for *Aloe* species for which the colourful thorns can be white, red, black, or yellow, and for thorny African *Euphorbia* species, many of which have colourful thorns or colourful markings associated with thorns along the ribs of the stems (Lev-Yadun 2001). Anthocyanins are known to contribute to the markings in *Agave* and *Euphorbia*, though the carotenoid rhodoxanthin is involved in leaf colouration in *Aloe*. Colourful thorns advertise their defensive quality directly, unlike poisonous aposematic organisms in which the poison is advertised indirectly (Ruxton et al. 2004; Speed and Ruxton 2005).

Since the colouration and markings of thorns in plants are so widespread, they are probably neither a neutral nor a random phenomenon. Lev-Yadun (2001) proposed that, like aposematic colouration in animals, the conspicuousness of thorns would be of adaptive value since herbivorous animals would remember the signal, and subsequently tend to avoid tasting such plants. Annual and perennial plants usually survive damage caused by herbivores (Williamson 1982; Crawley 1983; Ohgushi 2005), so an herbivore reacting to aposematic colouration would be of direct benefit to the individual plant, which would probably suffer fewer repeated attacks. Hence, as with animals (Sillén-Tullberg and Bryant 1983), there is no need to propose kin/group selection, or altruism, as the evolutionary drive for the spread of this character. Indeed, there are several probable reasons for the quick and easy route to aposematism in plants in general. Thorny or poisonous plants are already well-defended, even without aposematic colouration. Plants recover much better than most animals from herbivore (predator) damage (Crawley 1983; Ohgushi 2005). Thus, the original thorny, aposematic mutant would have had a good chance of survival and producing offspring, even despite the risks associated with being conspicuous to herbivores and being partly eaten. Furthermore, an herbivore might pass over an aposematic individual and eat its non-aposematic neighbour, thereby reducing the competition between the aposematic and neighbouring plants. A recent hypothesis proposed that since thorns harbour many types of pathogenic microbes, their aposematism serves to signal both about their biological risks as well as the mechanical ones (Halpern et al. 2007). It is possible, therefore, that the evolution and spread of aposematism progressed even more swiftly in plants than in animals.

2.8 Defensive Mimicry

Defensive mimicry is said to occur if a plant gains protection against herbivory by resembling a noxious or unpalatable model species. Williamson (1982) proposed that because plants are sessile organisms, mimicry is less likely to be successful in plants than in animals; plant mimics are less likely to be mistaken for their models than are the mobile animal mimics. Moreover, because plants that have been partially eaten by herbivores can often regenerate new organs, defensive mimicry would in theory provide less advantage to plants than animals. In contrast, Augner and Bernays (1998) studied the theoretical conditions for an evolutionary stable equilibrium of defended, signalling plants and of plants mimicking these signals. Their model showed that mimicry of plant defence signals could well be a common occurrence; even imperfect mimics had the potential to invade a population of defended model plants. Theoretically, natural selection would allow the success of even imperfect mimics (Edmunds 2000). Wiens (1978) estimated that about 5% of the land plants are mimetic, listing various types of protective plant mimicry. Several types of anthocyanin-related defensive mimicry have been proposed, especially in recent years, and these are discussed below.

2.8.1 Mimicry of Dead Leaves

An interesting hypothesis in relation to anthocyanins was published in a brief note by Stone (1979), who proposed that young leaves of understory palms mimic old or dead leaves. He observed that the combination of chlorophylls and anthocyanins in the developing leaves of several species produced a dull brown colour which "strikingly mimics the drab color of dying or withered dead leaves". The leaves, moreover, showed no evidence of browsing damage, possibly because they appeared unpalatable to potential herbivores.

Juniper (1994) suggested a similar explanation for the abundant red, brown and even blue flushes of young leaves in tropical trees and shrubs from the families Annonaceae, Fabaceae subfam. Caesalpinioideae, Guttiferae, Lauraceae, Meliaceae, Rutaceae, and Sapindaceae. It was postulated that to a phytophagous insect, such leaves would appear dull, like advanced senescent foliage. "If you do not look like a normal leaf … if you do not feel like a leaf to a palpating insect, because you are soggy; if you do not have the posture of a proper leaf, i.e. you are hanging down like wet facial tissue; if you do not smell like a leaf because you have no photosynthesis and, if on the first suck or bite, you do not taste like a leaf because you have no sugars … you might escape being eaten" (Juniper 1994).

Young foliage that acquires a protective advantage by resembling older, non-green leaves would be an example of cryptic mimicry (Pasteur 1982). The idea seems to have been generally accepted (Juniper 1994; Dominy et al. 2002; Gould 2004; Karageorgou and Manetas 2006; Lev-Yadun 2006; Manetas 2006) although there is little in the way of supporting experimental data. As is the case for many of these anthocyanin defence hypotheses, the chemical nature of the pigments in many such red leaves has not been tested, the anatomical location of the pigmentation is not always known, and possible physiological roles of the colouration are unclear.

2.8.2 Defensive Thorn Automimicry

Since aposematic thorns are so common, it is not surprising that mimics of this type of defence exist. Lev-Yadun (2003b) described two variations of thorn mimicry: (i) thorn-like imprints on lamina margins (a type of weapon automimicry, until recently believed to be exclusive to animals), and (ii) mimicry of aposematic colourful thorns by colourful, elongated, and pointed plant organs (buds, leaves and fruit) which, despite their appearances, are not sharp (Fig. 2.1B).

Weapon automimicry (in which the mimic has part of its structure resembling some other part) of horns and canine teeth has been observed in several mammalian species (Guthrie and Petocz 1970) and has been proposed to be of greatest value in intraspecific defence. This intriguing idea has now been suggested to occur in plants. More than 40 *Agave* species show what Lev-Yadun (2003b) described as "thorn automimicry". In those *Agave*, the developing leaves press strongly against one another as they grow. Teeth along the leaf margins press against the lamina surfaces of other leaves, leaving an impression of the teeth along the non-spiny parts. This phenomenon is particularly conspicuous in *A. impressa*, for which the teeth impressions are white. However, because many of the real thorns in other *Agave* species are red, brown and black, probably as the outcome of expression of anthocyanins, some of the imprints contain reddish colouration. Similar impressions of colourful teeth are obvious in fronds of the palm *Washingtonia filifera*, a common ornamental and a feral tree in Israel, as well as in certain *Aloe* species. It remains to be demonstrated that such leaf patterns effectively contribute to herbivore deterrence. If they do, then the imprints, which give the visual impression of a more extensive system of thorns than actually exists, should be considered a transitional type between Batesian and Müllerian mimicry; the mimicking of real thorns by non-thorny plants would be Batesian, and the illusion of more thorns than occurs in reality would be a Müllerian-Batesian intermediate (Lev-Yadun 2003b).

Colourful, thorn-like appendages have been observed in several species that grow wild in Israel (Lev-Yadun 2003b). They are especially prominent in the fruit of various *Erodium* species, an annual in the family Geraniaceae. *E. laciniatum* subsp. *laciniatum*, *E. crassifolium*, and *E. arborescens* all have elongated, beak-like fruits with pointed apices (Fig. 2.1C). These self-dispersing structures are usually pigmented red from anthocyanins, and although they look like thorns, their tissues are soft. In *Limonium angustifolium*, a wild and domesticated perennial of the Plumbaginaceae, the apical portions of its large leaves are red; again, this resembles a spine though its tissues are soft (Lev-Yadun 2003b). Such non-thorny plants that mimic thorny ones are cases of Batesian mimicry. There are two possible evolutionary routes toward mimicry of colourful thorns. In the first, an aposematic thorny plant may have lost its thorny characteristics but retained the shape and aposematic signal. In the second, a non-aposematic and non-thorny plant may acquire the signal, becoming a primary mimic. Alternatively, the structure and colouration may have a different, unknown function. The mode of evolution of these mimics has not been studied.

2.8.3 Defensive Animal Mimicry by Plants

Some plants utilise anthocyanins (and other pigments) to mimic insects, the presence of which would likely discourage visits by potential herbivores. It is widely accepted that many animals visually masquerade as parts of plants, thereby gaining protection from predators, or crypsis from their prey (Cott 1940; Wickler 1968; Edmunds 1974; Ruxton et al. 2004). Classic examples listed in Cott (1940) include: fish and crabs that resemble algae; geckos and moths that look like lichens; many insects, amphibians and reptiles that resemble leaves; spiders, caterpillars, moths, beetles, amphibians, lizards and birds that mimic tree bark; stick-insects that resemble branches; and Thomisid spiders that are disguised as flowers. Flowers, too, are known to mimic animals to attract pollinators (Dafni 1984). However, the possibility of animal mimicry as a defence mechanism of plants against herbivores has been largely overlooked. Evidence for the involvement of anthocyanins in plant mimicry of ants, aphids and caterpillars is discussed below.

2.8.4 Ant Mimicry

Ants are well known to defend plants from invertebrate herbivory. Indeed, in certain cases, the relationship between ants and their plant hosts has been recognized as mutualistic (Madden and Young 1992; Jolivet 1998). The potential benefit from ant-attendance, and therefore from mimicry of ant presence, is obvious. Ants bite, sting and are aggressive, and most insectivorous animals and herbivores will avoid them. Thus, ants have become models for a variety of arthropods, which have evolved to mimic them (Wickler 1968; Edmunds 1974). The importance of ants in defending plants was demonstrated in a field experiment in which the removal of ants and aphids resulted in a 76% increase in the abundance of other herbivores on narrow-leaf cottonwoods (Wimp and Whitham 2001). Not surprisingly, therefore, many plant species invest resources in attracting ants, providing them with shelter, food bodies and extrafloral nectaries (Huxley and Cutler 1991). Certain plants even tolerate aphid infestation to gain anti-herbivore protection from aphid-attending ants (Bristow 1991; Dixon 1998).

Ant mimicry has been observed in several plant families. The mimicry takes the form of conspicuous, darkly-coloured spots and flecks, usually 2–10 mm long, on the plant's epidermis. These markings resemble ants in size, shape and in the direction of their spatial patterns, which look like columns of ants. Dots predominate in some individual plants, flecks in others (Lev-Yadun and Inbar 2002). This phenomenon has so far been found on the stems and petioles of *Xanthium trumarium* (Asteraceae) and *Arisarum vulgare* (Araceae) growing in Israel, and in several other plant species growing in Eastern USA and Northern Greece (Lev-Yadun unpublished). Real ant swarms, as seen from a distance, comprise many moving dark flecks, each varying in size from several mm to over 1 cm. The swaying of leaves, stems and branches in the wind, in combination with the dark spots and flecks, many of which are arranged in lines, give the illusion that the ant mimics are moving. Olfactory components may also be involved, though this has not been tested. The ant-mimicking colouration has not been analysed for pigment composition, but it is likely that anthocyanins are involved since they are abundant in the flowers and/or foliage of these species. In any case, it seems evident that ant mimicry would be highly beneficial to plants, since they could acquire herbivore deterrence without paying the cost of feeding or housing real ants (Lev-Yadun and Inbar 2002).

2.8.5 Aphid Mimicry

A phenomenon similar to ant mimicry is aphid mimicry. Lev-Yadun and Inbar (2002) described aphid mimicry in *Paspalum paspaloides* (= *P. distichum*), for which the dark-pigmented anthers are the size, shape and colour of aphids. The stems of *Alcea setosa* are also covered with dark flecks that look like aphids, and similar morphological features have been found in several wild grasses growing in North Carolina (Lev-Yadun, unpublished). Plants which look infested would likely be left untouched both by grazers and insects (Lev-Yadun and Inbar 2002). Several studies have shown that early infestation by aphids and other homopterans has a negative impact on host plant preferences and larval performance of other insect herbivores. Finch and Jones (1989) reported that large colonies of the cabbage aphid *Brevicoryne brassicae* and of the peach aphid *Myzus persicae* deterred ovipositioning by the root fly *Delia radicum*. Inbar et al. (1999) demonstrated that the presence of Homopterans (whiteflies) not only altered adult cabbage looper (*Trichoplusia ni*) host selection, but also actually reduced the feeding efficiency of their offspring. Aphids respond to crowding by enhanced dispersal (Dixon 1998), and it is probable, therefore, that they avoid previously infested or infestation-mimicked hosts. The clear zoological data are consistent with a potential defensive value in aphid mimicry by plants, but the hypothesis requires testing experimentally.

2.8.6 Mimicry of Aposematic Poisonous Caterpillars

The final example of red colouration from anthocyanins mimicking insects for defence is that of immature legume pods, which often resemble aposematic poisonous caterpillars. Lev-Yadun and Inbar (2002) described pods of several wild annual legumes (*Lathyrus ochrus, Pisum elatius, Pisum humile*, and *Vicia peregrina*) that were comparable in general shape, size and colour to those of Lepidopteran caterpillars. The pods were ornamented with apparent "spiracles" or other conspicuous spots in various shades of red and purple. In one of the species (*V. peregrina*), two different phenotypes were observed. The first had red spots along the length of the pods, similar to those of *L. ochrus, P. elatius*, and *P. humile*. The second phenotype was characterized by red circles with green centres along the pods. Lev-Yadun and Inbar (2002) proposed that these morphological traits may serve as herbivore-repellence cues, and are part of the defence system of the plants. Caterpillars employ a large array of defences that reduce predation. Unpalatable caterpillars armed with stinging and irritating hairs, functional osmeteria (scent glands) or body-fluid toxins often advertise their presence by aposematic colouration and aggregation (Cott 1940; Bowers 1993; Eisner et al. 2005). The usual warning colours are yellow, orange, red, black, and white, often in stripes along the body and/or spots, especially around the abdominal spiracles. By mimicking aposematic caterpillars with red "spiracle spots", wild legumes may reduce immature pod predation (Lev-Yadun and Inbar 2002). It has been shown that ungulates actively select leaves in the field by shape and colour, and avoid eating spotted ones (e.g., Cahn and Harper 1976) but there seems to be no published data on the response of vertebrate herbivores to aposematic (or cryptic) caterpillars. Again, the possible involvement of olfactory deterrence has not been studied.

The examples of ant, aphid and caterpillar mimicry may signal unpalatability to more than one group of animals in two ways (Lev-Yadun and Inbar 2002). First, insect mimicry would reduce attacks by insect herbivores, which refrain from colonizing or feeding on infested plants because of perceived competition and/or induced plant defences. Second, where the insect mimicked is aposematic, this could deter larger herbivores from eating the plants. None of these hypotheses concerning the various types of defensive insect mimicry has been tested directly, though there is good indirect evidence from studies on insects, and the hypotheses appear reasonable.

2.9 Camouflage

2.9.1 Whole Plants and Seeds

There is a remarkable dearth of scientific information on the use and effectiveness of plant pigments to camouflage plant organs from potential herbivores. Given that plants are largely sessile and often produce seeds that are dispersed on bare ground or on plant litter, an ability to blend in with their general surroundings (eucrypsis) would present them with an obvious advantage. To our knowledge, however, not a single monograph has been published on this topic, though several workers have alluded to the possibility of camouflage. In the New Zealand tree *Pseudopanax crassifolius*, for example, the combination of chlorophylls and anthocyanins produces a brown colour in seedling leaf laminae (Gould 1993), effectively concealing them (at least to human eyes) amongst the background leaf litter. *Lithops*, too, appears to be well camouflaged amidst the rocks and gravel in the desert (Cole 1970; Cole and Cole 2005), although betalains, rather than anthocyanins, are involved in that taxon.

Camouflage has been postulated to act as defence in seeds of various *Pinus* species (Saracino et al. 1997, 2004; Lanner 1998); in a field experiment involving various combinations of soil types and seed colours, light grey seeds, which were rich in polyphenols, were observed to be predated less if the seeds were on a similar coloured substrate, suggesting that eucrypsis functioned as a protective strategy against predation by granivorous birds. The predation of black seeds, in contrast, did not vary consistently with substrate colour.

From years of field work it has become evident to one of the authors (Lev-Yadun) that anthocyanins play a significant role in plant camouflage, yet there has been no systematic study on this phenomenon. Since seed survival in particular is critical for plants, and camouflage of seeds is so obvious, it is therefore difficult to understand why such studies, which are commonplace in zoology have not featured similarly in botany.

2.9.2 Variegation in Understory Herbs

The leaves of many understory herbs in certain temperate and tropical floras are variegated. Their colours are diverse; many of these plants are variegated green and white, others are mottled with red, purple or even black, often the result of anthocyanin accumulation. Givnish (1990) proposed that variegation in evergreen understory herbs in the forests of New England serves as camouflage. He argued that in that dappled light environment, variegation would disrupt the leaf outlines as perceived by colour-blind vertebrate herbivores such as deer. Unfortunately, this appealing hypothesis lacks empirical data. As stated by Allen and Knill (1991), "Givnish's hypothesis, elegant in its simplicity, demands to be tested experimentally."

2.10 Undermining Crypsis of Invertebrate Herbivores

Another recent hypothesis holds that red and yellow pigments in both vegetative and reproductive shoots can undermine the camouflage of invertebrate herbivores (Lev-Yadun et al. 2004; Lev-Yadun 2006). Thus exposed, the invertebrates would be vulnerable to predation. Moreover, potential herbivores would likely avoid settling on plant organs with unsuitable colouration in the first place, thereby compounding the benefit to the plants.

Plants provide the habitat and food for many animals. Intuitively, the common optimal camouflage for herbivorous insects would be green, since many (e.g., aphids, caterpillars, grasshoppers) have evolved green colouration (Cott 1940; Purser 2003). It has been claimed that the considerable variation in the colours of leaves and stems, as well as those of flowers and fruits, could serve to undermine the camouflage of invertebrate herbivores, especially insects (Lev-Yadun et al. 2004; Lev-Yadun 2006). For example, colour differences between the upper and lower surfaces of leaves, or between the petiole, veins and leaf lamina, are common across diverse plant taxonomic groups. They occur across a range of plant forms, from short annuals to tall trees, and in habitats ranging from deserts to rain forests, and from the tropics to temperate regions. When a given leaf has two different colours, such as green on its upper (adaxial) surface and blue, brown, pink, red, white, yellow or simply a different shade of green on its lower (abaxial) surface, a green insect (or otherwise coloured one) that is camouflaged on one of the leaf surfaces will not be camouflaged on the other. The same is true for vein, petiole, branch, stem, flower, or fruit colouration. Moreover, when a green herbivore moves from one green region to another, passing through a non-green region, it would immediately become more conspicuous to its predators (Lev-Yadun et al. 2004). The foliage of many plants is simply too colourful to allow the universal camouflage of folivorous insects, and it compels small animals to cross "killing zones" of colours that do not match their camouflage (Fig. 2.2). This is a special case of "the enemy of my enemy is my friend", and a visual parallel of the chemical signals that are emitted by plants to signal wasps when attacked by caterpillars (Kessler and Baldwin 2001). It is also a natural parallel to the well-known phenomenon of industrial melanism (e.g., Kettlewell 1973; Majerus 1998), which illustrated the importance of plant-based camouflage for herbivorous insect survival.

Fig. 2.2 Variegated red autumn leaf of *Acer* may undermine the crypsis of herbivorous insects. See Plate 1 for a colour version of this photograph

Interestingly, the antithesis of this, that background chromatic heterogeneity *promotes* herbivore crypsis, has also been argued by some workers (Merilaita 2003; Merilaita et al. 1999; Schaefer and Rolshausen 2006). Using a mathematical model to simulate the evolution of cryptic colouration against different backgrounds, Merilaita (2003) concluded that the risk of prey being detected was appreciably lower in the more visually complex (more colourful) habitats. Similarly, Schaefer and Rolshausen (2006) showed that a variegated leaf comprising red primary veins and pink or white secondary veins on a green lamina could effectively camouflage a diverse assortment of computer-generated insects, many more than could be accommodated by a green-only leaf.

This dichotomy of opinion is, perhaps, resolvable if the ratio between the size of the herbivore and that of the colour patch on the plant be taken into consideration. Green insects that are smaller than, say, the average red patch on a leaf would run the greater risk of being exposed to their predators. Conversely, those types of variegation that consist of small-scale mosaics, wherein each patch is smaller than an insect, are unlikely to facilitate the undermining of the animal's camouflage.

There is a rich literature on the effects of background matching on crypsis in animals (e.g., Cott 1940; Endler 1984; Ruxton et al. 2004). However, the role of anthocyanins in undermining crypsis on plant surfaces awaits experimental confirmation.

2.11 Red Young Leaves Divert Herbivores from More Costly Old Ones

The young, expanding leaves of many woody plant species in the tropics are red (Stone 1979; Lee et al. 1987; Juniper 1994; Richards 1996; Dominy et al. 2002). Lüttge (1997) proposed that the colourful young leaves may attract herbivores such as primates, diverting them from the metabolically more expensive and photosynthetically active older leaves. The hypothesis, which was not elaborated even by Lüttge (1997), remains to be tested in the field. However, young leaves are usually chemically and physiologically less defended than old ones (Coley and Barone 1996). This might cause difficulties in interpreting data from field observations.

2.12 Signalling by Red Autumn Leaves

2.12.1 General

The spectacular phenomenon of red autumn foliage of deciduous shrubs and trees, especially in eastern USA and Canada (Matile 2000; Hoch et al. 2001) has received broad scientific attention in recent years, especially in relation to the potential for defensive colouration by anthocyanins. For decades, many scientists believed that these colours simply appeared after the degradation of chlorophyll that masked these pigments, and that they served no function. Recently, however, it has been shown that for many plants, anthocyanins are not simply unmasked, but rather are synthesized from precursors by leaves in mid senescence (Hoch et al. 2001; Lee 2002; Lee et al. 2003; Ougham et al. 2005). Thus, the question of the possible physiological and ecological benefits of this colouration has attracted considerable scientific attention. There is very good evidence for physiological benefits of autumn leaf colouration, such as an enhanced recovery of foliar nitrogen owing to the protection of degrading chloroplasts by anthocyanins from the effects of photooxidation (Hoch et al. 2001; Yamasaki 1997; Chalker-Scott 1999; Matile 2000; Hoch et al. 2003; Feild et al. 2001; Lee and Gould 2002a, 2002b; Close and Beadle 2003; Schaefer and Wilkinson 2004; Gould 2004). These physiological advantages notwithstanding, certain hypotheses for non-physiological functions of autumn leaf colouration also merit consideration.

An early hypothesis held that autumn leaves function as a fruit flag (Stiles 1982). The reddening of leaves adjacent to ripe fruits was postulated to attract frugivorous birds, thereby enhancing the chances of seed dispersal. The hypothesis was tested experimentally by Facelli (1993), who tied plastic, leaf-shaped "flags" of different colours and sizes to the infructescences of *Rhus glabra*; birds removed the most fruit from the shrub when the flags were large and red. Other experiments, however, failed to confirm the hypothesis (e.g. Willson and Hoppes 1986), and this may have stimulated the search for alternative ecological explanations. Since the year 2000, several variations on possible signalling roles for anthocyanins in defence have been proposed in relation to autumn foliage.

2.12.2 Signalling of Defensive Potential

It has been proposed that the red and yellow colours of autumn leaves signal to potential herbivores that the plants are chemically well defended (Archetti 2000, 2007a, 2007b; Hamilton and Brown 2001; Hagen et al. 2003, 2004; Archetti and Brown 2004; Archetti and Leather 2005; Brown 2005). The hypothesis is a classic example of Zahavi's handicap principle (Zahavi 1975), which states that because signalling is metabolically expensive, it is therefore likely to be reliable. Organisms that operate under the handicap principle send *honest* signals for the receiver to evaluate (Zahavi 1975, 1977, 1987; Grafen 1990; Zahavi and Zahavi 1997). Accordingly, Hamilton and Brown (2001) argued that trees which face the more intense pressure from herbivory would likely present the more ostentatious displays of colourful foliage; those displays would correlate to, and be perceived by insects as, a greater metabolic investment by the plant into active defence compounds. Archetti (2000) theorised that the production of red foliage as a warning signal would benefit both parties, the tree and the parasitic insects; the tree gains from a reduction in browsing, and the insects gain information on where not to lay their eggs. Thus, far from being "a kind of extravagancy without a vital function" (Matile 2000), autumn foliage was considered to be an adaptive phenomenon resulting from the coevolution between insects and trees.

There is evidence, experimental as well as theoretical, both for and against this signalling hypothesis (Archetti 2000; Hamilton and Brown 2001; Holopainen and Peltonen 2002; Wilkinson et al. 2002; Hagen et al. 2003, 2004; Archetti and Brown 2004; Schaefer and Wilkinson 2004; Archetti and Leather 2005; Brown 2005; Ougham et al. 2005; Sinkkonen 2006a, 2006b; Schaefer and Rolshausen 2006, 2007; Archetti 2007a, 2007b; Rolshausen and Schaefer 2007; Lev-Yadun and Gould 2007). Among those papers consistent with the hypothesis, Hamilton and Brown (2001) noted that tree species for which autumn foliage was especially rich in red or yellow pigmentation were among those that potentially risked colonisation by the greatest diversity of specialist aphid species. Hagen et al. (2003) found that in mountain birch, the first trees to develop autumn colouration were the least likely to show insect damage in the following season. Similarly, Archetti and Leather (2005) found a negative correlation between aphid counts and the proportions of red or yellow leaves in a population of bird cherry (*Prunus padus*).

Arguing against the signalling hypothesis are compelling data recently published by Schaefer and Rolshausen (2007). These authors experimentally manipulated the colour of leaves in mountain ash, then monitored visits to the leaves by winged aphids. Contrary to the predictions of Archetti's (2000) coevolutionary hypothesis, aphid counts were similar for both green and red leaves. Noting a positive correlation between aphid numbers and fruit production, however, Schaefer and Rolshausen (2007) concluded that aphids probably do select their hosts non-randomly, but they do not use leaf colour as a cue.

Other workers have argued that rather than serving as a signal to potential herbivores, the synthesis of anthocyanins in autumn shoots might better be explained in terms of the possible benefits to senescent leaf physiology (e.g. Wilkinson et al. 2002; Schaefer and Wilkinson 2004; Ougham et al. 2005). This leads to an important point: as far as we are aware, none of the proponents of the leaf signalling hypothesis have adequately addressed the question of the timing of autumn leaf colouration. Why would a plant preferentially protect from herbivores a dying leaf soon to detach from the branch, yet leave its younger, productive, acyanic leaves vulnerable to herbivory? There are possible explanations for this, of course, though none has been tested experimentally. For example, in autumn the phloem sap is particularly rich with the nutrients remobilized from senescing leaves, and signalling by anthocyanins could well benefit the tree by deterring phloem sap feeders, such as aphids. (Incidentally, Holopainen and Peltonen (2002) argued the opposite of this –

that autumn foliage *attracts* aphids because they associate the bright colours with the availability of a nitrogen feast! The authors' reasoning was based largely on the well-known attractive effects of yellow colours found in carotenoids, however, rather than of the red anthocyanins). Other possibilities are that the timing of autumn red leaves coincides with visits of a particularly aggressive herbivore, or else a key stage in the animal's lifecycle; again, these require empirical testing. Plant physiologists, on the other hand, have demonstrated a function of anthocyanins that is unique to the senescing foliage: by reducing levels of photooxidative stress on the degenerating chloroplasts, anthocyanins improve rates of nitrogen recovery from autumn leaves (Hoch et al. 2001, 2003). In terms of explaining the phenology of anthocyanin production, therefore, evidence for physiological benefits of autumn leaf colouration currently appears to outweigh that for defensive signalling.

2.12.3 The "Defence Indication Hypothesis"

Schaefer and Rolshausen (2006) proposed a variation on the signalling hypothesis which goes some way towards bridging the gap between the putative physiological and defensive functions of anthocyanins in autumn leaves. Elaborating on previous ideas by Willson and Whelan (1990) on the evolution of fruits colours, and by Fineblum and Rausher (1997) concerning the common biochemical pathways for fruit and flower colour and defensive molecules, Schaefer and Rolshausen (2006) formulated the "defence indication hypothesis". The hypothesis holds that foliar anthocyanins evolved in response to abiotic stressors such as drought, cold, and strong light. However, because anthocyanins share the same phenylpropanoid biosynthetic pathway with many defensive phenolic compounds (such as condensed tannins), the upregulation of anthocyanin biosynthesis inevitably leads to elevated defensive strengths. Fewer herbivorous insects would feed on plants that have strong anthocyanic pigmentation because it correlates with the strength of a chemical defence. Thus, anthocyanins in leaves at once serve to mitigate the effects of environmental stress and to indicate to potential herbivores their level of investment in defensive compounds. The hypothesis is an attractive alternative to Archetti's (2000) coevolutionary model, and warrants experimental testing.

2.12.4 Aposematism of Red Autumn Leaves

If the "defence indication hypothesis" is accepted, it follows that plant parts rich in anthocyanins might serve as aposematic (warning) colouration for chemical-based unpalatability. Even if red autumn leaves are not chemically well-defended, but have a low nutritive value (another case of unpalatability) they could still be considered aposematic. The possibility of aposematism in chemically defended plants has been appraised in previous studies (Cook et al. 1971; Hinton 1973; Harborne 1982; Lev-Yadun and Ne`eman 2004; Lev-Yadun 2006); red autumn leaves are simply another case of such phenomena. We emphasize that aposematism does not exclude the possible simultaneous operation of any other types of visual or non-visual defence.

As in other cases of aposematism (Cott 1940; Wickler 1968; Lev-Yadun 2003b), it seems likely that mimics of aposematic autumn leaves also exist. Indeed, the

Döring, T.F. and Chittka, L. (2007) Visual ecology of aphids – a critical review on the role of colours in host finding. Arthropod Plant Interact. 1, 3–16.

Edmunds, M. (1974) *Defence in Animals. A Survey of Anti-Predator Defences.* Longman Press, New York.

Edmunds, M. (2000) Why are there good and poor mimics? Biol. J. Linn. Soc. 70, 459–466.

Eisner, T. and Grant, R.P. (1981) Toxicity, odor aversion, and "olfactory aposematism". Science 213, 476.

Eisner, T., Eisner, M. and Siegler, M. (2005). *Secret Weapons. Defenses of Insects, Spiders, Scorpions, and Other Many-Legged Creatures.* Harvard University Press, Cambridge.

Endler, J.A. (1984) Progressive background matching in moths, and a quantitative measure of crypsis. Biol. J. Linn. Soc. 22, 187–231.

Facelli, J.M. (1993) Experimental evaluation of the foliar flag hypothesis using fruits of *Rhus glabra* (L.). Oecologia 93, 70–72.

Faegri, K. and van der Pijl, L. (1979) *The Principles of Pollination Ecology,* 3rd Edn. Pergamon Press, Oxford.

Feild, T.S., Lee, D.W. and Holbrook, N.M. (2001) Why leaves turn red in autumn. The role of anthocyanins in senescing leaves of red-osier dogwood. Plant Physiol. 127, 566–574.

Finch, S. and Jones, T.H. (1989) An analysis of the deterrent effect of aphids on cabbage root fly (*Delia radicum*) egg-laying. Ecol. Entomol. 14, 387–391.

Fineblum, W.L. and Rausher, M.D. (1997) Do floral pigmentation genes also influence resistance to enemies? The W locus in *Ipomoea purpurea.* Ecology 78, 1646–1654.

Gittleman, J.L. and Harvey, P.H. (1980) Why are distasteful prey not cryptic? Nature 286, 149–150.

Givnish, T.J. (1990) Leaf mottling: relation to growth form and leaf phenology and possible role as camouflage. Funct. Ecol. 4, 463–474.

Gould, K.S. (1993) Leaf heteroblasty in *Pseudopanax crassifolius*: functional significance of leaf morphology and anatomy. Ann. Bot. 71, 61–70.

Gould, K.S. (2004) Nature's Swiss army knife: the diverse protective roles of anthocyanins in leaves. J. Biomed. Biotechnol. 2004, 314–320.

Gould, K.S., Neill, S.O. and Vogelmann, T.C. (2002) A unified explanation for anthocyanins in leaves? Adv. Bot. Res. 37, 167–192.

Grafen, A. (1990) Biological signals as handicaps. J. Theor. Biol. 144, 517–546.

Gronquist, M., Bezzerides, A., Attygalle, A., Meinwald, J., Eisner, M. and Eisner, T. (2001) Attractive and defensive functions of the ultraviolet pigments of a flower (*Hypericum calycinum*). Proc. Natl. Acad. Sci. USA 98, 13745–13750.

Grubb, P.J. (1992) A positive distrust in simplicity – lessons from plant defences and from competition among plants and among animals. J. Ecol. 80, 585–610.

Guthrie, R.D. and Petocz, R.G. (1970) Weapon automimicry among animals. Am. Nat. 104, 585–588.

Hagen, S.B., Folstad, I. and Jakobsen, S.W. (2003) Autumn colouration and herbivore resistance in mountain birch (*Betula pubescens*). Ecol. Lett. 6, 807–811.

Hagen, S.B., Debeauss, S., Yoccoz, N.G. and Folstad, I. (2004) Autumn coloration as a signal of tree condition. Proc. Roy. Soc. Lond., Ser. B: Biol. Sci. 271 (Suppl.), S184-S185.

Halpern, M., Raats, D. and Lev-Yadun, S. (2007) Plant biological warfare: thorns inject pathogenic bacteria into herbivores. Environ. Microbiol. 9, 584–592.

Hamilton, W.D. and Brown, S.P. (2001) Autumn tree colours as a handicap signal. Proc. Roy. Soc. Lond., Ser. B: Biol. Sci. 268, 1489–1493.

Harborne, J.B. (1982) *Introduction to Ecological Biochemistry.* Academic Press, London.

Harborne, J.B. (1997) Biochemical plant ecology. In: Dey, P.M. and Harborne, J.B. (Eds.), *Plant Biochemistry.* Academic Press, London, pp. 503–516.

Harper, J.L. (1977) *Population Biology of Plants.* Academic Press, London.

Harvey, P.H. and Paxton, R.J. (1981) The evolution of aposematic coloration. Oikos 37, 391–396.

Hatier, J.-H. and Gould, K.S. (2007) Black coloration in leaves of *Ophiopogon planiscapus* "Nigrescens". Leaf optics, chromaticity, and internal light gradients. Funct. Plant Biol. 34, 130–138.

Herrera, C.M., Medrano, M., Rey, P.J., Sánchez-Lafuente, A.M., Garcia, M.B., Guitián, J. and Manzaneda, A.J. (2002) Interaction of pollinators and herbivores on plant fitness suggests a pathway for correlated evolution of mutualism- and antagonism-related traits. Proc. Natl. Acad. Sci. USA 99, 16823–16828.

Hinton, H.E. (1973) Natural deception. In: Gregory, R.L. and Gombrich, E.H. (Eds.), *Illusion in Nature and Art*. Duckworth, London, pp. 97–159.

Hoch, W.A., Zeldin, E.L. and McCown, B.H. (2001) Physiological significance of anthocyanins during autumnal leaf senescence. Tree Physiol. 21, 1–8.

Hoch, W.A., Singsaas, E.L. and McCown, B.H. (2003) Resorption protection. Anthocyanins facilitate nutrient recovery in autumn by shielding leaves from potentially damaging light levels. Plant Physiol. 133, 1296–1305.

Hoekstra, H.E. (2006) Genetics, development and evolution of adaptive pigmentation in vertebrates. Heredity 97, 222–234.

Holopainen, J.K. and Peltonen, P. (2002) Bright autumn colours of deciduous trees attract aphids: nutrient retranslocation hypothesis. Oikos 99, 184–188.

Huxley, C.R. and Cutler, D.F. (1991) *Ant-Plant Interactions*. Oxford University Press, Oxford.

Ichiishi, S., Nagamitsu, T., Kondo, Y., Iwashina, T., Kondo, K. and Tagashira, N. (1999) Effects of macro-components and sucrose in the medium on in vitro red-color pigmentation in *Dionaea muscipula* Ellis and *Drosera spathulata* Laill. Plant Biotechnol. 16, 235–238.

Inbar, M. and Lev-Yadun, S. (2005) Conspicuous and aposematic spines in the animal kingdom. Naturwiss. 92, 170–172.

Inbar, M., Doostdar, H. and Mayer, R.T. (1999) Effects of sessile whitefly nymphs (Homoptera: Aleyrodidae) on leaf-chewing larvae (Lepidoptera: Noctuidae). Environ. Entomol. 28, 353–357.

Janzen, D.H. (1986) Chihuahuan desert nopaleras: defaunated big mammal vegetation. Annu. Rev. Ecol. Syst. 17, 595–636.

Janzen, D.H. and Martin, P.S. (1982) Neotropical anachronisms: the fruits the gomphotheres ate. Science 215, 19–27.

Jolivet, P. (1998) *Interrelationship Between Insects and Plants*. CRC Press, Boca Raton.

Juniper, B.E. (1994) Flamboyant flushes: a reinterpretation of non-green flush colours in leaves. Int. Dendrol. Soc. Yrbk. 1993, 49–57.

Jürgens, A. (2004) Flower scent composition in diurnal *Silene* species (Caryophyllaceae): phylogenetic constraints or adaption to flower visitors? Biochem. Syst. Ecol. 32, 841–859.

Jürgens, A., Witt, T. and Gottsberger, G. (2002) Flower scent composition in night-flowering *Silene* species (Caryophyllaceae). Biochem. Syst. Ecol. 30, 383–397.

Jürgens, A., Witt, T. and Gottsberger, G. (2003) Flower scent composition in *Dianthus* and *Saponaria* species (Caryophyllaceae) and its relevance for pollination biology and taxonomy. Biochem. Syst. Ecol. 31, 345–357.

Karageorgou, P. and Manetas, Y. (2006) The importance of being red when young: anthocyanins and the protection of young leaves of *Quercus coccifera* from insect herbivory and excess light. Tree Physiol. 26, 613–621.

Karban, R. and Baldwin, I.T. (1997) *Induced Responses to Herbivory*. University of Chicago Press, Chicago.

Kelber, A. (2001) Receptor based models for spontaneous colour choices in flies and butterflies. Entomol. Exp. et Applic. 99, 231–244.

Kelber, A., Vorobyev, M. and Osorio, D. (2003) Animal colour vision – behavioural tests and physiological concepts. Biol. Rev. 78, 81–118.

Kessler, A. and Baldwin, I.T. (2001) Defensive function of herbivore induced plant volatile emissions in nature. Science 291, 2141–2144.

Kettlewell, B. (1973) *The Evolution of Melanism*. Clarendon Press, Oxford.

Kirchner, S.M., Döring, T.F. and Saucke, H. (2005) Evidence for trichromacy in the green peach aphid *Myzus persicae* (Homoptera: Aphididae). J. Insect Physiol. 51, 1255–1260.

Knight, R.S. and Siegfried, W.R. (1983) Inter-relationships between type, size and color of fruits and dispersal in Southern African trees. Oecologia 56, 405–412.

Komárek, S. (1998) *Mimicry, Aposematism and Related Phenomena in Animals and Plants: Bibliography 1800–1990*. Vesmir, Prague.

Konczak, I. and Zhang, W. (2004) Anthocyanins – more than nature's colours. J. Biomed. Biotechnol. 2004, 239–240.

Lanner, R.M. (1998) Seed dispersal in *Pinus*. In: Richardson, D.M. (Ed.), *Ecology and Biogeography of* Pinus. Cambridge University Press, Cambridge, pp. 281–295.

Launchbaugh, K.L. and Provenza, F.D. (1993) Can plants practice mimicry to avoid grazing by mammalian herbivores? Oikos 66, 501–504.

Lee, D.W. (2002) Anthocyanins in autumn leaf senescence. Adv. Bot. Res. 37, 147–165.

Lee, D.W. and Gould, K.S. (2002a) Why leaves turn red. Am. Sci. 90, 524–531.

Lee, D.W. and Gould, K.S. (2002b) Anthocyanins in leaves and other vegetative organs: an introduction. Adv. Bot. Res. 37, 2–16.

Lee, D.W. and Lowry, J.B. (1980) Young-leaf anthocyanin and solar ultraviolet. Biotropica 12, 75–76.

Lee, D.W., Brammeier, S. and Smith, A.P. (1987) The selective advantages of anthocyanins in developing leaves of mango and cacao. Biotropica 19, 40–49.

Lee, D.W., O`Keefe, J., Holbrook, N.M. and Feild, T.S. (2003) Pigment dynamics and autumn leaf senescence in a New England deciduous forest, eastern USA. Ecol. Res. 18, 677–694.

Lev-Yadun, S. (2001) Aposematic (warning) coloration associated with thorns in higher plants. J. Theor. Biol. 210, 385–388.

Lev-Yadun, S. (2003a) Why do some thorny plants resemble green zebras? J. Theor. Biol. 244, 483–489.

Lev-Yadun, S. (2003b) Weapon (thorn) automimicry and mimicry of aposematic colorful thorns in plants. J. Theor. Biol. 244, 183–188.

Lev-Yadun, S. (2006) Defensive coloration in plants: a review of current ideas about anti-herbivore coloration strategies. In: Teixeira da Silva, J.A. (Ed.), *Floriculture, Ornamental and Plant Biotechnology: Advances and Topical Issues,* Volume IV. Global Science Books, London, pp. 292–299.

Lev-Yadun, S. and Inbar, M. (2002) Defensive ant, aphid and caterpillar mimicry in plants. Biol. J. Linn. Soc. 77, 393–398.

Lev-Yadun, S. and Gould, K.S. 2007. What do red and yellow autumn leaves signal? Bot. Rev. 73, 279–289.

Lev-Yadun, S. and Ne`eman, G. (2004) When may green plants be aposematic? Biol. J. Linn. Soc. 81, 413–416.

Lev-Yadun, S., Dafni, A., Inbar, M., Izhaki, I. and Ne`eman, G. (2002) Colour patterns in vegetative parts of plants deserve more research attention. Trends Plant Sci. 7, 59–60.

Lev-Yadun, S., Dafni, A., Flaishman, M.A., Inbar, M., Izhaki, I., Katzir, G. and Ne`eman, G. (2004) Plant coloration undermines herbivorous insect camouflage. BioEssays 26, 1126–1130.

Lüttge, U. (1997) *Physiological Ecology of Tropical Plants*. Springer-Verlag, Berlin.

Madden, D. and Young, Y.P. (1992) Symbiotic ants as an alternative defense against giraffe herbivory in spinescent *Acacia drepanolobium*. Oecologia 91, 235–238.

Majerus, M.E.N. (1998) *Melanism. Evolution in Action.* Oxford University Press, Oxford.

Manetas, Y. (2006) Why some leaves are anthocyanic and why most anthocyanic leaves are red? Flora 201, 163–177.

Matile, P. (2000) Biochemistry of Indian summer: physiology of autumnal leaf coloration. Exp. Gerontol. 35, 145–158.

Mendez, M., Gwynn-Jones, D. and Manetas, Y. (1999) Enhanced UV-B radiation under field conditions increases anthocyanin and reduces the risk of photoinhibition but does not affect growth in the carnivorous plant *Pinguicula vulgaris*. New Phytol. 144, 275–282.

Merilaita, S. (2003) Visual background complexity facilitates the evolution of camouflage. Evolution 57, 1248–1254.

Merilaita, S., Tuomi, J. and Jormalainen, V. (1999) Optimization of cryptic coloration in heterogeneous habitat. Biol. J. Linn. Soc. 67, 151–161.

Moran, J.A. and Moran, A.J. (1998) Foliar reflectance and vector analysis reveal nutrient stress in prey-deprived pitcher plants (*Nepenthes rafflesiana*). Int. J. Plant Sci. 159, 996–1001.

Myers, J.H. and Bazely, D. (1991). Thorns, spines, prickles, and hairs: are they stimulated by herbivory and do they deter herbivores? In: Tallamy, D.W. and Raupp, M.J. (Eds.), *Phytochemical Induction by Herbivores*. John Wiley and Sons, New York, pp. 325–344.

Neill, S.O. and Gould, K.S. (1999) Optical properties of leaves in relation to anthocyanin concentration and distribution. Can. J. Bot. 77, 1777–1782.

Nottingham, S.F., Hardie J. and Tatchell, G.M. (1991) Flight behaviour of the bird cherry aphid, *Rhopalosiphum padi*. Physiol. Entomol. 16, 223–229.

Ohgushi, T. (2005) Indirect interaction webs: herbivore-induced effects through trait change in plants. Annu. Rev. Ecol. Syst. 36, 81–105.

Ougham, H.J., Morris, P. and Thomas, H. (2005) The colors of autumn leaves as symptoms of cellular recycling and defenses against environmental stresses. Curr. Top. Dev. Biol. 66, 135–160.

Padmavati, M., Sakthivel, N., Thara, K.V. and Reddy, A.R. (1997) Differential sensitivity of rice pathogens to growth inhibition by flavonoids. Phytochemistry 46, 499–502.

Pasteur, G. (1982) A classification review of mimicry systems. Annu. Rev. Ecol. Syst. 13, 169–199.

Pichersky, E. and Gershenzon, J. (2002) The formation and function of plant volatiles: perfumes for pollinator attraction and defense. Curr. Opin. Plant Biol. 5, 237–243.

Pichersky, E. and Dudareva, N. (2007) Scent engineering: toward the goal of controlling how flowers smell. Trends Biotech. 25, 105–110.

Provenza, F.D., Kimball, B.A. and Villalba, J.J. (2000) Roles of odor, taste, and toxicity in the food preferences of lambs: implications for mimicry in plants. Oikos 88, 424–432.

Purser, B. (2003) *Jungle Bugs: Masters of Camouflage and Mimicry*. Firefly Books. Toronto.

Rebollo, S., Milchunas, D.G., Noy-Meir, I. and Chapman, P.L. (2002) The role of spiny plant refuge in structuring grazed shortgrass steppe plant communities. Oikos 98, 53–64.

Reichelt, G. and Wilmanns, O. (1973) *Vegetationsgeographie. Das Geographisce Seminar.* Westermann, Braunschweig.

Richards, P.W. (1996) *The Tropical Rain Forest: An Ecological Study*, 2nd Edn. Cambridge University Press, Cambridge.

Ridley, H.N. (1930) *The Dispersal of Plants Throughout the World.* L. Reeve and Co., Ashford.

Rolshausen, G. and Schaefer, H.M. (2007) Do aphids paint the tree red (or yellow) – can herbivore resistance or photoprotection explain colourful leaves in autumn? Plant Ecol. 191, 77–84.

Rothschild, M. (1980) Remarks on carotenoids in the evolution of signals. In: Gilbert, L.E. and Raven, P.H. (Eds.), *Coevolution of Animals and Plants.* University of Texas Press, Austin, pp. 20–51.

Rubino, D.L. and McCarthy, B.C. (2004) Presence of aposematic (warning) coloration in vascular plants of southeastern Ohio. J. Torrey Bot. Soc. 131, 252–256.

Ruxton, G.D., Sherratt, T.N. and Speed, M.P. (2004) *Avoiding Attack. The Evolutionary Ecology of Crypsis, Warning Signals and Mimicry.* Oxford University Press, Oxford.

Saracino, A., Pacella, R., Leone, V. and Borghetti, M. (1997) Seed dispersal and changing seed characteristics in a *Pinus halepensis* Mill. forest after fire. Plant Ecol. 130, 13–19.

Saracino, A., D'Alessandro, C.M. and Borghetti, M. (2004) Seed colour and post-fire bird predation in a Mediterranean pine forest. Acta Oecol. 26, 191–196.

Schaefer, H.M. and Rolshausen, G. (2006) Plants on red alert: do insects pay attention? BioEssays 28, 65–71.

Schaefer, H.M. and Rolshausen, G. (2007) Aphids do not attend to leaf colour as visual signal, but to the handicap of reproductive investment. Biol. Lett. 3, 1–4.

Schaefer, H.M. and Wilkinson, D.M. (2004) Red leaves, insects and coevolution: a red herring? Trends Ecol. Evol. 19, 616–618.

Schaefer, H.M., Schaefer, V. and Levey, D.J. (2004) How plant-animal interactions signal new insights in communication. Trends Ecol. Evol. 19, 577–584.

Schaefer, H.M., Levey, D.J., Schaefer, V. and Avery, M.L. (2006) The role of chromatic and achromatic signals for fruit detection by birds. Behav. Ecol. 17, 784–789.

Schiestl, F.P., Ayasse, M., Paulus, H.F., Löfstedt, C., Hansson, B.S., Ibarra, F. and Francke, W. (2000) Sex pheromone mimicry in the early spider orchid (*Ophrys sphegodes*): patterns of hydrocarbons as the key mechanism for pollination by sexual deception. J. Comp. Physiol., A 186, 567–574.

Schulze, E.-D., Beck, E. and Müller-Hohenstein, K. (2002) *Plant Ecology.* Springer-Verlag, Berlin.

Schwinn, K.E. and Davies, K.M. (2004) Flavonoids. In: Davies, K.M. (Ed.), *Plant Pigments and their Manipulation.* Annual Plant Reviews, Volume 14. Blackwell Publishing, Oxford, pp. 92–149.

Shimohigashi, M. and Tominaga, Y. (1991) Identification of UV, green and red receptors, and their projection to lamina in the cabbage butterfly, *Pieris rapae.* Cell Tiss. Res. 263, 49–59.

Sillén-Tullberg, B. and Bryant, E.H. (1983) The evolution of aposematic coloration in distasteful prey: an individual selection model. Evolution 37, 993–1000.

Sinkkonen, A. (2006a) Sexual reproduction advances autumn leaf colours in mountain birch (*Betula pubescens* ssp. *czerepanovii*). J. Evol. Biol. 19, 1722–1724.

Sinkkonen, A. (2006b) Do autumn leaf colours serve as a reproductive insurance against sucking herbivores? Oikos 113, 557–562.

Smith, A.P. (1986) Ecology of leaf color polymorphism in a tropical forest species: habitat segregation and herbivory. Oecologia 69, 283–287.

Speed, M.P. and Ruxton, G.D. (2005) Warning displays in spiny animals: one (more) evolutionary route to aposematism. Evolution 59, 2499–2508.

Stafford, H.A. (1994) Anthocyanins and betalains: evolution of the mutually exclusive pathways. Plant Sci. 101, 91–98.

Stiles, E.W. (1982) Fruit flags: two hypotheses. Am. Nat. 120, 500–509.

Stone, B.C. (1979) Protective coloration of young leaves in certain Malaysian palms. Biotropica 11, 126.

Tuomi, J. and Augner, M. (1993) Synergistic selection of unpalatability in plants. Evolution 47, 668–672.

Weiss, M.R. (1995) Floral colour change: a widespread functional convergence. Am. J. Bot. 82, 167–195.

Werlein, H.-D., Kutemeyer, C., Schatton, G., Hubbermann, E.M. and Schwarz, K. (2005) Influence of elderberry and blackcurrant concentrates on the growth of microorganisms. Food Control 16, 729–733.

Wickler, W. (1968) *Mimicry in Plants and Animals*. Weidenfeld and Nicolson, London.

Wiens, D. (1978) Mimicry in plants. Evol. Biol. 11, 365–403.

Wiklund, C. and Järvi T. (1982) Survival of distasteful insects after being attacked by naive birds: a reappraisal of the theory of aposematic coloration evolving through individual selection. Evolution 36, 998–1002.

Wilkinson, D.M., Sherratt, T.N., Phillip, D.M., Wratten, S.D., Dixon, A.F.G. and Young, A.J. (2002) The adaptive significance of autumn leaf colours. Oikos 99, 402–407.

Williamson, G.B. (1982) Plant mimicry: evolutionary constraints. Biol. J. Linn. Soc. 18, 49–58.

Willson, M.F. and Hoppes, W.G. (1986) Foliar "flags" for avian frugivores: signal or serendipity? In: Estrada, A. and Fleming, T.H. (Eds.), *Frugivores and Seed Dispersal.* Springer, Dordrecht, pp. 55–69.

Willson, M.F. and Whelan, C.J. (1990) The evolution of fruit color in fleshy-fruited plants. Am. Nat. 136, 790–809.

Wimp, G.M. and Whitham, T.G. (2001) Biodiversity consequences of predation and host plant hybridization on an aphid-ant mutualism. Ecology 82, 440–452.

Wrolstad, R.E. (2004). Symposium 12: Interaction of natural colors with other ingredients. Anthocyanin pigments - Bioactivity and coloring properties. J. Food Sci. 69, C419-C421.

Yamasaki, H. (1997) A function of colour. Trends Plant Sci. 2, 7–8.

Zahavi, A. (1975) Mate selection - a selection for a handicap. J. Theor. Biol. 53, 205–214.

Zahavi, A. (1977) The cost of honesty (further remarks on the handicap principle). J. Theor. Biol. 67, 603–605.

Zahavi, A. (1987) The theory of signal selection and some of its implications. In: Delfino, V.P. (Ed.), *International Symposium of Biological Evolution.* Adriatica Editrica, Bari, pp. 305–327.

Zahavi, A. and Zahavi, A. (1997) *The Handicap Principle: A Missing Piece of Darwin's Puzzle.* Oxford University Press, New York.

3

Modifying Anthocyanin Production in Flowers

Kevin M. Davies

New Zealand Institute for Crop & Food Research Ltd, Private Bag 11600, Palmerston North, New Zealand, daviesk@crop.cri.nz

Abstract. Anthocyanin biosynthesis is a key aspect of flower development for many angiosperms, providing one of the major influences on the choice of potential pollinators. In some species evolution has resulted in complex anthocyanin structures that provide bright flower colours, whereas in other species sophisticated combinations of pigment patterning and floral shape have developed to attract pollinators. There is now a good understanding of the molecular biology of both the genes encoding the biosynthetic enzymes for anthocyanins and copigments, and the temporal and spatial regulation of anthocyanin production. The availability of genes relating to anthocyanin biosynthesis has allowed for the molecular breeding of flower colour in several ornamental species. Since the first publication detailing the generation of new flower colours using recombinant DNA techniques (approximately 20 years ago) there have been many notable advances in the gene technologies available for genetic modification of anthocyanin biosynthesis. Transgenic carnation cultivars that produce delphinidin-derived anthocyanins and that have novel mauve-violet colours are now available commercially, and it is anticipated that these will be followed to market by many more genetically modified ornamental crops during the next 10 to 15 years.

3.1 Introduction

The major pigments that cause flower colour are carotenoids, flavonoids and betalains. Although other pigment types such as chlorophylls, phenylphenalenones and quinochalcones can generate flower colours, they are rare examples (Davies 2004). The flavonoids are phenylpropanoid compounds of great variation in structure and function (Bohm 1998). Those involved in flower colour are water-soluble and generally located in the vacuole, with by far the most common type being the anthocyanins. Anthocyanins are the basis for nearly all pink, red, orange, scarlet, purple, blue and blue-black flower colours. Although flavonoids such as the yellow aurones also give rise to flower colours, they are comparatively rare, and only the anthocyanin pigments are discussed in detail in this review. Recent reviews of the biosynthesis of other pigment types include those of Strack et al. (2003) and Zrÿd and Christinet (2004) on betalains, Cuttriss and Pogson (2004) and Fraser and Bramley (2004) on carotenoids, and that of Davies (2004) on some of the less common pigment types.

K. Gould et al. (eds.), *Anthocyanins*, DOI: 10.1007/978-0-387-77335-3_3,

Flavonoid biosynthesis is part of the larger phenylpropanoid pathway, which produces a range of secondary metabolites from the aromatic amino acid phenylalanine. There are many branches to the flavonoid-specific pathway, producing coloured and colourless compounds with diverse biological functions. As mentioned above, anthocyanins are the most significant flavonoid pigments, with aurones, chalcones and some flavonols playing a limited role in flower colour. The generally colourless (or weakly coloured) flavones and flavonols also have a role in flower colour for their function as co-pigments. They stabilise and maintain anthocyanins in their coloured forms, in a process of complex molecular interactions known as co-pigmentation (Brouillard and Dangles 1993). Flavones and flavonols, strong absorbers of UV-light, are also the basis for some floral insect nectar guides.

This review focuses on the current status of gene technologies for manipulating anthocyanin production in flowers. To help with understanding of the various genetic modification (GM) approaches discussed, a brief overview is also given of the biosynthetic pathway for anthocyanins, the regulation of anthocyanin production, and the character of anthocyanins as flower pigments.

3.2 Anthocyanin Biosynthesis in Flowers

The anthocyanin biosynthetic pathway is well defined at the genetic and enzymatic level, with gene sequences available for all the key biosynthetic steps to the primary anthocyanins and also for many of the secondary modification activities. Extensive reviews have been published on the molecular biology of flavonoid biosynthesis (e.g. Springob et al. 2003; Schwinn and Davies 2004; Davies and Schwinn 2006; Grotewold 2006), and only a brief overview is given here. There are also reviews available on some of the specific biosynthetic enzyme groups, including acyltransferases (Nakayama et al. 2003), glycosyltransferases (Vogt 2000), methyltransferases (Ibrahim and Muzac 2000) and flavonoid dioxygenases (Gebhardt et al. 2005).

The base pigments are the anthocyanidins, which are then glycosylated to form the anthocyanins. In all examples to date, except for one recent report of C-glycosylation (Saito et al. 2003), only O-glycosylation occurs for anthocyanins in plants. The core of the anthocyanidin is a 15-carbon (C_{15}) structure of two aromatic rings (the A and B rings) joined by a third ring of C_3O_1 (the C-ring; Fig. 3.1). The degree of oxidation of the C-ring defines the various flavonoid types (Fig. 3.2). Anthocyanidins have two double bonds in the C-ring – and hence carry a positive charge.

The core anthocyanidin structure is modified by the addition of a wide range of chemical groups, in particular through hydroxylation, acylation and methylation. Hydroxylation and methylation usually, but not exclusively, occur on the anthocyanidin prior to further modifications. Thus, there are a small number of anthocyanidin types that have been identified as the basis of the subsequent large number of known anthocyanins with differing glycosylation and acylation patterns. Table 3.1 lists some of the 31 known naturally occurring anthocyanidin types. Not listed are some of the more recent structures identified, which include anthocyanidins with additional rings incorporated, for example the pyranoanthocyanidins and riccionidin A. The structure of violet rosacyanin B (5-carboxypyranoanthocyanidin), which has been isolated from rose (*Rosa hybrida*) petals and red onion bulbs (*Allium cepa*) (Fukui et al. 2002; Andersen and Jordheim 2006), is shown in Fig. 3.1.

Table 3.1 Structures of some of the naturally occurring anthocyanidins. The numbering of the relevant carbons for the base anthocyanidin structure is shown in Fig. 3.1

	Substitution pattern at numbered position						
	3	5	6	7	3'	4'	5'
Common anthocyanidins							
Pelargonidin (Pg)	OH	OH	H	OH	H	OH	H
Cyanidin (Cy)	OH	OH	H	OH	OH	OH	H
Delphinidin (Dp)	OH	OH	H	OH	OH	OH	OH
Peonidin	OH	OH	H	OH	OMe	OH	H
Petunidin	OH	OH	H	OH	OMe	OH	OH
Malvidin	OH	OH	H	OH	OMe	OH	OMe
3-Deoxyanthocyanidins							
Apigeninidin	H	OH	H	OH	H	OH	H
Luteolinidin	H	OH	H	OH	OH	OH	H
Tricetinidin	H	OH	H	OH	OH	OH	OH
6-Hydroxyanthocyanidins							
6-Hydroxypelargonidin	OH	OH	OH	OH	H	OH	H
6-Hydroxycyanidin	OH	OH	OH	OH	OH	OH	H
6-Hydroxydelphinidin	OH	OH	OH	OH	OH	OH	OH
Rare methylated anthocyanidins							
5-Methoxycyanidin	OH	OMe	H	OH	OH	OH	H
5-Methoxydelphinidin	OH	OMe	H	OH	OH	OH	OH
5-Methoxypetunidin (Europinidin)	OH	OMe	H	OH	OMe	OH	OH
5-Methoxymalvidin (Capensinidin)	OH	OMe	H	OH	OMe	OH	OMe
7-Methoxypeonidin (Rosinidin)	OH	OH	H	OMe	OMe	OH	H
7-Methoxymalvidin (Hirsutidin)	OH	OH	H	OMe	OMe	OH	OMe

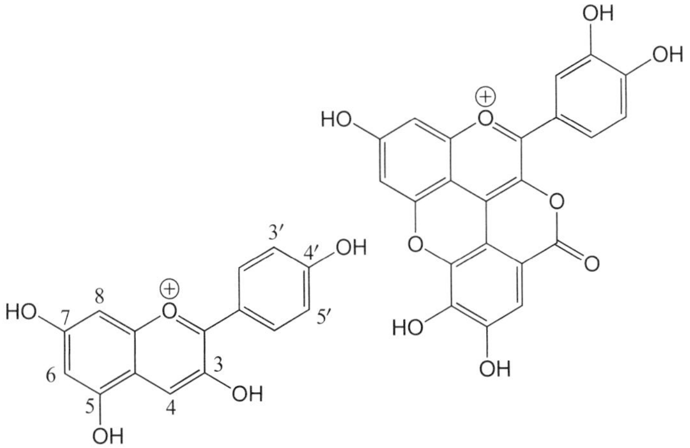

Fig. 3.1 Anthocyanidin structures. The structure on the left (pelargonidin) shows the numbering of some of the carbons for the common anthocyanidins and anthocyanins. The structure of rosacyanin B, a pyranoanthocyanidin, is shown on the right

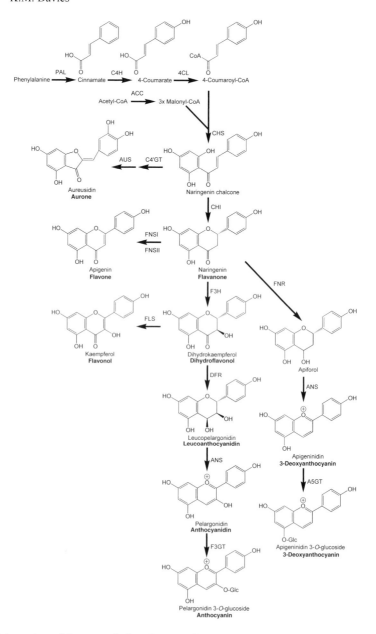

Fig. 3.2 A section of the general phenylpropanoid and flavonoid biosynthetic pathways leading to the anthocyanins and other flavonoids found in flowers. For ease of presentation, generally only the route for flavonoids with 4′-hydroxylation of the B-ring is shown. For formation of anthocyanins from leucoanthocyanidins only the simplified scheme via the anthocyanidin is shown. Enzyme abbreviations are defined in the text except for PAL (phenylalanine ammonia lyase) and 4CL (4-coumaroyl CoA:ligase)

Of particular relevance to this chapter are the six common anthocyanidins that have C-3,5,7 hydroxylation of the A- and C-rings, as these account for about 90% of the anthocyanins that have been identified to date (Andersen and Jordheim 2006), and the 3-deoxyanthocyanidins. Anthocyanin hydroxylation patterns are of prime importance in flower colour, as they have a major effect on the colour resulting from the pigment. A comparison of the colours of the common 3-hydroxyanthocyanidins to two of the equivalent 3-deoxyanthocyanidins is shown in Fig. 3.3.

The key flavonoid precursors are phenylalanine and malonyl-CoA, derived from the shikimate/arogenate pathway and the TCA cycle, respectively. The first flavonoids are the C_{15} chalcones, which are formed by chalcone synthase (CHS), a member of the polyketide synthase group of enzymes. CHS takes a hydroxycinnamic acid-CoA ester unit, usually p-coumaroyl-CoA, and carries out three sequential additions of the "extender" molecule malonyl-CoA.

The chalcones are the first coloured flavonoid, and provide yellow colouration to petals of a few plant species, most notably carnation (*Dianthus caryophyllus*). They

Fig. 3.3 Top row: Flowers of lisianthus cultivars pigmented predominately by pelargonidin- (left), cyanidin- (centre) or delphinidin-based (right) anthocyanins. Centre row left: Solutions of, from left to right, pelargonidin, cyanidin, delphinidin, apigeninidin (3-deoxypelargonidin) and luteolinidin (3-deoxycyanidin). Centre row right: 3-Deoxyanthocyanin pigmented flowers of *Sinningia cardinalis*. Bottom row left: Floral organs of the Black Bat Flower. Bottom row centre and right: The floral organs of calla lily Treasure and a close up of the epidermis of the spathe of the same cultivar. See Plate 2 for colour version of these photographs

may also be converted to the brighter yellow aurones by the vacuolar-located aure-usidin synthase (AUS), via a glycosylated intermediate generated by uridine diphos-phate-glucose (UDPG): chalcone 4'-O-glucosyltransferase (C4'GT) (Ono et al. 2006a, 2006b). The identification of cDNAs for both of the required aurone biosyn-thetic enzymes (Nakayama et al. 2000; Ono et al. 2006a) should enable the introduc-tion of yellow flower colours into the wide range of ornamentals species that currently lack them.

More commonly the chalcones are converted by stereospecific isomerisation to the corresponding (2S)-flavanones by chalcone isomerase (CHI), establishing the heterocyclic C-ring. Two types of CHI have been identified; CHI-I can use only 6'-hydroxychalcone substrates whereas CHI-II, which is found mostly in legumes, can catalyse isomerisation of both 6'-hydroxy- and 6'-deoxychalcones. With 6'-hydroxychalcones, but not 6'-deoxychalcones due to the intramolecular hydrogen bond between the 2'-hydroxyl and the carbonyl group, the isomerisation reaction can occur non-enzymically to form racemic (2R,2S) flavanone, and this may occur *in vivo* to enable formation of anthocyanins in plants with mutations for CHI.

However, the requirement for CHI for full flavonoid biosynthesis *in vivo* is evi-dent from, for example, the reduction in anthocyanin biosynthesis in transgenic to-bacco in which CHI gene activity is suppressed (Nishihara et al. 2005).

The (2S)-flavanones are converted stereospecifically to the respective (2R,3R)-dihydroflavonols (DHFs) by flavanone 3ß-hydroxylase (F3H), a 2-oxoglutarate de-pendent dioxygenase (2OGD). The next step in the pathway to anthocyanins is the conversion of DHFs to the respective (2R,3S,4S)-flavan-2,3-*trans*-3,4-*cis*-diols (leu-coanthocyanidins) by dihydroflavonol 4-reductase (DFR), a member of the Single-Domain-Reductase/Epimerase/Dehydrogenase (RED) protein family. Of interest with regard to the genetic control of flower colour is the varying preference shown by DFR of some species towards the three common DHFs. For example, DFR in cymbidium orchids (*Cymbidium hybrida*) and *Petunia* cannot efficiently reduce dihydrokaempferol (DHK), so that pelargonidin-based anthocyanins rarely accumu-late in these species (Meyer et al. 1987; Johnson et al. 1999). DFR substrate usage is also relevant to the engineering of blue flower colours, as the presence of a DFR with a strong preference for dihydromyricetin (DHM) will encourage accumulation of delphinidin-derived anthocyanins. The endogenous DFR of two leading target ornamentals for the introduction of blue colours, chrysanthemum (*Dendranthema*) and carnation, can use DHM even though delphinidin derivatives do not naturally occur in these species (Heller and Forkmann 1994; Davies and Schwinn 1997).

The flavanones and dihydroflavonols may be converted to flavones and fla-vonols, respectively, by flavone synthase I or II (FNSI, FNSII) and the flavonol synthase (FLS). Indeed, in most flowers studied, flavones or, more commonly, fla-vonols are found in amounts exceeding that of the anthocyanins. FLS and FNSI, which has been characterised from species in the Apiaceae, are 2OGD enzymes, whereas FNSII is a membrane bound cytochrome P450-dependent mono-oxygenase (P450). Flavones and flavonols generally have little colour in themselves, but they influence the colour of the anthocyanins through a process called co-pigmentation, and also are the basis for some UV-absorbing insect nectar guides.

Although the enzymes involved in the final steps to the formation of the initial anthocyanins were characterised some time ago as anthocyanidin synthase (ANS) and (most commonly) UDPG:flavonoid 3-*O*-glucosyltransferase (F3GT), it is only more recently that the nature of their *in vivo* mechanism has been clarified. It is thought that the ANS reaction proceeds through a series of intermediates, starting with the stereospecific hydroxylation of the leucoanthocyanidin at the C-3, to yield an anthocyanidin pseudobase (3-flaven-2,3-diol). The pseudobase is then the substrate for addition of glucose at the 3-hydroxyl by F3GT. Interestingly, recent evidence indicates that oxidation at the C-3 by ANS is actually a minor or "side reaction", with regard to the quantum mechanics of the various possible enzymatic reactions of ANS, and recombinant ANS exhibit primarily other activities (e.g. FLS) with negligible ANS activity (Wilmouth et al. 2002; Nakajima et al. 2006; Wellmann et al. 2006). Furthermore, recombinant ANS enzyme may use (+)-catechin as a substrate to form cyanidin (Wellmann et al. 2006).

Regarding glycosylation, 3-*O*-glucosylation is the most common initial activity in plants, but 3-*O*-galactosylation is the first reaction in some species and there are rare examples of other patterns, such as 5-*O*-glucosylation with 3-deoxyanthocyanins, B-ring *O*-glucosylation, and 3,5-*O*-diglucosylation (Ogata et al. 2005; Andersen and Jordheim 2006).

In what could be viewed as the final step of anthocyanin biosynthesis, they are transported to their site of accumulation – the vacuole. There is evidence for a number of alternative intracellular transport systems for the flavonoids, with the best characterised for anthocyanins being those related to xenobiotic detoxification processes. In brief, the components identified for anthocyanins that are shared with the transport of toxins are the action of glutathione *S*-transferases (GSTs) and ATP-binding cassette (ABC) transmembrane transporters (Marrs et al. 1995; Winefield 2002). The formation of prevacuolar vesicles at the site of anthocyanin biosynthesis on the endoplasmic reticulum may also be part of the transport process, with these vesicles migrating to fuse with the primary vacuole. There are comparatively rare examples of anthocyanins forming collated bodies in the vacuole, often termed Anthocyanic Vacuolar Inclusions (AVIs, Markham et al. 2000). First characterised in detail from sweet potato (*Ipomoea batatas*) cell cultures (Nozue et al. 1993) they have since been found in intact petals (Markham et al. 2000; Gonnet 2003) and fruit skin (Bae et al. 2006). The mechanism of their formation and function, if any, is unknown, but it has been suggested that they may be associated with intensifying flower colour (Markham et al. 2000; Gonnett 2006). Examples of AVIs of carnation and lisianthus (*Eustoma grandiflorum*) are shown in Fig. 3.4.

Variations in the secondary modifications of anthocyanins, particularly the glycosylation and acylation, account for the enormous range of known anthocyanin structures, which was at 539 when Andersen and Jordheim (2006) prepared their recent review. The majority of these anthocyanin structures characterised to date contain two or more monosaccharide units and at least one acyl group. DNA sequences are available for several of the enzymes that catalyse these reactions. In addition to F3GT from several species, cDNAs are available for UDP-rhamnose:anthocyanidin 3-*O*-glucoside-6″-*O*-rhamnosyltransferase (Brugliera et al. 1994; Kroon et al. 1994), UDPG:anthocyanin 5-*O*-glucosyltransferase (Yamazaki et al. 1999, 2002), UDPG:anthocyanin

Fig. 3.4 Top row: Anthocyanic vacuolar inclusions in petals of carnation (left) and lisianthus (centre and right). Centre row and bottom row right: Examples of floral pigmentation pattern-ing in painted tongue (*Salpiglossis sinuate*, centre left), *Paphiopedilum* orchid (centre right), and the snapdragon cultivar Venosa (bottom right). Bottom row left: 35S:LC transgenic petu-nia plants with purple-leaf phenotypes growing in field trials. See Plate 3 for colour version of these photographs

3′-*O*-glucosyltransferase (Fukuchi-Mizutani et al. 2003), UDPG:anthocyanin 3′,5′-*O*-glucosyltransferase (Noda et al. 2004) and UDPG:anthocyanin 3-*O*-glucoside-2″-*O*-glucosyltransferase (Morita et al. 2005). Of particular note is RhGT1 from rose that catalyses the sequential addition of glucose to the hydroxyls at the C-5 and C-3 posi-tions (Ogata et al. 2005), and so is UDPG:anthocyanidin 5,3-*O*-glycosyltransferase (A3,5GT).

With regard to acyltransferases, at present cDNAs have been obtained for two aromatic and six aliphatic anthocyanin acyltransferases (AATs) (Nakayama et al. 2003; Davies and Schwinn 2006). In all cases, they transfer the acyl group from a CoA-donor molecule to hydroxyl residues of sugars attached to the anthocyanin. For both the FGTs and the AATs, narrow substrate specificity may be shown. Thus, although recombinant 5AT enzyme of gentian (*Gentiana trifolia*) can use either caffeoyl-CoA or 4-coumaroyl-CoA as the donor and will accept pelargonidin, cya-nidin or delphinidin derivatives, specificity is shown with regard to the glycosylation and acylation pattern of the substrate anthocyanin (Fujiwara et al. 1998). This pattern of specificity, with flexibility for the number of hydroxyl groups on the B-ring but specificity to the other substitution patterns, is typical of AATs studied to date. Thus, in an example for the aliphatic AATs, the recombinant cineraria (*Senecio cruentus*) malonyl-CoA:anthocyanidin 3-*O*-glucoside-6″-*O*-malonyltransferase protein accepts pelargonidin-, cyanidin- or delphinidin-3-*O*-glucosides but does not use the anthocyanin diglycosides (Suzuki et al. 2003).

There is a range of other anthocyanin biosynthetic enzymes relevant to the modification of anthocyanin production in flowers, for example flavanone 4-reductase (FNR), flavonoid 3′-hydroxylase (F3′H) and flavonoid 3′,5′-hydroxylase (F3′,5′H). These are discussed in the appropriate sections later in the review.

3.3 Anthocyanins as Flower Pigments

The six common anthocyanidins are the basis for a large range of anthocyanin-based flower colours. Evidently, the hydroxylation pattern of the B-ring is a major component of colour variation. The impact on the colour of pelargonidin, cyanidin and delphinidin can be seen from the solutions in Fig. 3.3, and the subsequent effect on flower pigmentation is apparent from the lisianthus cultivars in Fig. 3.3. In comparison, methylation of the B-ring hydroxyls has a comparatively minor effect on the colour, causing a small shift towards red. However, a second major contribution to the natural range of anthocyanin colours observed comes from the other secondary modifications of the anthocyanins, due to the resultant change in the interaction of the anthocyanin with the vacuolar environment.

In aqueous solutions, anthocyanins can undergo pH-dependent changes in chemical form, which alter colour intensity and hue and can even result in loss of colour. Anthocyanins are usually presented as the coloured flavylium cation form in diagrams such as Fig. 3.2. However, anthocyanins may also assume various quinonoidal base, water adduct and chalcone forms, which are referred to as anthocyanin secondary structures. At the mildly acidic pHs typically of vacuoles (pH 3–6), the flavylium cation is vulnerable to hydration, which triggers a reaction that converts them to colourless hemiacetal and (later) chalcone forms. A schematic of the structural transformations of cyanidin 3-*O*-glucoside in aqueous solutions in response to pH is presented in Chapter 9. The anthocyanin secondary modifications influence these chemical changes though altering intra- or inter-molecular interactions of the anthocyanins (including with co-pigments and/or metal cations), which in turn affect the formation of pigment tertiary structures that may prevent hydration and stabilise the anthocyanin in a coloured form. An example of such a tertiary structure would be the vertical stacking of molecules, with an anthocyanin sandwiched between two flavone co-pigments. Tertiary structure also influences flower colour through causing changes in the amount of light and the specific light wavelengths absorbed by the pigment. Many of the brilliant blue flower colours studied are based on pigments with complex tertiary structures.

Thus, it is principally the interaction of anthocyanin structure and concentration, co-pigment structure and concentration and vacuolar pH that determines the final pigmentation. Excellent reviews of the possible structural forms of anthocyanins, the impact these have on colour, and the role of co-pigmentation can be found in Goto and Kondo (1991), Brouillard and Dangles (1993) and Andersen and Jordheim (2006).

It should not be forgotten that pigmentation is an interaction with light, so the light must pass through at least the epidermal cell wall, interact with the pigment and then pass out of the cell to be perceived by the observer. Thus, changes in the structure of the cell, for example the formation of conical cell shape, can alter the resulting flower colour (Gorton and Vogelmann 1996). At least one gene controlling formation of conical cells has been cloned (Noda et al. 1994), raising the prospect of genetic modification of this aspect of flower colour.

3.4 Regulation of Anthocyanin Production in Flowers

The anthocyanin biosynthetic pathways of flowers are regulated in response to a range of environmental, spatial and developmental/temporal signals. The most evident developmental regulation is the production of anthocyanins during flower opening, usually to attract pollinators coincident with flower fertility. In at least some species, there is a direct linkage between attainment of fertility and induction of the biosynthetic genes. In petunia, this may be achieved through the release of gibberellic acid (GA) from the anthers, as removal of the anthers will prevent normal anthocyanin pigment formation and GA_1, GA_3 or GA_4 application will compensate for anther excision and promote gene transcription for several anthocyanin biosynthetic enzymes (reviewed in Weiss 2000). The other endogenous signals that are involved in controlling anthocyanin production in petals are not well characterised. However, it is probable that MADS box genes of the A, B and E class that determine petal identity also have a role in later petal function, including pigmentation. In antirrhinum (*Antirrhinum majus*), loss of the function of the B-class protein DEFICIENS causes reduced pigmentation and lower transcript abundance for the CHS and F3H (Bey et al. 2004), and in petunia, plants inhibited for the activity of the E class gene *floral binding protein2* have small, aberrant green corollas that fail to produce anthocyanins or express CHS (Angenent et al. 1994; Ferrario et al. 2003).

Anthocyanin biosynthesis may be regulated by both abiotic and biotic environmental stimuli. Light is the principal environmental signal, and is required for full floral colouration in many species, either through direct exposure of the flowers or via the leaves (Weiss 2000; Meng and Wang 2004). However, despite extensive knowledge of the mechanisms of light regulation of pigment production in vegetative tissues, there are few data for petals. A further level of temporal control of anthocyanin biosynthetic gene expression, that is likely to be mediated by light, has been observed for anthurium (*Anthurium andraeanum*), in which DFR transcript levels show a diurnal rhythm of abundance (Collette et al. 2004). The notable biotic interaction for flowers is with the target pollinators, and it is thus no surprise that pigmentation can be linked to pollinator activity. In addition to pigments appearing coincident with fertility, flower colour may also change in response to pollination, as the result of changes in petal cell pH, degradation of pigments, or *de novo* pigment biosynthesis (Weiss 1995; Bohm 1998; Farzad et al. 2002, 2003).

The "highest" level of spatial regulation is the association of anthocyanins with the flower itself. In most angiosperms petals are the major coloured organ in the flower. However, there are species in which the sepals are the main coloured organ (e.g. *Hydrangea*, *Limonium*), others in which petals and sepals have become fused as tepals (e.g. *Sandersonia* and many orchids), and a few in which the bracts provide the major colour source. For example, in the Araceae family (e.g. anthurium and calla lily, *Zantedeschia*; Fig. 3.3) the flowers make no contribution to pigmentation, being tiny colourless organs on a cylindrical inflorescence (the spadix), and the major pigmented organ is a modified leaf (the spathe). Within the petal, anthocyanins are most often epidermally located. Figure 3.3 shows an example for calla lily spathe, in which the red anthocyanins in the epidermis are seen above the yellow carotenoids in the mesophyll cells. The patchy production of anthocyanins shown in the calla lily image, with a mix of weakly- and strongly-pigmented cells, is seen for many species.

In most species there is further spatial regulation of anthocyanin biosynthesis to generate floral pigment patterning. These patterns may be as simple as separation into inner (or throat) and outer petal regions. However, they may also be comprised of spots, stripes, irregular blotches, or combinations of all of these (Fig. 3.4). These may be accompanied by morphological changes to assist in pollinator attraction. Providing some of the most dramatic examples are flowers of the orchid genus *Ophrys*, which use a combination of colour, scent, and shape to mimic female bees, causing the male bee to attempt copulation, and coincidentally, pollination (Paxton and Tengo 2001). There are now data on the molecular mechanism behind at least one type of floral pigment pattern, stripes associated with the flower veins (Fig. 3.4), which in antirrhinum is controlled by the *Venosa* gene (see below and Schwinn et al. 2006).

Although data on these higher-level developmental or environmental signals is limited, there is now a good understanding of the direct regulation of the biosynthetic genes. The evidence to date suggests that changes in the transcription rates for the biosynthetic genes are the key regulatory target for anthocyanin biosynthesis in flowers, and that this is mediated by transcription factors (TFs) of the R2R3-MYB and basic Helix-Loop-Helix (bHLH) (or MYC) type (reviewed in Davies and Schwinn 2003; Vom Endt et al. 2002; Grotewold 2006). The involvement of MYB and/or bHLH factors in regulating anthocyanin biosynthesis has been shown for several species, both dicot and monocot, and is likely to be a mechanism that is conserved for most plants. Furthermore, species for which detailed studies are available, specifically antirrhinum, arabidopsis, maize (*Zea mays*), morning glory (*Ipomea tricolor*) and petunia, have been found to contain anthocyanin-regulating gene families for both types of TF (Table 3.2). The function of the gene families has recently been characterised for antirrhinum (Schwinn et al. 2006), in which three MYBs genes and two bHLH genes regulating anthocyanin biosynthesis have been identified (Goodrich et al. 1992; Schwinn et al. 2006). The bHLH genes are *Delila* and *Mutabilis* and the MYB genes are *Rosea1*, *Rosea2* and *Venosa*. The combined action of these genes, through variation in both transcript abundance and biochemical activity of the encoded proteins, allows for the complex spatial variations in pigmentation that can occur in the *Antirrhinum* genus. Thus, *Delila* is required for petal tube pigmentation, while both *Delila* and *Mutabilis* contribute in the lobes. *Rosea1* enables

Table 3.2 Anthocyanin-related transcription factor gene families in some ornamental and/or model plant species

Species	R2R3-MYB	BHLH	WD-Repeat	References
Antirrhinum	ROSEA1	DELILA,	–	Goodrich et al. (1992)
majus	ROSEA2	MUTABILIS		Schwinn et al. (2006)
	VENOSA			
Arabidopsis	PAP1/AtMYB75	GL3	TTG1	Walker et al. (1999)
thaliana	PAP2/AtMYB90	EGL3		Borevitz et al. (2000)
	(MYB113)			Payne et al. (2000)
	(MYB114)			Ramsay et al. (2003)
				Zimmermann et al. (2004)
Ipomoea nil	INMYB1	INBHLH1	INWDR1	Park et al. (2004)
	INMYB3	INBHLH2	INWDR2	Morita et al. (2006)
	INMYB3	INBHLH3		
Petunia hybrida	AN2	AN1	AN11	de Vetten et al. (1999)
				Quattrocchio et al. (1999)
				Spelt et al. (2000)
Zea mays	C1	B-PERU	PAC1	Chandler et al. (1989)
	PL	LC		Ludwig and Wessler (1990)
		IN1		Cone et al. (1993)
		R		Carey et al. (2004)

strong, "wild-type" pigmentation, and *Rosea2* is a closely linked gene that only colours the inner epidermis weakly. As mentioned earlier, *Venosa* gives a venation phenotype, and *in situ* RNA analysis reveals that *Venosa* is expressed only in small groups of cells between the flower veins and adaxial epidermis (the author and co-workers' unpublished data). Analysis of a range of *Antirrhinum* species showed that differences in anthocyanin pigmentation between at least six species were attributable to variations in the activity of the *Rosea* and *Venosa* loci. This observation fits with data on anthocyanin regulation in other species to suggest that the activity of the MYB genes is probably the major determinant of the natural variation of anthocyanin production observed in plants.

Although it is the MYB and bHLH TFs that have been characterised in detail for regulation of anthocyanin biosynthesis in petals, there will likely be several other proteins that are also involved either as members of the transcriptional complex for the biosynthetic genes, or as part of the transduction pathway from the developmental and environmental signal to the MYB and bHLH factors. One example, now identified for several species (de Vetten et al. 1997; Walker et al. 1999; Sompornpailin et al. 2002; Carey et al. 2004), is the WD-repeat (or WD40) protein, which may assist in stabilising the MYB-bHLH complex.

3.5 Genetic Modification of Anthocyanin Biosynthesis

It is now almost 20 years since the first publication detailing the generation of new flower colours using GM (Meyer et al. 1987). During the intervening years there have been many notable successes in molecular breeding of flower colour, the most recent being the introduction of aurone production for yellow flower colours in torenia (Ono et al. 2006a). However, GM has yet to make a significant impact on the ornamentals industry. The only GM flowers commercialised to date are carnation engineered to produce delphinidin-derived anthocyanins and novel mauve-violet colours. Thus, there are clearly major barriers that must be overcome, principally regulatory approval costs and freedom to operate for intellectual property, before the new GM cultivars produced can make it into commercial trade. Nevertheless, given the demand for novelty in the world ornamental trade, in the long term, it would be surprising if GM technologies did not become a major component of new cultivar development. A partial survey reveals that applications have been made to date in the USA, Europe, Japan or Australasia, for the field trial of GM cultivars of almost 20 different ornamental species.

There are significant opportunities for application of GM in most of the leading cut flower, potted plant and bedding crops. For example, chrysanthemum, gerbera and rose lack cultivars with production of delphinidin-derived anthocyanins (and associated blue/mauve colours), and cyclamen (*Cyclamen persicum*), impatiens, pelargonium and saintpaulia lack cultivars with bright yellow colours. These are just some of the more obvious opportunities that can be identified. To give an idea of the scope available for application of GM technologies for anthocyanins in ornamental crops, Table 3.3 lists a range of species that lack one or more of the main anthocyanin types. Furthermore, this list does not include examples where a major colour group is lacking from an important crop but no gene technology is yet available to address the reasons for lack of the colour – such as species like tulip, in which delphinidin production occurs but blue colours do not result. Nor does it include crops such as African marigold (*Tagetes erecta*), daffodil (*Narcissus* sp.) and *Sandersonia aurantiaca*, which generally lack any significant anthocyanin-pigmented cultivars.

The following sections cover some of the current approaches to modifying anthocyanin biosynthesis in flowers. Also discussed are some of the anthocyanin types for which there has been no success to date with regard to GM of their production, such as 3-deoxyanthocyanins and 6-hydroxyanthocyanins, but for which there is an increasing understanding of their biosynthetic mechanism. The final sections of the review address the use of transcription factor genes for GM of anthocyanin production in ornamentals.

Table 3.3 Examples of commercial crops with opportunities for introduction of new anthocyanin types. Species are listed for which the available non-GM cultivars do not produce one or more of the major groups of anthocyanins, with reference to the three principal anthocyanidin types, pelargonidin, cyanidin and delphinidin

Species common name	Species binomial	Anthocyanidin type lacking
Calla lily	*Zantedeschia* sp.	Pelargonidin, delphinidin
Carnation	*Dianthus caryophyllus*	Delphinidin
Chrysanthemum	*Dendranthema*	Pelargonidin, delphinidin
Cyclamen	*Cyclamen persicum*	Pelargonidin
Cymbidium orchid	*Cymbidium hybrida*	Pelargonidin, delphinidin
Dendrobium orchid	*Dendrobium* sp.	Delphinidin
Gerbera	*Gerbera hybrida*	Delphinidin
Nierembergia	*Nierembergia* sp.	Pelargonidin, delphinidin
Petunia	*Petunia* sp.	Pelargonidin
Rose	*Rosa hybrida*	Delphinidin
Snapdragon	*Antirrhinum majus*	Delphinidin

3.5.1 Preventing Anthocyanin Production

There are a number of reasons why preventing anthocyanin production may be advantageous in molecular breeding for ornamental crops. Firstly, it may enable development of white-flowered cultivars. Secondly, it may be used to remove anthocyanins to allow carotenoid pigments to be seen more clearly, for example for better yellow colours. Thirdly, experiments to date have shown that, unexpectedly, some approaches to gene inhibition may generate novel floral pigmentation patterning. There are two basic approaches to preventing anthocyanin production; inhibiting production of a key biosynthetic enzyme or introducing an enzyme that competes with an anthocyanin biosynthetic enzyme for substrate.

Inhibiting production of an anthocyanin biosynthetic enzyme may be achieved by targeting the enzyme itself (e.g. through single-chain antibodies), targeting the RNA transcript (e.g. by RNAi) or by preventing the transcription of the encoding gene. These three approaches have all been used in transgenic plants. By far the most widely used approach has been the targeting of the encoding RNA. Inhibition of gene transcription using repressor proteins has only recently been demonstrated for anthocyanin production (see Section 3.5.7), and the targeting of DFR with single-chain antibodies has been only partially successful (e.g. Santos et al. 2004). It is worth noting that in several instances anthocyanin production has been used as a convenient reporter system, rather than through a direct interest in generating new flower colours.

The first published examples of prevention of anthocyanin biosynthesis by gene inhibition in transgenic plants were for CHS, using initially antisense RNA (van der Krol et al. 1988) and subsequently sense RNA (Napoli et al. 1990; van der Krol et al. 1990a). Since then there have been numerous published examples of using antisense or sense RNA inhibition for the anthocyanin pathway, mostly targeting CHS or DFR. The species targeted for CHS and/or DFR include carnation (Gutterson 1995), chrysanthemum (Courtney-Gutterson et al. 1994), cyclamen (Tanaka et al. 2005), gentian (Nishihara et al. 2003), gerbera (*Gerbera hybrida*, Elomaa et al. 1993), lisianthus (Deroles et al. 1998), petunia (*Petunia hybrida*, van der Krol et al. 1988, 1990a; Napoli et al. 1990; Tanaka et al. 1995; Jorgensen et al. 2002), rose (Firoozabady et al. 1994) and torenia species (*T. fournieri* or *T. hybrida*, which is a derivative of *T. fournieri* x *T. concolor*) (Aida et al. 2000a, 2000b; Suzuki et al. 2000; Fukusaki et al. 2004). There are also published examples of inhibiting F3H (in carnation, Zuker et al. 2002), ANS (in torenia, Nakamura et al. 2006) and CHI (in tobacco, Nishihara et al. 2005) production. A listing of publications reporting inhibition of flavonoid biosynthesis in transgenic plants is given in Davies and Schwinn (2006).

In all cases, lines with the expected pale or white flower colour phenotypes were obtained. However, there were also some unexpected phenotypes. In some, but not all species, inhibiting flavonoid production resulted in male (e.g. Taylor and Jorgensen 1992; Fischer et al. 1997) or female (Jorgensen et al. 2002) sterility. In carnation, inhibition of F3H production resulted in an increase in methylbenzoate levels and more fragrant flowers (Zuker et al. 2002). With regard to flower colour, both ordered and erratic corolla pigmentation patterns have been obtained in addition to a general reduction in anthocyanin production, but to date only for species that naturally have patterned varieties. Thus, patterns were found for lisianthus, petunia and torenia but not carnation, chrysanthemum, gerbera and rose. This suggests that the variations seen in the effectiveness of the gene inhibition process are determined by predetermined morphological signals within the petal. However, such morphological signals, for example related to the petal junctions or midribs, can also interact with environmental signals influencing pigmentation (van der Krol et al. 1990b). Figure 3.5 shows examples of both white and patterned flowers of torenia produced by inhibition of ANS gene activity in a purple-flowered line.

With regard to the commercial usefulness of the GM-generated patterns, some of the more novel patterns show instability, not only within a particular plant but also in their inheritance (e.g. Jorgensen 1995; Bradley et al. 2000). Stability may also vary

Moonaqua™   Moonlite™   Moonshade™   Moonvista™

Fig. 3.5 Top row left: Flowers of a non-transgenic line of torenia (left) and two lines containing a transgene for inhibition of ANS gene activity. Top row right: A flower of a transgenic rose cultivar engineered to accumulate delphinidin-derived anthocyanins. (Top row photographs courtesy of Suntory Ltd, http://www.suntory.co.jp). Centre row: Examples of transgenic carnation cultivars engineered to accumulate delphinidin-derived anthocyanins (photographs courtesy of Suntory Ltd and Florigene Pty Ltd, http://www.florigene.com). Bottom row left: Flowers of a non-transgenic line of lisianthus (top) and a line containing a transgene for inhibition of F3′,5′H gene activity. Bottom row right: Flowers of a line of lisianthus containing a transgene for inhibition of CHS gene activity, with all flowers shown being on a single plant. See Plate 4 for colour version of these photographs

for the non-patterned white varieties. Nakamura et al. (2006) studied the impact of the choice of target gene and the method used for its down-regulation. White flowered torenia (*T. hybrida*) were originally obtained through sense suppression of CHS or DFR in a purple-flowered variety (Suzuki et al. 2000), but the phenotypes were not stable and the loss of CHS activity was thought detrimental to plant health. As an alternative, ANS was targeted in the same torenia cultivar using gene constructs for antisense suppression, sense suppression/cosuppression or RNAi hairpins. Around half of the RNAi transgenics had white flowers, but few or no white flowers were found for antisense or sense suppression transgenics, respectively. The RNAi suppression phenotype was shown to be stable for at least three years.

The type of patterning seen suggests that some of the patterning observed in non-GM cultivars might be due to endogenous RNA inhibition processes. With advances in the understanding of RNA-dependent silencing in plants it has become easier to test for endogenous RNA silencing, in particular by detection of the small interfering RNAs or the use of viruses that suppress the silencing process. These methods have been used to show that inhibitory alleles of CHS trigger RNA silencing in both *C2-Inhibitor diffuse* lines of maize (Della Vedova et al. 2005) and the Red Star cultivar of petunia (Koseki et al. 2005). Although these mechanisms may account for some of the patterning of pigmentation seen in flowers, it is still likely that the majority of patterns observed are due to the programmed, cell-specific transcription of regulatory factors such as the MYB and bHLH TFs (see Section 3.4). However, it is possible that RNA silencing plays a role in determining these "ordered" pigment patterns too, as microRNAs have recently been shown to be part of the spatial regulatory process for some MYB genes in plants (Millar and Gubler 2005).

The over-production of enzymes that compete for substrate with anthocyanin-biosynthetic enzymes has been successful in reducing pigment levels, generating plants with paler flower colours, but has not been shown to reduce the pathway sufficiently to result in white flowers. Examples to date include the introduction of stilbene synthase (Fischer et al. 1997) or polyketide reductase to compete with CHS (Davies et al. 1998; Joung et al. 2003), or anthocyanidin reductase to compete with F3GT (Xie et al. 2003). The expected products – stilbenes, 6′-deoxychalcones and flavan-3-ols (or their derivatives) – accumulated in the flowers, and in the case of 6′-deoxychalcones imparted pale yellow pigmentation to the flowers (Davies et al. 1998).

3.5.2 Increasing Anthocyanin Production by Altering Biosynthetic Enzyme Activity

If either a rate-determining biosynthetic step or key flux points can be identified, then introducing new biosynthetic activities may permit increased anthocyanin levels. Although it was initially thought that over-production of CHS might direct more substrate into the flavonoid pathway, and thus increase anthocyanin levels, this has not proven to be the case when transgenics have been produced containing a CHS transgene driven by the strong Cauliflower Mosaic Virus 35S promoter (*35S:CHS*). The successful examples of using over-production of biosynthetic enzymes to increase anthocyanin production have been with DFR, for petunia and tobacco (Davies et al. 2003; Polashock et al. 2002), and DFR and ANS together into forsythia (*Forsythia x intermedia*) (Rosati et al. 1997, 2003). These examples were probably successful because DFR is a competing flux step with FLS in many species, and the target species or cultivars had low (or no) anthocyanin levels to begin with. This also indicates the alternative approach for increasing anthocyanin biosynthesis of redirection of pathway flux by reducing FLS production. Thus, inhibition of FLS production in lisianthus, petunia or tobacco resulted in an increase in anthocyanin content in the flowers (Holton et al. 1993b; Nielsen et al. 2002; Davies et al. 2003).

Overall transgenes for biosynthetic enzymes have been only partially successful in increasing anthocyanin biosynthesis in target species. Their most promising application is to complement missing enzyme steps, as accomplished with forsythia. An alternative is the use of TF transgenes to increase levels of multiple biosynthetic activities (see Section 3.5.7).

3.5.3 Anthocyanins with Unusual Patterns of A- or C-Ring Hydroxylation

The 3-deoxyanthocyanins provide colours in the orange to bright red range to flowers of some species of the Gesneriaceae, including *Columena hybrida, Gesneria cuneifolia, Kohleria eriantha* and *Sinningia cardinalis* (Bohm 1998; Fig. 3.3). Various 3-deoxyflavonoids are also produced in grass species such as maize and sorghum (*Sorghum bicolor*). The absence of the 3-hydroxyl arises through the action of a variant DFR termed the flavanone 4-reductase (FNR). FNR can use both DHFs and flavanones as substrates, producing either leucoanthocyanidins (flavan-3,4-diols) or flavan-4-ols. FNR activity has been demonstrated for recombinant DFR/FNR proteins from species that commonly produce 3-deoxyanthocyanins (e.g. maize, *S. cardinalis*), and species that can produce 3-deoxyflavonoids under some circumstances (*Malus domestica* and *Pyrus communis*) (Fischer et al. 2003; Halbwirth et al. 2003; Winefield et al. 2005). As the enzymes studied to date have both FNR and DFR activity, it is likely that other conditions need to be met before 3-deoxyanthocyanin biosynthesis occurs. One aspect is the need for reduction in the potentially competitive F3H activity (Lo and Nicholson 1998; Winefield et al. 2005). Another is the requirement for an A5GT that can accept the 3-deoxyanthocyanidin substrates. With the exception of the rose A3,5GT, recombinant A5GTs studied to date (from 3-hydroxyanithocyanin producing species) require, at a minimum, prior 3-*O*-glucosylation of the substrate (Yamazaki et al. 1999; 2002). It is not certain whether a specific ANS is also required, as the 3-hydroxyl that is thought to be important in the ANS reaction mechanism (Wilmouth et al. 2002; Nakajima et al. 2006) is lacking from the substrates (e.g. apiforol). Thus, GM for introduction of 3-deoxyanthocyanin biosynthesis into additional species may require not only introduction of a transgene for FNR into a cultivar with low F3H activity, but potentially additional transgenes for downstream enzymes able to use the 3-deoxy substrates.

The most common hydroxylation pattern of the A-ring of anthocyanins is for the C-5 and C-7, which are introduced during the formation of chalcones by CHS. A small number of species produce anthocyanins with 6- or 8-hydroxylation, in particular *Alstromeria* (Saito et al. 1988; Tatsuzawa et al. 2003). The additional hydroxyl groups shift the spectrum of the pigment further towards the orange-red colours. There is little information on the biosynthesis of 6-hydroxyanthocyanins, but there are some details for other 6- and 8-hydroxyflavonoids. Latunde-Dada et al. (2001) identified a P450 cDNA whose recombinant product had flavonoid 6-hydroxylase (F6H) activity with flavanone and DHF substrates, including liquiritigenin. It is thought the F6H is part of the biosynthetic route to isoflavonoids with 6,7-dihydroxylation, with the aryl migration of the B-ring occurring subsequently to the 6-hydroxylation. Halbwirth and Stich (2006) have characterised a NADPH and FAD dependent enzyme that catalyzes hydroxylation of flavonols and flavones at the C-8 in petals of *Chrysanthemum segetum*.

3.5.4 Generating New Flower Colours by Altering Anthocyanin B-Ring Hydroxylation

The 4′-hydroxyl group of anthocyanins is incorporated during formation of the precursor 4-coumarate by the cinnamate 4-hydroxylase (C4H). The addition of the 3′ and 5′ hydroxyl groups is usually catalyzed by F3′H and F3′,5′H. Thus, in general, the presence of the different anthocyanidin types can be accounted for by the activity of these two enzymes. Given the impact of B-ring hydroxylation on flower colour, an obvious approach for flavonoid biotechnology is to direct the accumulation of pelargonidin-, cyanidin- or delphinidin-derived anthocyanins by inhibiting or adding F3′H and/or F3′,5′H. DNA sequences for both enzymes are now available from many species (Holton et al. 1993a; Brugliera et al. 1999; Seitz et al. 2006).

Blue flower colours are lacking from some of the leading floriculture species, such as rose, carnation, chrysanthemum, cyclamen, daffodil, gerbera, impatiens, lisianthus, pelargonium, poinsettia and tulip. From the known anthocyanin biosynthetic pathway it might be assumed that introduction of a *35S:F3′,5′H* transgene would be sufficient to direct synthesis of the delphinidin precursors and generate blue flower colours in target species. However, biotechnology for blue flowers has not proved that simple. The most successful examples are the range of carnations with mauve and violet flower colours developed by Florigene Ltd (Melbourne, Australia), which are now in commercial release (Lu et al. 2003; Fig. 3.5). However, the attractive and novel flower colours now available were only achieved after much research to enhance the impact of the F3′,5′H transgene. Initial experiments introducing F3′,5′H into cyanidin-accumulating carnation or tobacco lines generated transgenic plants with relatively low levels of delphinidin accumulation and only small changes in flower colour (Shimada et al., 1999; Tanaka et al. 2005). For the petunia F3′,5′H, it was subsequently discovered that it required the alternative electron donor Cyt *b* (*difF*) for full activity (de Vetten et al. 1999), and co-introduction of a *difF* transgene markedly increased the level of delphinidin in flowers of transgenic carnation and a significant shift towards purple colours (Tanaka et al. 2005). However, more dramatic changes in flower colour were obtained when genetic and transgenic approaches were combined. Competition from F3′H and a preference of the endogenous DFR for DHK or DHQ were identified as problems for maximizing delphinidin production, so the F3′,5′H transgene was co-introduced with a transgene for a DFR that preferred DHM into a plant background that accumulates DHK (a *DFR/F3′H* double mutant) (International Patent Application WO96/36716; Tanaka et al. 2005). The resultant cultivars have been marketed as the Moon series in Australia, Japan and the USA (Tanaka et al. 2005). Fukui et al. (2003) examined in detail the flavonoids present in the petals of the Moonshadow™ cultivar. The major pigment was delphinidin 3,5-O-diglucoside-6″-O-4,6‴-O-1-cyclic-malyl diester, the same pigment as commonly accumulates in carnation petals but with the addition of an hydroxyl at the C-5′. Thus, in keeping with studies of other species (see Section 3.2), the secondary modification enzymes of carnation do not seem to be inhibited by the presence of the additional hydroxyl group. Besides carnation, a transgene for lisianthus F3′,5′H has been used to introduce delphinidin production, and blue flower colour, into a pink-flowered cultivar of lobelia (*Lobelia erinus*, Tanaka et al. 2005). Furthermore, in research yet to be published in detail, rose cultivars have been engineered to produce delphinidin using a F3′,5′H transgene, resulting in cultivars with novel colours suitable for commercial release (Suntory Press Release 8826, 30 June 2004; Fig. 3.5).

A further improvement in the 'blue gene' technology has been suggested from analysis of the efficiency of different F3′,5′H proteins for directing production of delphinidin precursors. Thus, a transgene for the Canterbury bells (*Campanula medium*) F3′,5′H generated a higher level of delphinidin production in tobacco (99% of total anthocyanins) than when transgenes for the petunia or lisianthus F3′,5′H were used (Okinaka et al. 2003).

Despite the success with generation of novel, delphinidin-based colours in carnation, and more recently rose (Tanaka et al. 2005), the highly sought-after true blue flower colours have not been achieved. Evidently, the presence of delphinidin-derived anthocyanins is not sufficient in itself for blue flower colour. A good illustration of this is that some of the ornamental species listed earlier as lacking blue colours can already produce delphinidin-derived anthocyanins in the flowers, such as cyclamen, impatiens, lisianthus, pelargonium and tulip. In addition to the presence of delphinidin-derived anthocyanins, the vacuolar pH (>pH 5.5 is preferable), and intra- or inter-molecular (co-pigmentation) interactions need to be appropriate to enable blue colours to result. Fukui et al. (2003) review the role of these factors in formation of the mauve carnation colours. Although the petals of the transgenic carnation produce acylated delphinidins, acylated flavones and have a pH around 5.5, all of which favour blue flower colours, the actual colours are mauve-violet rather than true blue.

Some of the most attractive blue flower colours are based on complex anthocyanins, which may have features such as the presence of multiple monosaccharide and acyl units or the covalent attachment of co-pigments. Furthermore, there is an increasing number of reports of anthocyanin-metal ion complexes. A good example of a metalloanthocyanin is the blue pigment commelinin from *Commelina* flowers, which consists of six molecules of a malonated delphinidin 3,5-O-diglucoside co-pigmented with six molecules of a methylated derivative of an apigenin 4′-O-glucoside-6-C-glucoside and complexed with two magnesium atoms. Excellent reviews of the role of anthocyanin structure and interactions in generating blue flower colours have been presented by Goto and Kondo (1991), and more recently Andersen and Jordheim (2006). Shiono et al. (2005) recently used X-ray crystallography to determine how the cyanidin-based anthocyanins in cornflower (*Centaurea cyanus*) generate blue flower colours, and demonstrated the formation of a tetrametal complex containing one ferric ion, one magnesium and two calcium ions. With regard to the use of GM technology for imparting production of "blue anthocyanins" to additional species, DNA sequences are not available for several of the required enzyme activities. Even if all the necessary sequences were available, successful GM would still likely entail the introduction of multiple gene constructs into the target species. Nevertheless, the examples of blue flower colours from cyanidin-based anthocyanins, such as cornflower and morning glory, do allow for the development of GM approaches to engineering blue flower colours without the use of F3′,5′H transgenes.

Despite the critical role of pH in determining anthocyanin-based flower colours, there are no reports of engineering significant changes in pH in ornamental crops. This is despite knowledge of the genetics and molecular biology for control of petal vacuolar pH in petunia and morning glory (Fukada-Tanaka et al. 2000; Spelt et al. 2002). The *Purple* gene of morning glory encodes a Na^+/H^+ antiporter that is expressed coincident with flower opening, when the pH changes from 6.5 to 7.5 and flower colour from purple to blue, and mutant analysis suggests it plays a key role in this process (Fukada-Tanaka et al. 2000). Similar sequences are present in petunia, nierembergia (*Nierembergia* sp.) and torenia (Yamaguchi et al. 2001), suggesting that the transporter may provide useful gene technology for controlling pH in target species. However, given the importance of maintaining an appropriate cellular pH, it is likely that mechanisms exist that may counterbalance changes induced by a single transporter transgene. Interestingly, some of the anthocyanin regulatory MYBs also modify vacuolar pH, probably via interaction with a different group of bHLH factors than those involved in anthocyanin regulation (Spelt et al. 2002). The identification of the target genes for these TFs may enable a better understanding of the regulation of vacuolar pH in flowers, and provide more effective gene technology.

Several leading ornamental species, such as those listed in Table 3.3, lack pelargonidin-accumulating commercial cultivars. Inhibition of F3′H and/or F3′,5′H gene activity to encourage production of pelargonidin- or cyanidin-derived anthocyanins in species that lack them should, in theory, be more straightforward than generating transgenics accumulating delphinidin-derivatives. Indeed, there are some successful examples of the switching of the type of anthocyanin accumulated by manipulating only F3′H and/or F3′,5′H gene activity, with the most complete studies perhaps being for torenia. Suzuki et al. (2000) used inhibition of F3′,5′H in transgenics of a commercial, purple-flowered male and female sterile line of torenia (cv. Summerwave Blue) to generate a pink-flowered line that produced pelargonidin-based anthocyanins. Subsequently to this, Ueyama et al. (2002) introduced a sense F3′H transgene into the F3′,5′H inhibited Summerwave Blue transgenic lines to increase the amount of cyanidin-based anthocyanins and generate red flower colours. Figure 3.5 illustrates a similar example to that of Suzuki et al. (2000) but for lisianthus (K. Nielsen and S. Deroles, Crop & Food Research, New Zealand, unpublished data). An antisense gene construct to F3′,5′H was introduced into a lisianthus cultivar that accumulates delphinidin-based anthocyanins, and the transgenic plants produced pelargonidin-based anthocyanins and had pink flower colours.

Experiments have been also been conducted for both the F3′H and F3′,5′H in petunia. Shimada et al. (2001) used sense-suppression constructs for F3′,5′H inhibition in petunia lines that normally accumulate delphinidin-based anthocyanins, resulting in accumulation of a mix of cyanidin- and delphinidin-based anthocyanins in the transgenics. Jorgensen et al. (2002) also inhibited F3′,5′H production in petunia, but in conjunction with DFR cosuppression, so the resultant plants lacked anthocyanin biosynthesis. Dramatic changes in flower colour have been produced through engineering pelargonidin biosynthesis into petunia, which normally lacks commercial cultivars with significant accumulation of pelargonidin-based anthocyanins. Tsuda et al. (2004) used inhibition of F3′H in petunia to generate target lines for subsequent introduction of DFR transgenes, to generate pelargonidin production and novel flower colours. Previously, in what was the first published example of GM for novel flower colours, Meyer et al. (1987) had used a petunia line with mutations in the F3′H and F3′,5′H genes as the target for a *35S:maizeDFR* transgene. The subsequent crosses of these lines into commercial cultivars produced F_2 plants with attractive orange and bright red colours novel to this species (Meyer 1991; Griesbach 1993). Nierembergia is another Solanaceous ornamental that lacks commercial cultivars with pelargonidin-based anthocyanins. As part of determining the reasons for lack of pelargonidin production in this species, Ueyama et al. (2006) used an antisense construct to inhibit F3′,5′H in a purple-flowered cultivar. The transgenic plants had reduced levels of delphinidin-based anthocyanins but accumulated kaempferol, rather than cyanidin- or pelargonidin-based anthocyanins, suggesting that the endogenous DFR has low activity with DHK.

The conditions necessary for generating orange or bright red colours from pelargonidin are less well defined than those for blue colours. For example, lisianthus flowers can produce pelargonidin-, cyanidin- and delphinidin-based anthocyanins, yet the colour range is limited to a range of shades from pink to purple. A low vacuolar pH is certainly one factor that promotes red-orange colours. There is also the suggestion that loss of co-pigments encourages a red shift (Asen et al. 1971; Arisumi et al. 1990). It is common for flowers to contain a mix of anthocyanin types. Developing cultivars that contain only pigments based on one anthocyanidin may provide for more attractive flower colours.

3.5.5 Changing Flower Colour by Altering Anthocyanin Secondary Modifications

The importance of secondary modification for generating blue flower colours has been discussed in the previous sections. For GT and AAT activities, their introduction into other species has resulted in changes in the anthocyanin types produced (Schwinn et al. 1997; Suzuki et al. 2002; Fukuchi-Mizutani et al. 2003), but there is only one example of a colour change resulting (Brugliera et al. 1994). In this case, altering the glycosylation pattern also affected the methylation pattern of the anthocyanin, and it was this that produced the colour change. The common anthocyanin methylation pattern at the 3′- and 5′-hydroxyls, generating the anthocyanidins peonidin, petunidin and malvidin, has a noticeable reddening effect on the resulting flower colours. This methylation is carried out by *S*-adenosyl-l-methionine (SAM)-dependent *O*-methyltransferases (OMTs), and they have been characterised for petunia, being encoded by the genes *Mt1, Mt2, Mf1* and *Mf2*. However, the cloning and analysis of the corresponding cDNAs has been reported in detail only in the patent literature to date (International Patent Application WO03/062428). The patent also describes cDNAs for anthocyanin OMTs from *Fuchsia, Plumbago* and torenia. Anthocyanins with methylation at the 5- or 7-hydroxyls are also known, although there are few data on the enzymes involved.

3.5.6 Black Flower Colours

There are few studies on so-called black flowers. There are many species with very dark flower colours, but which are clearly dark purple rather than black. Figure 3.3 includes an example of such a flower, the Black Bat Flower (*Tacca chantrieri*). True black flowers would absorb light in all of the visible wavelengths, and to be black to some birds and insects would need to absorb in the UV wavelengths also. In a survey of over 500 species by Chittka et al. (1994) no flowers were found that absorbed wavelengths in the four spectral domains chosen to cover the range between 300 and 700 nm, indicating the rarity of black colours. However, there are examples of flowers that appear almost black. Two species with black flower colours that are more commonly grown in gardens are pansy (*Viola tricolor*) and tulip. For both species, where black-flowered cultivars have been characterised, the major anthocyanin pigments are delphinidin-based (Goto et al. 1978; Markham et al. 2004). In at least the case of pansy, it seems likely that the background presence of yellow carotenoids in the sub-epidermal cells contributes considerably to the depth of colour and the black appearance. *Lisianthus nigerescens*, a native of Central America, has been termed "the blackest flower in the world". Markham et al. (2004) studied the chemistry behind the black colour. Spectral analysis confirmed that the petals absorbed almost all wavelengths of both UV and visible light, suggesting it indeed qualifies as a black flower. The anthocyanins, which as for pansy and tulip were delphinidin-based, were produced to an extraordinary level of 24% of the petal dry weight, and were present in an approximately 1:1 ratio with flavone co-pigments. In comparison, purple-flowered cultivars of lisianthus (*E. grandiflorum*), such as those shown in Fig. 3.3, accumulate anthocyanins to only around 1 to 1.4% dry weight (Markham et al. 2000).

The molecular basis of the production of uncommonly high levels of delphinidin pigments in flowers such as *L. nigerescens* is not known, and no GM approaches to generating black flowers have been published. To date, although transgenic plants have been produced that have greatly increased anthocyanin levels from over-expression of anthocyanin regulatory genes (discussed in the following section), they have not generated black flower colours. It is possible that the pathways supplying precursors to the flavonoid pathway will also need to be up-regulated. Furthermore, it has been suggested that AVI structures may assist in the accumulation of high concentrations of anthocyanins and darker flower colours (Markham et al. 2000; 2004).

3.5.7 GM Application of Anthocyanin-related Transcription Factors

Although it is not yet clear what role metabolic channeling may have in regulating the activity of the flavonoid pathway, the evidence to date suggests that anthocyanin production is primarily dependent on the transcriptional activity of the biosynthetic genes, with little impact from post-transcriptional regulation. As outlined in Section 3.4, transcription of the biosynthetic genes can be induced by the activity of a few TFs, principally of the MYB and bHLH type. It was first shown in maize that ectopic expression of one MYB and one bHLH from the anthocyanin-related gene families was necessary and sufficient to induce anthocyanin production in a wide range of maize cell types, indicating that the WD40 factor was either constitutively present or induced by the MYB and bHLH (Ludwig et al. 1990). Thus, the TF transgenes offer the prospect of powerful gene technology for modulating the amount as well as the temporal and spatial occurrence of anthocyanins in plants. This has subsequently been found to hold true in experiments with several target species, although differences in the efficacy of TFs from different species have become apparent. In some of the successful experiments the power of TF transgenes to up-regulate whole pathways has been apparent. For example in petunia, a transgene for the maize bHLH factor LC up-regulated at least eight biosynthetic genes in the leaves, resulting in plants with deep purple foliage (Bradley et al. 1998; Fig. 3.4).

Many of transgenic experiments using anthocyanin-related TF transgenes have been in non-ornamental species. Commonly, model species such as arabidopsis, tobacco and tomato have been used to test either the function of the TF or the utility of TF transgene technology (Lloyd et al. 1992; Mooney et al. 1995; Goldsbrough et al. 1996; Bradley et al. 1998; Elomaa et al. 2003; Gong et al. 1999; Borevitz et al. 2000; Mathews et al. 2003). Other experiments have targeted either proanthocyanidin biosynthesis for improved agricultural characters (Damiani et al. 1999; de Majnik et al. 2000; Robbins et al. 2003), or food crops for increasing flavonoid content for human health benefit (Bovy et al. 2002; Mathews et al. 2003). Results from these experiments have been varied. In some examples, especially with Solanaceous species, dramatic increases in anthocyanin production occurred. There are also examples of no apparent change in flavonoid biosynthesis occurring. Interestingly, in some cases, the over-expression of anthocyanin-related TFs also enhanced production of other flavonoids, such as proanthocyanidins, flavonols and isoflavonoids. It is not clear why what are thought to be non-target pathways can be up-regulated. Possible alternatives are that the TFs increase levels of required precursor compounds, that the high levels of TF protein produced in transgenics result in recognition of additional promoter sites or interference with the endogenous regulatory environment, or that they have previously unidentified regulatory roles for other branches of the phenylpropanoid pathway as part of their normal function.

Surprisingly, there are few published accounts of using TF transgenes for increasing anthocyanin production with the intent of generating new ornamental cultivars. Early experiments in non-commercial lines of petunia demonstrated the potential for generating cultivars with attractive vegetative pigmentation (Bradley et al. 1998). The other published results to date for stable transgenics are for the introduction of a *35S:LC* transgene into lisianthus and pelargonium (Bradley et al. 1999), *Caladium bicolor* (Li et al. 2005) and chrysanthemum (Boase et al. 1998). Transgenic *C. bicolor* plants had enhanced anthocyanin accumulation in the roots, leaves and stems, including the vascular bundles. However, no change in phenotype was observed for the other species, although in the case of chrysanthemum no transgene expression was detected. In addition, gene expression analysis for the lisianthus and pelargonium transgenics failed to reveal any change in transcript levels for anthocyanin biosynthetic genes tested. The simplest explanation for the lack of phenotype in, for example, the lisianthus transgenics, is that a suitable MYB to interact with the LC transgene product is lacking. This has been tested by generating *35S:LC+35S:ROSEA* double transgenics, and comparing them with transgenics containing *35S:ROSEA* or *35S:LC* alone (the author and coworkers' unpublished data). *35S:ROSEA* transgenics do indeed have some enhancement of anthocyanin biosynthesis, most notably in the sepals and young buds, suggesting that a suitable MYB factor is lacking in some tissues. However, the LC+ROSEA double transgenics display no increase in anthocyanin biosynthesis beyond that observed for the *35S:ROSEA* alone transgenics.

It is not yet clear why the co-expression of an anthocyanin-related MYB and bHLH is sufficient to induce anthocyanin biosynthesis in species such as maize and tobacco but not in others such as lisianthus. With regard to the lisianthus experiments, it is known from petunia transgenics that the LC and ROSEA factors, produced from the same transgenes as used with lisianthus, can interact to induce anthocyanin biosynthesis (Dr K. Schwinn, Crop & Food Research, New Zealand, unpublished data). It is possible that a required WD40 factor is lacking, but data from other species suggest that the WD40 genes are expressed in most tissues. Other possibilities are that there is a transcriptional repressor present in some plants, or that other TFs are lacking, perhaps those required to up-regulate the early genes for supply of flavonoid precursors. It is unlikely that LC+ROSEA cannot recognise the target promoter sequences, as some anthocyanin biosynthesis is induced in *35S:ROSEA* lisianthus transgenics. Another interesting result from the transgenic studies is that apparently equivalent TFs can have different effects in transgenics of the same species, for example alfalfa (*Medicago sativa*) (Ray et al. 2003). This suggests that the TFs from different species, and indeed different members of the gene families of one species, may differ in their interaction with other TFs or in their promoter recognition or activation strengths. In this regard, recent studies of the anthocyanin-related MYB gene family of antirrhinum provide evidence supporting varying activation strengths between family members (Schwinn et al. 2006).

A further aspect of the use of TF transgenes for controlling anthocyanin biosynthesis in plants that has come to light is the impact of the environment on the resultant transgenic phenotypes. *35S:LC* transgenics of alfalfa and petunia showed markedly different vegetative pigmentation phenotypes dependent on the environmental conditions (Ray et al. 2003; Boase et al. 2006). *35S:LC* petunia with weak pigmentation when grown in a greenhouse developed intense anthocyanin pigmentation of the leaves within a few days of planting in field conditions, due to the much higher white light levels (Fig. 3.4). Clearly, light can induce an endogenous factor required for anthocyanin production, either to partner the LC protein in up-regulation of the anthocyanin biosynthetic genes or to increase precursor supply to the flavonoid pathway.

From the TF transgenic experiments to date it can be concluded that for generation of the desired phenotypes in target ornamental species, appropriate TFs may need to be identified for each species and/or combinations of TF transgenes may be required. One of the best prospects is the generation of new cultivars with vegetative pigmentation. In theory, it should also be possible to use TF transgenes to generate novel spatial pigmentation patterning in flowers, as it has been shown that at least some patterns in nature are based on cell-specific TF expression. However, this may require both promoters with the required spatially regulated expression and appropriate target cultivars e.g. mutant lines that lack the activity of endogenous anthocyanin-related TFs that can be complemented by the TF transgene.

TF transgenes can also be used to inhibit anthocyanin biosynthesis, for example to generate white-flowered cultivars. There are some naturally occurring TF alleles that have an inhibitory effect on anthocyanin biosynthetic gene transcription, with both MYB and bHLH types being identified (Paz-Ares et al. 1990; Burr et al. 1996; Aharoni et al. 2001; Chen et al. 2004). Although some of these may be mutant alleles of loci that encode activators of anthocyanin biosynthesis, there is growing evidence for the general involvement of inhibitors of transcription in the normal regulation of secondary metabolism in plants (e.g. Jin et al. 2000; Aharoni et al. 2001). Artificial repressors of anthocyanin biosynthesis can also be constructed, by modifying anthocyanin-related activator TFs through the addition of repression domains from other proteins, such as the ERF-associated amphiphilic repression (EAR)-motif (Hiratsu et al. 2003; Matsui et al. 2004).

A further development of the application of anthocyanin-related TFs for modifying flower colour is their use to generate variegated floral patterns (Liu et al. 2001). By inserting the arabidopsis transposon *Tag1* between the 35S promoter and the maize bHLH factor R, and introducing this construct into tobacco, plants were generated that showed variegation of flower pigmentation.

3.6 Concluding Comments

A wealth of knowledge is now available on the molecular biology of anthocyanin biosynthesis. This provides a range of gene technology approaches for modifying flower colour in ornamental crops. The potential for GM to contribute significantly to breeding of new ornamental cultivars was first demonstrated almost 20 years ago (Meyer et al. 1987), but as yet there are few commercial GM ornamental products on the market. However, it would be expected that more products would appear over the next 10 to 15 years, as GM plants (particularly non-food crops) become more acceptable to the public and regulatory authorities.

Acknowledgments

My thanks to Dr Yoshi Tanaka (Suntory Ltd, Japan), Dr John Mason (Florigene Pty Ltd, Australia), Ms Yang Shaoyong (SouthWest Forestry College, China), Dr David Lewis (Crop & Food Research, New Zealand) and Dr Huaibi Zhang (Crop & Food Research) for some of the photographs in Figs. 3.3, 3.4 and 3.5.

References

Aharoni, A., De Vos, C.H., Wein, M., Sun, Z., Greco, R., Kroon, A., Mol, J.N. and O'Connell, A.P. (2001) The strawberry FaMYB1 transcription factor suppresses anthocyanin and flavonol accumulation in transgenic tobacco. Plant J. 28, 319–332.

Aida, R., Kishimoto, S., Tanaka, Y. and Shibata, M. (2000a) Modification of flower colour in torenia (*Torenia fournieri* Lind.) by genetic transformation. Plant Sci. 153, 33–42.

Aida, R., Yoshida, K., Kondo, T., Kishimoto, S. and Shibata, M. (2000b) Copigmnetation gives bluer flowers on transgenic torenia plants with the antisense dihydroflavonol-4-reductase gene. Plant Sci. 160, 49–56.

Andersen, Ø.M. and Jordheim, M. (2006) The anthocyanins. In: Anderson, Ø.M. and Markham, K.R. (eds) *Flavonoids: Chemistry, Biochemistry, and Applications*, CRC Press, Boca Raton, Florida, USA, pp. 471–553.

Angenent, G.C., Franken, J., Busscher, M., Weiss, D. and van Tunen, A.J. (1994) Co-suppression of the petunia homeotic gene *fbp2* affects the identity of the generative meristem. Plant J. 5, 33–44.

Arisumi, K., Sakata, Y. and Takeshita, S. (1990) The pigment constitution of *R. griersonianum*. Am. Rhod. Soc. J. 44, 15–17.

Asen, S., Stewart, R.N and Norris, K.H. (1971) Co-pigmentation effect of quercetin glycosides on absorption characteristics of cyanidin glycosides and color of Red Wing azalea. Phytochemistry 10, 171–175.

Bae, R.N., Kim, K.W., Kim, T.C. and Lee, S.K. (2006) Anatomical observations of anthocyanin rich cells in apple skins. HortScience 41, 733–736.

Bey, M., Stuber, K., Fellenberg, K., Schwarz-Sommer, Z., Sommer, H., Saedler, H. and Zachgo, S. (2004) Characterization of antirrhinum petal development and identification of target genes of the class B MADS box gene *DEFICIENS*. Plant Cell 16, 3197–3215.

Boase, M.R., Bradley, J.M. and Borst, N.K. (1998) Genetic transformation mediated by *Agrobacterium tumefaciens* of florists' chrysanthemum (*Dendranthema* × *grandiflorum*) cultivar "Peach Margaret". In Vitro Cell. Dev. Biol. Plant 34, 46–51.

Boase, M.R., Lill, T.R., Rains, R.S., Lewis, D.H., Schwinn, K.E., King, I.S. and Davies, K.M. (2006) Impact of the environment on phenotypes of petunia plants carrying transgenes for flavonoid regulatory factors or the ROLC protein. In: Mercer, C.F. (ed.) 13th Australasian Plant Breeding Conference Proceedings, CD-ROM IBSN 978-0-86476-167-8. Grains Research & Development Corporation, New Zealand.

Bohm, B.A. (1998) *Introduction to Flavonoids*. Harwood Academic Publishers, Amsterdam, The Netherlands.

Borevitz, J.O., Xia, Y., Blount, J., Dixon, R.A. and Lamb, C. (2000) Activation tagging identifies a conserved MYB regulator of phenylpropanoid biosynthesis. Plant Cell 12, 2383–2394.

Bovy, A., de Vos, R., Kemper, M., Schijlen, E., Pertejo, M.A., Muir, S., Collins, G., Robinson, S., Verhoeyen, M., Hughes, S., Santos-Buelga, C. and van Tunen, A. (2002) High-flavonol tomatoes resulting from the heterologous expression of the maize transcription factor genes LC and C1. Plant Cell 14, 2509–2526.

Bradley, J.M., Davies, K.M., Deroles, S.C., Bloor, S.J. and Lewis D.H. (1998) The maize *Lc* regulatory gene up-regulates the flavonoid biosynthetic pathway of *Petunia*. Plant J. 13, 381–392.

Bradley, J.M., Deroles, S.C., Boase, M.R., Bloor, S., Swinny, E. and Davies, K.M. (1999) Variation in the ability of the maize *Lc* regulatory gene to upregulate flavonoid biosynthesis in heterologous systems. Plant Sci. 140, 31–39.

Bradley, J.M., Rains, R.S., Manson, J.L. and Davies, K.M. (2000) Flower pattern stability in genetically modified lisianthus (*Eustoma grandiflorum*) under commercial growing conditions. New Zealand J. Crop Hort. Sci. 28, 175–184.

Brouillard, R. and Dangles, O. (1993) Flavonoids and flower colour. In: Harborne, J.B. (ed.) *The Flavonoids: Advances in Research Since 1986*. Chapman & Hall, London, UK, pp. 565–587.

Brugliera, F., Holton, T.A., Stevenson, T.W., Farcy, E., Lu, C.Y. and Cornish, E.C. (1994) Isolation and characterization of a cDNA clone corresponding to the *Rt* locus of *Petunia hybrida*. Plant J. 5, 81–92.

Brugliera, F., Barri-Rewell, G., Holton, T.A. and Mason, J.G. (1999) Isolation and characterization of a flavonoid 3'-hydroxylase cDNA clone corresponding to the *Ht1* locus of *Petunia hybrida*. Plant J. 19, 441–451.

Burr, F.A., Burr, B., Scheffler, B.E., Blewitt, M., Wienand, U. and Matz, E.C. (1996) The maize repressor-like gene *intensifier1* shares homology with the *R1/B1* multigene family of transcription factors and exhibits missplicing. Plant Cell 8, 1249–1259.

Carey, C.C., Strahle, J.T., Selinger, D.A. and Chandler, V.L. (2004) Mutations in the *pale aleurone color1* regulatory gene of the *Zea mays* anthocyanin pathway have distinct phenotypes relative to the functionally similar *TRANSPARENT TESTA GLABRA1* gene in *Arabidopsis thaliana*. Plant Cell 16, 450–464.

Chandler, V.L., Radicella, J.P., Robbins, T.P., Chen, J. and Turks, D. (1989) Two regulatory genes of the maize anthocyanin pathway are homologous: isolation of *B* utilizing *R* genomic sequences. Plant Cell, 1, 1175–1183.

Chen, B., Wang, X., Hu, Y., Wang, Y. and Lin, Z. (2004) Ectopic expression of a *c1-I* allele from maize inhibits pigment formation in the flower of transgenic tobacco. Mol. Biotech. 26, 187–192.

Chittka, L., Shmida, A., Troje, N. and Menzel, R. (1994). Ultraviolet as a component of flower reflections, and the colour perception of Hymenoptera. Vision Res. 34, 1489–1508.

Collette, V.E., Jameson, P.E., Schwinn, K.E., Umaharan, P. and Davies, K.M. (2004) Temporal and spatial expression of flavonoid biosynthetic genes in flowers of *Anthurium andraeanum*. Physiol. Plant. 122, 297–304.

Cone, K.C., Cocciolone, S.M., Burr, F.A. and Burr, B. (1993) Maize anthocyanin regulatory gene *pl* is a duplicate of *c1* that functions in the plant. Plant Cell, 5, 1795–1805.

Courtney-Gutterson, N., Napoli, C., Lemieuz, C., Morgan, A., Firoozabady, E. and Robinson, K.E.P. (1994) Modification of flower color in florist's chrysanthemum: production of a white-flowering variety through molecular-genetics. Biotechnology 12, 268–271.

Cuttriss, A. and Pogson, B. (2004) Carotenoids. In: Davies, K. (ed.) *Plant Pigments and their Manipulation*. Blackwell Publishing, Oxford, UK, pp. 57–91.

Damiani, F., Paolocci, F., Cluster, P.D., Arcioni, S., Tanner, G.J., Joseph, R.J., Li, Y.G., de Majnik, J. and Larkin, P.J. (1999) The maize transcription factor *Sn* alters proanthocyanidin synthesis in transgenic *Lotus corniculatus* plants. Aust. J. Plant Physiol. 26, 159–169.

Davies, K.M. (2004) Important rare plant pigments. In: Davies, K.M. (ed.) *Plant Pigments and their Manipulation*. Annual Plant Reviews, Blackwell Press, UK, pp. 214–247.

Davies, K.M. and Schwinn, K.E. (1997) Flower colour. In: Geneve, R.L., Preece, J.E. and Merkle, S.A. (eds) *Biotechnology of Ornamental Plants*. CAB International, Wallingford, UK, pp. 259–294.

Davies, K.M. and Schwinn, K.E. (2003) Transcriptional regulation of secondary metabolism. Funct. Plant Biol. 30, 913–925.

Davies, K.M. and Schwinn, K.E. (2006) Molecular biology and biotechnology of flavonoid biosynthesis. In: Anderson, Ø.M. and Markham, K.R. (eds) *Flavonoids: Chemistry, Biochemistry, and Applications*, CRC Press, Boca Raton, Florida, USA, pp. 143–218.

Davies, K.M., Bloor, S.J., Spiller, G.B. and Deroles, S.C. (1998) Production of yellow colour in flowers: redirection of flavonoid biosynthesis in *Petunia*. Plant J. 13, 259–266.

Davies, K.M., Schwinn, K.E., Deroles, S.C., Manson, D.G., Lewis, D.H., Bloor, S.J. and Bradley, J.M. (2003) Enhancing anthocyanin production by altering competition for substrate between flavonol synthase and dihydroflavonol 4-reductase. *Euphytica* 131, 259–268.

Della Vedova, C.B., Lorbiecke, R., Kirsch, H., Schulte, M.B., Scheets, K., Borchert, L.M., Scheffler, B.E., Wienand, U., Cone, K.C. and Birchler, J.A. (2005) The dominant inhibitory chalcone synthase allele *C2-Idf* (*inhibitor diffuse*) from *Zea mays* (L.) acts via an endogenous RNA silencing mechanism. Genetics 170, 1989–2002.

de Majnik, J., Weinman, J.J., Djordjevic, M.A., Rolfe, B.G., Tanner, G.J., Joseph, R.G. and Larkin, P.J. (2000) Anthocyanin regulatory gene expression in transgenic white clover can result in an altered pattern of pigmentation. Aust. J. Plant Physiol. 27, 659–667.

Deroles, S.C., Bradley, J.M., Schwinn, K.E., Markham, K.R., Bloor, S.J., Manson, D.G. and Davies, K.M. (1998) An antisense chalcone synthase gene leads to novel flower patterns in lisianthus (*Eustoma grandiflorum*). Mol. Breed. 4, 59–66.

de Vetten, N., Quattrocchio, F., Mol, J. and Koes, R. (1997) The *an11* locus controlling flower pigmentation in petunia encodes a novel WD-repeat protein conserved in yeast, plants, and animals. Genes Dev. 11, 1422–1434.

de Vetten, N., ter Horst, J., van Schaik, H.-P., de Boer, A., Mol, J. and Koes, R. (1999) A cytochrome b_5 is required for full activity of flavonoid 3',5'-hydroxylase, a cytochrome P450 involved in the formation of blue flower colours. Proc. Natl. Acad. Sci. USA 96, 778–783.

Elomaa, P., Honkanen, J., Puska, R., Seppanen, P., Helariutta, Y., Mehto, M., Kotilainen, M., Nevalainen, L. and Teeri, T.H. (1993) *Agrobacterium*-mediated transfer of antisense chalcone synthase cDNA to *Gerbera hybrida* inhibits flower pigmentation. Biotechnology 11, 508–511.

Elomaa, P., Uimari, A., Mehto, M., Albert, V.A., Laitinen, R.A.E. and Teeri, T.H. (2003) Activation of anthocyanin biosynthesis in *Gerbera hybrida* (Asteraceae) suggests conserved protein-protein and protein-promoter interactions between the anciently diverged monocots and eudicots. Plant Physiol. 133, 1831–1842.

Farzad, M., Griesbach, R. and Weiss, M.R. (2002) Floral color change in *Viola cornuta* L. (Violaceae): a model system to study regulation of anthocyanin production. Plant Sci. 162, 225–231.

Farzad, M., Griesbach, R., Hammond, J., Weiss, M.R. and Elmendorf, H.G. (2003) Differential expression of three key anthocyanin biosynthetic genes in a color-changing flower, *Viola cornuta* cv. Yesterday, Today and Tomorrow. Plant Sci. 165, 1333–1342.

Ferrario, S., Immink, R.G., Shchennikova, A., Busscher-Lange, J. and Angenent, G.C. (2003) The MADS box gene FBP2 is required for SEPALLATA function in petunia. Plant Cell 15, 914–925.

Firoozabady, E., Moy, Y., Courtney-Gutterson, N. and Robinson, K. (1994) Regeneration of transgenic rose (*Rosa hybrida*) plants from embryogenic tissue. Biotechnology 12, 609–613.

Fischer, R., Budde, I. and Hain, R. (1997) Stilbene synthase gene expression causes changes in flower colour and male sterility in tobacco. Plant J. 11, 489–498.

Fischer, T.C., Halbwirth, H., Meisel, B., Stich, K. and Forkmann, G. (2003) Molecular cloning, substrate specificity of the functionally expressed dihydroflavonol 4-reductases from *Malus domestica* and *Pyrus communis* cultivars and the consequences for flavonoid metabolism. Arch. Biochem. Biophys. 412, 223–230.

Fraser, P.D. and Bramley, P.M. (2004) The biosynthesis and nutritional uses of carotenoids. Prog. Lipid Res. 43, 228–265.

Fujiwara, H., Tanaka, Y., Yonekura-Sakakibara, K., Fukuchi-Mizutani, M., Nakao, M., Fukui, Y., Yamaguchi, M., Ashikari, T. and Kusumi, T. (1998) cDNA cloning, gene expression and subcellular localization of anthocyanin 5-aromatic acyltransferase from *Gentiana triflora*. Plant J. 16, 421–431.

Fukada-Tanaka, S., Inagaki, Y., Yamaguchi, T., Saito, N. and Iida, S. (2000) Colour-enhancing protein in blue petals. Nature 407, 581.

Fukuchi-Mizutani, M., Okuhara, H., Fukui, Y., Nakao, M., Katsumoto, Y., Yonekura-Sakakibara, K., Kusumi, T., Hase, T. and Tanaka, Y. (2003) Biochemical and molecular characterization of a novel UDP-glucose:anthocyanin 3'-*O*-glucosyltransferase, a key enzyme for blue anthocyanin biosynthesis, from gentian. Plant Physiol. 132, 1652–1663.

Fukui, Y., Kusumi, T., Matsuda, K., Iwashita, T. and Nomoto, K. (2002) Structure of rosacyanin B, a novel pigment from the petals of *Rosa hybrida*. Tetrahedron Lett. 43, 2637–2639.

Fukui, Y., Tanaka, Y., Kusumi, T., Iwashita, T. and Nomoto, K. (2003) A rationale for the shift in colour towards blue in transgenic carnation flowers expressing the flavonoid 3',5'-hydroxylase gene. Phytochemistry 63, 15–23.

Fukusaki, E., Kawasaki, K., Kajiyama, S., An, C.I., Suzuki, K., Tanaka, Y., Kobayashi, A. (2004) Flower color modulations of *Torenia hybrida* by downregulation of chalcone synthase genes with RNA interference. J. Biotech. 111, 229–240.

Gebhardt, Y., Witte, S., Forkmann, G., Lukačin, R., Marten, U. and Martens, S. (2005) Molecular evolution of flavonoid dioxygenases in the family Apiaceae. Phytochemistry 66, 1273–1284.

Goldsbrough, A., Tong, Y. and Yoder, J.I. (1996). *Lc* as a non-destructive visual reporter and transposition excision marker gene for tomato. Plant J. 9, 927–933.

Gong, Z.Z., Yamagishi, E., Yamazaki, M. and Saito, K. (1999) A constitutively expressed *Myc*-like gene involved in anthocyanin biosynthesis from *Perilla frutescens*: molecular characterization, heterologous expression in transgenic plants and transactivation in yeast cells. Plant Mol. Biol. 41, 33–44.

Gonnet, J.F. (2003) Origin of the color of Cv. rhapsody in blue rose and some other so-called "blue" roses. J. Agric. Food. Chem. 51, 4990–4994.

Goodrich, J., Carpenter, R. and Coen, E.S. (1992) A common gene regulates pigmentation pattern in diverse plant species. Cell 68, 955–964.

Gorton, H.L. and Vogelmann, T.C. (1996) Effects of epidermal cell shape and pigmentation on optical properties of Antirrhinum petals at visible and ultraviolet wavelengths. Plant Physiol 112, 879–888.

Goto, T. and Kondo, T. (1991) Structure and molecular stacking of anthocyanins – flower color variation. Angew. Chem. Int. Ed. Eng. 30, 17–33.

Goto, T., Takase, S. and Kondo, T. (1978) PMR spectra of natural acylated anthocyanins. Determination of stereostructure of awobanin, shisonin and violanin, Tetrahedron Lett. 2413–2416.

Griesbach, R.J. (1993) Characterization of the flavonoids from *Petunia* x*hybrida* flowers expressing the A1 gene of *Zea mays*. HortScience 28, 659–660.

Grotewold, E. (2006) The genetics and biochemistry of floral pigments. Annu. Rev. Plant Biol. 57, 761–780.

Gutterson, N. (1995) Anthocyanin biosynthetic genes and their application to flower colour modification through sense suppression. HortScience 30, 964–966.

Halbwirth, H. and Stich, K. (2006) An NADPH and FAD dependent enzyme catalyzes hydroxylation of flavonoids in position 8. Phytochemistry 67, 1080–1087.

Halbwirth, H., Martens, S., Wienand, U., Forkmann, G. and Stich, K. (2003) Biochemical formation of anthocyanins in silk tissue of *Zea mays*. Plant Sci. 164, 489–495.

Heller, W. and Forkmann, G. (1994) Biosynthesis of flavonoids. In: Harborne, J.B. (ed.) *The Flavonoids: Advances in research since 1986*. Chapman & Hall, London, pp. 499–535.

Hiratsu, K., Matsui, K., Koyama, T. and Ohme-Takagi, M. (2003) Dominant repression of target genes by chimeric repressors that include the EAR motif, a repression domain, in *Arabidopsis*. Plant J. 34, 733–739.

Holton, T.A., Brugliera, F., Lester, D.R., Tanaka, Y., Hyland, C.D., Menting, J.G.T., Lu, C.-Y., Farcy, E., Stevenson, T.W. and Cornish, E.C. (1993a) Cloning and expression of cytochrome P450 genes controlling flower colour. Nature 366, 276–279.

Holton, T.A., Cornish, E.C. and Tanaka, Y. (1993b) Cloning and expression of flavonol synthase from *Petunia hybrida*. Plant J. 4, 1003–1010.

Ibrahim, R.K. and Muzac, I. (2000) The methyltransferase gene superfamily: a tree with multiple branches. In: Romeo, J.T., Ibrahim, R., Varin, L. and De Luca, V. (eds) *Evolution of Metabolic Pathways*. Elsevier Science Ltd, Oxford, UK, pp. 349–384.

Jin, H., Cominelli, E., Bailey, P., Parr, A., Mehrtens, F., Jones, J., Tonelli, C., Weisshaar, B. and Martin, C. (2000) Transcriptional repression by AtMYB4 controls production of UV-protecting sunscreens in *Arabidopsis*. EMBO J. 19, 6150–6161.

Johnson, E.T., Yi, H., Shin, B., Oh, B.-J., Cheong, H. and Choi, G. (1999) *Cymbidium hybrida* dihydroflavonol 4-reductase does not efficiently reduce dihydrokaempferol to produce orange pelargonidin-type anthocyanins. Plant J. 19, 81–85.

Jorgensen, R. (1995) Cosuppression, flower color patterns, and metastable gene expression states. Science 268, 686–691.

Jorgensen, R.A., Que, Q.D. and Napoli, C.A. (2002) Maternally-controlled ovule abortion results from cosuppression of dihydroflavonol-4-reductase or flavonoid-3',5'-hydroxylase genes in *Petunia hybrida*. Funct. Plant Biol. 29, 1501–1506.

Joung, J.Y., Kasthuri, G.M., Park J.Y., Kang, W.J., Kim, H.S., Yoon, B.S., Joung, H. and Jeon, J.H. (2003) An overexpression of chalcone reductase of *Pueraria montana* var. lobata alters biosynthesis of anthocyanin and 5'-deoxyflavonoids in transgenic tobacco. Biochem. Biophys. Res. Comm. 303, 326–331.

Koseki, M., Goto, K., Masuta, C. and Kanazawa, A. (2005) The star-type color pattern in *Petunia hybrida* "Red Star" flowers is induced by sequence-specific degradation of chalcone synthase RNA. Plant Cell Physiol. 46, 1879–1883.

Kroon, J., Souer, E., de Graaff, A., Xue, Y., Mol, J. and Koes, R. (1994) Cloning and structural analysis of the anthocyanin pigmentation locus *Rt* of *Petunia hybrida*: characterization of insertion sequences in two mutant alleles. Plant J. 5, 69–80.

Latunde-Dada, A.O., Cabello-Hurtado, F., Czittrich, N., Didierjean, L., Schopfer, C., Hertkorn, N., Werck-Reichhart, D. and Ebel, J. (2001) Flavonoid 6-hydroxylase from soybean (*Glycine max* L.), a novel plant P-450 monooxygenase. J. Biol. Chem. 276, 1688–1695.

Li, S.J., Deng, X.M., Mao, H.Z. and Hong, Y. (2005) Enhanced anthocyanin synthesis in foliage plant *Caladium bicolor*. Plant Cell Rep. 23, 716–720.

Liu, D., Galli, M. and Crawford, N.M. (2001) Engineering variegated floral patterns in tobacco plants using the arabidopsis transposable element *Tag1*. Plant Cell Physiol. 42, 419–423.

Lloyd, A.M., Walbot, V. and Davis, R.W. (1992) *Arabidopsis* and *Nicotiana* anthocyanin production activated by maize regulators *R* and *C1*. Science 258, 1773–1775.

Lo, S.C. and Nicholson, R.L. (1998) Reduction of light-induced anthocyanin accumulation in inoculated sorghum mesocotyls. Implications for a compensatory role in the defense response. Plant Physiol. 116, 979–989.

Lu, C.Y., Chandler, S.F., Mason, J.G. and Brugliera, F. (2003) Florigene flowers: from laboratory to market. In: Vasil, I.K. (ed.) *Plant Biotechnology 2002 and Beyond*. Kluwer Academic Publishers, Dordrecht, The Netherlands, pp. 333–336.

Ludwig, S.R. and Wessler, S.R. (1990) Maize *R* gene family: tissue-specific helix-loop-helix proteins. Cell, 62, 849–851.

Ludwig, S.R., Bowen, B., Beach, L. and Wessler, S.R. (1990) A regulatory gene as a novel visible marker for maize transformation. Science 24, 449–450.

Markham, K.R., Gould, K.S., Winefield, C.S., Mitchell, K.A., Bloor, S.J. and Boase, M.R. (2000) Anthocyanic vacuolar inclusions - their nature and significance in flower colouration. Phytochemistry 55, 327–336.

Markham, K.R., Bloor, S.J., Nicholson, R., Rivera, R., Shemluck, M., Kevan, P.G. and Michener, C. (2004) Black flower coloration in wild *Lisianthius nigrescens*: its chemistry and ecological consequences. Z. Naturforsch. 59, 625–630.

Marrs, K.A., Alfenito, M.R., Lloyd, A.M. and Walbot, V. (1995) A glutathione S-transferase involved in vacuolar transfer encoded by the maize gene *Bronze-2*. Nature 375, 397–400.

Mathews, H., Clendennen, S.K., Caldwell, C.G., Liu, X.L., Connors, K., Matheis, N., Schuster, D.K., Menasco, D.J., Wagoner, W., Lightner, J. and Wagner, D.R. (2003) Activation tagging in tomato identifies a transcriptional regulator of anthocyanin biosynthesis, modification, and transport. Plant Cell 15, 1689–1703.

Matsui, K., Tanaka, H. and Ohme-Takagi, M. (2004) Suppression of the biosynthesis of proanthocyanidin in *Arabidopsis* by a chimeric PAP1 repressor. Plant Biotech. J. 2, 487–493.

Meng, X. and Wang, X. (2004) Regulation of flower development and anthocyanin accumulation in *Gerbera hybrida*. J. Hort. Sci. Biotech. 79, 131–137.

Meyer, P. (1991) Engineering of novel flower colours. In: Harding, J., Singh, F. and Mol, J.N.M. (eds) *Genetics and Breeding of Ornamental Species*. Kluwer Academic Publishers, Dordrecht, The Netherlands pp. 285–307.

Meyer, P., Heidmann, I., Forkmann, G. and Saedler, H. (1987) A new petunia flower colour generated by transformation of a mutant with a maize gene. Nature 330, 677–678.

Millar, A.A. and Gubler F. (2005) The arabidopsis *GAMYB-Like* genes, *MYB33* and *MYB65*, are microRNA-regulated genes that redundantly facilitate anther development. Plant Cell 17, 705–721.

Mooney, M., Desnos, T., Harrison, K., Jones, J., Carpenter, R. and Coen, E. (1995) Altered regulation of tomato and tobacco pigmentation genes caused by the *delila* gene of *Antirrhinum*. Plant J. 7, 333–339.

Morita, Y., Hoshino, A., Kikuchi, Y., Okuhara, H., Ono, E., Tanaka, Y., Fukui, Y., Saito, N., Nitasaka, E., Noguchi, H. and Iida, S. (2005) Japanese morning glory *dusky* mutants displaying reddish-brown or purplish-gray flowers are deficient in a novel glycosylation enzyme for anthocyanin biosynthesis, UDP-glucose:anthocyanidin 3-*O*-glucoside-2"-*O*-glucosyltransferase, due to 4-bp insertions in the gene. Plant J. 42, 353–63.

Morita, Y., Saitoh, M., Hoshino, A., Nitasaka, E. and Iida, S. (2006) Isolation of cDNAs for R2R3-MYB, bHLH and WDR transcriptional regulators and identification of *c* and *ca* mutations conferring white flowers in the Japanese morning glory. Plant and Cell Physiology 47, 457–470.

Nakajima, J.-I., Sato, Y., Hoshino, T., Yamazaki, M. and Saito, K. (2006) Mechanistic study on the oxidation of anthocyanidin synthase by quantum mechanical calculation. J. Biol. Chem. 281, 21387–21398.

Nakamura, N., Fukuchi-Mizutani, M., Miyazaki, K., Suzuki, K. and Tanaka, Y. (2006) RNAi suppression of the anthocyanidin synthase gene in *Torenia hybrida* yields white flowers with higher frequency and better stability than antisense and sense suppression. Plant Biotech. 23, 13–17.

Nakayama, T., Yonekura-Sakakibara, K., Sato, T., Kikuchi, S., Fukui, Y., Fukuchi-Mizutani, M., Ueda, T., Nakao, M., Tanaka, Y., Kusumi, T. and Nishino, T. (2000) Aureusidin synthase: a polyphenol oxidase homolog responsible for flower coloration. Science 290, 1163–1166.

Nakayama, T., Suzuki, H. and Nishino, T. (2003) Anthocyanin acyltransferases: specificities, mechanism, phylogenetics, and applications. J. Mol. Cat. 23, 117–132.

Napoli, C., Lemieux, C. and Jorgensen, R. (1990) Introduction of a chimeric chalcone synthase gene into Petunia results in reversible co-suppression of homologous genes in trans. Plant Cell 2, 279–289.

Nielsen, K.M., Deroles, S.C., Markham, K.R., Bradley, J.M., Podivinsky, E. and Manson, D. (2002) Antisense flavonol synthase alters copigmentation and flower color in lisianthus. Mol. Breed. 9, 217–229.

Nishihara, M., Nakatsuka, T., Mishiba, K., Kikuchi, A. and Yamamura, S. (2003) Flower color modification by suppression of chalcone synthase gene in gentian. Plant Cell Physiol. 44s1, 59.

Nishihara, M., Nakatsuka, T. and Yamamura, S. (2005) Flavonoid components and flower color change in transgenic tobacco plants by suppression of chalcone isomerase gene. FEBS Lett. 579, 6074–6078.

Noda, K.-I., Glover, B.J., Linstead, P. and Martin, C. (1994) Flower colour intensity depends on specialized cell shape controlled by a Myb-related transcription factor. Nature 369, 661–664.

Noda, N., Kato, N., Kogawa, K., Kazuma, K. and Suzuki, M. (2004) Cloning and characterization of the gene encoding anthocyanin 3',5'-O-glucosyltransferase involved in ternatin biosynthesis from blue petals of butterfly pea (Clitoria ternatea). Plant Cell Physiol. 45s1, 32.

Nozue, M., Kubo, H., Nishimura, M., Katou, A., Hattori, C., Usuda, N., Nagata, T., Yasuda, H. (1993) Characterization of intravacuolar pigmented structures in anthocyanin-containing cells of sweet-potato suspension-cultures. Plant Cell Physiol. 34, 803–808.

Okinaka, Y., Shimada, Y., Nakano-Shimada, R., Ohbayashi, M., Kiyokawa, S. and Kikuchi, Y. (2003) Selective accumulation of delphinidin derivatives in tobacco using a putative flavonoid 3',5'-hydroxylase cDNA from Campanula medium. BioSci. Biotech. Biochem. 67, 161–165.

Ogata, J., Kanno, Y., Itoh, Y., Tsugawa, H. and Suzuki, M. (2005) Plant biochemistry: anthocyanin biosynthesis in roses. Nature 435, 757–758.

Ono, E., Fukuchi-Mizutani, M., Nakamura, N., Fukui, Y., Yonekura-Sakakibara, K., Yamaguchi, M., Nakayama, T., Tanaka, T., Kusumi, T. and Tanaka, Y. (2006a) Yellow flowers generated by expression of the aurone biosynthetic pathway. Proc. Natl. Acad. Sci. USA 103, 11075–11080.

Ono, E., Hatayama, M., Isono, Y., Sato, T., Watanabe, R., Yonekura-Sakakibara, K., Fukuchi-Mizutani, F., Tanaka, Y., Kusumi, T., Nishino T. and Nakayama, T. (2006b) Localization of a flavonoid biosynthetic polyphenol oxidase in vacuoles. Plant J. 45, 133–143.

Park, K.-I., Choi, J.-D., Hoshino, A., Morita, Y. and Iida, S. (2004) An intragenic tandem duplication in a transcriptional regulatory gene for anthocyanin biosynthesis confers pale-colored flowers and seeds with fine spots in Ipomoea tricolor. Plant J. 38, 840–849.

Paxton, R.J. and Tengo, J. (2001) Doubly duped males: the sweet and sour of the orchid's bouquet. Trends Ecol. Evol. 16, 167–169.

Payne, C.T., Zhang, F. and Lloyd, A.M. (2000) GL3 encodes a bHLH protein that regulates trichome development in Arabidopsis through interaction with GL1 and TTG1. Genetics 156, 1349–1362.

Paz-Ares, J., Peterson, P.A. and Saedler, H. (1990) Molecular analysis of the C1–I allele from Zea mays: a dominant mutant of the regulatory C1 locus. EMBO J. 9, 315–321.

Polashock, J.J., Griesbach, R.J., Sullivan, R.F. and Vorsa, N. (2002) Cloning of a cDNA encoding the cranberry dihydroflavonol-4-reductase (DFR) and expression in transgenic tobacco. Plant Sci. 163, 241–251.

Quattrocchio, F., Wing, J., van der Woude, K., Souer, E., de Vetten, N., Mol, J. and Koes, R. (1999) Molecular analysis of the *anthocyanin2* gene of petunia and its role in the evolution of flower color. Plant Cell 11, 1433–1444.

Ramsay, N.A., Walker, A.R., Mooney, M. and Gray, J.C. (2003) Two basic-helix-loop-helix genes (*MYC-146* and *GL3*) from *Arabidopsis* can activate anthocyanin biosynthesis in a white-flowered *Matthiola incana* mutant. Plant Molecular Biology 52, 679–688.

Ray, H., Yu, M., Auser, P., Blahut-Beatty, L., McKersie, B., Bowley, S., Westcott, N., Coulman, B., Lloyd, A. and Gruber, M.Y. (2003) Expression of anthocyanins and proanthocyanidins after transformation of alfalfa with maize *Lc*. Plant Physiol. 132, 1448–1463.

Robbins, M.P., Paolocci, F., Hughes, J.W., Turchetti, V., Allison, G., Arcjoni, S., Morris, P. and Damiani, F. (2003) Sn, a maize *bHLH* gene, modulates anthocyanin and condensed tannin pathways in *Lotus corniculatus*. Journal of Experimental Botany 54, 239–248.

Rosati, C., Cadic, A., Duron, M., Renou, J.P. and Simoneau, P. (1997) Molecular cloning and expression analysis of dihydroflavonol 4-reductase gene in flower-organs of *Forsythia x intermedia*. Plant Mol. Biol. 35, 303–311.

Rosati, C., Simoneau, P., Treutter, D., Poupard, P., Cadot, Y., Cadic, A. and Duron, M. (2003) Engineering of flower colour in forsythia by expression of two independently-transformed dihydroflavonol 4-reductase and anthocyanidin synthase genes of flavonoid pathway. Mol. Breed. 12, 197–208.

Saito, N., Tatsuzawa, F., Miyoshi, K., Shigihara, A. and Honda, T. (2003) The first isolation of *C*-glycosylanthocyanin from the flowers of *Tricyrtis formosana*. Tetrahedron Letters 44, 6821–6823.

Saito, N., Yokoi, M., Ogawa, M., Kamijo, M. and Honda, T. (1988) 6-Hydroxyanthocyanidin glycosides in the flowers of Alstroemeria. Phytochemistry 27, 1399–1401.

Santos, M.O., Crosby, W.L. and Winkel, B.S.J. (2004) Modulation of flavonoid metabolism in Arabidopsis using a phage-derived antibody. Mol. Breed. 13, 333–343.

Schwinn, K.E. and Davies, K.M. (2004) Flavonoids. In: Davies, K. (ed.) *Plant Pigments and their Manipulation*. Blackwell Publishing, Oxford, UK, pp. 92–149.

Schwinn, K.E., Davies, K.M., Deroles, S.C., Markham, K., Miller, R.M., Bradley, M., Manson, D.G. and Given, N.K. (1997) Expression of an *Antirrhinum majus* UDP-glucose:flavonoid-3-*O*-glucosyltransferase transgene alters flavonoid glycosylation and acylation in lisianthus (*Eustoma grandiflorum* Grise.). Plant Sci. 125, 53–61.

Schwinn, K., Venail, J., Shang, Y., Mackay, S., Alm, V., Butelli, E., Oyama, R., Bailey, P., Davies, K. and Martin, C. (2006) A small family of MYB-regulatory genes controls floral pigmentation intensity and patterning in the genus Antirrhinum. Plant Cell 18, 831–851.

Seitz, C., Eder, C., Deiml, B., Kellner, S., Martens, S. and Forkmann, G. (2006) Cloning, functional identification and sequence analysis of flavonoid 3'-hydroxylase and flavonoid 3',5'-hydroxylase cDNAs reveals independent evolution of flavonoid 3',5'-hydroxylase in the Asteraceae family. Plant Mol. Biol. 61, 365–381.

Shimada, Y., Nakano-Shimada, R., Ohbayashi, M., Okinaka, Y., Kiyokawa, S. and Kikuchi, Y. (1999) Expression of chimeric P450 genes encoding flavonoid-3',5'-hydroxylase in transgenic tobacco and petunia plants. FEBS Lett. 461, 241–245.

Shimada, Y., Ohbayashi, M., Nakano-Shimada, R., Okinaka, Y., Kiyokawa, S. and Kikuchi, Y. (2001) Genetic engineering of the anthocyanin biosynthetic pathway with flavonoid-3',5'-hydroxylase: specific switching of the pathway in petunia. Plant Cell Rep. 20, 456–462.

Shiono, M., Matsugaki, N. and Takeda, K. (2005) Structure of the blue cornflower pigment. Nature 436, 791.

Sompornpailin, K., Makita, Y., Yamazaki, M. and Saito, K. (2002) A WD-repeat-containing putative regulatory protein in anthocyanin biosynthesis in *Perilla frutescens*. Plant Molecular Biology 50, 485–495.

Spelt, C., Quattrocchio, F., Mol, J.N.M. and Koes, R. (2000) *Anthocyanin1* of petunia encodes a basic helix-loop-helix protein that directly activates transcription of structural anthocyanin genes. Plant Cell 12, 1619–1631.

Spelt, C., Quattrocchio, F., Mol, J. and Koes, R. (2002) ANTHOCYANIN1 of petunia controls pigment synthesis, vacuolar pH, and seed coat development by genetically distinct mechanisms. Plant Cell 14, 2121–2135.

Springob, K., Nakajima, J., Yamazaki, M. and Saito, K. (2003) Recent advances in the biosynthesis and accumulation of anthocyanins. Nat. Prod. Rep. 20, 288–303.

Strack, D., Vogt, T. and Schliemann, W. (2003) Recent advances in betalain research. Phytochemistry 62, 247–269.

Suzuki, K., Xue, H., Tanaka, Y., Fukui, Y., Fukuchi-Mizutani, M., Katsumoto, Y., Tsuda, S. and Kusumi, T. (2000) Flower color modifications of *Torenia hybrida* by cosuppression of anthocyanin biosynthesis genes. Mol. Breed. 6, 239–246.

Suzuki, H., Nakayama, T., Yonekura-Sakakibara, K., Fukui, Y., Nakamura, N., Yamaguchi, M., Tanaka, Y., Kusumi, T. and Nishino, T. (2002) cDNA cloning, heterologous expressions, and functional characterization of malonyl-coenzyme A:anthocyanidin 3-*O*-glucoside-6"-*O*-malonyltransferase from dahlia flowers. Plant Physiol. 130, 2142–2151.

Suzuki, H., Sawada, K., Yonekura-Sakakibara, K., Nakayama, T., Yamaguchi, M.-A. and Nishino, T. (2003) Identification of a cDNA encoding malonyl-coenzyme A:anthocyanidin 3-*O*-glucoside-6"-*O*-malonyltransferase from cineraria (*Senecio cruentus*) flowers. Plant Biotech. 20, 229–234.

Tanaka, Y., Fukui, Y., Fukuchi-Mizutani, M., Holton, T.A., Higgins, E. and Kusumi, T. (1995) Molecular cloning and characterization of *Rosa hybrida* dihydroflavonol 4-reductase gene. Plant Cell Physiol. 36, 1023–1031.

Tanaka, Y., Katsumoto, Y., Brugliera, F. and Mason, J. (2005) Genetic engineering in floriculture. Plant Cell Tissue Organ Cult. 80, 1–24.

Tatsuzawa, F., Murata, N., Shinoda, K., Suzuki, R. and Saito, N. (2003) Flower colors and anthocyanin pigments in 45 cultivars of *Alstroemeria* L. J. Jap. Soc. Hort. Sci. 72, 243–251.

Taylor, L.P. and Jorgensen, R. (1992) Conditional male fertility in chalcone synthase-deficient petunia. J. Heredity 83, 11–17.

Tsuda, S., Fukui, Y., Nakamura, N., Katsumoto, Y., Yonekura-Sakakibara, K., Fukuchi-Mizutani, M., Ohira, K., Ueyama, Y., Ohkawa, H., Holton, T.A., Kusumi, T. and Tanaka, Y. (2004) Flower color modification of *Petunia hybrida* commercial varieties by metabolic engineering. Plant Biotech. 21, 377–386.

Ueyama, Y., Suzuki, K., Fukuchi-Mizutani, M., Fukui, Y., Miyazaki, K., Ohkawa, H., Kusumi, T. and Tanaka, Y. (2002) Molecular and biochemical characterization of torenia flavonoid 3'-hydroxylase and flavone synthase II and modification of flower color by modulating the expression of these genes. Plant Sci. 163, 253–263.

Ueyama, Y., Katsumoto, Y., Fukui, Y., Fukuchi-Mizutani, M., Ohkawa, H., Kusumi, T., Iwashita, T. and Tanaka, Y. (2006) Molecular characterization of the flavonoid biosynthetic pathway and flower color modification of *Nierembergia* sp. Plant Biotech. 23, 19–24.

van der Krol, A.R., Lenting, P.E., Veenstra, J.G., van der Meer, I.M., Koes, R.E., Gerats, A.G.M., Mol, J.N.M. and Stuitje, A.R. (1988) An antisense chalcone synthase gene in transgenic plants inhibits flower pigmentation. Nature 333, 866–869.

van der Krol, A.R., Mur, L.A., Beld, M., Mol, J.N.M. and Stuitje, A.R. (1990a) Flavonoid genes in petunia: addition of a limited number of additional copies may lead to a suppression of gene activity. Plant Cell 2, 291–299.

van der Krol, A.R., Mur, L.A., Delange, P., Gerats, A.G.M., Mol, J.N.M. and Stuitje, A.R. (1990b) Antisense chalcone synthase genes in petunia – visualization of variable transgene expression. Mol. Gen. Genet. 22, 204–212.

Vogt, T. (2000) Glycosyltransferases involved in plant secondary metabolism. In: Romeo, J.T., Ibrahim, R., Varin, L. and De Luca, V. (eds) *Evolution of Metabolic Pathways*. Elsevier Science Ltd, Oxford, UK, pp. 317–347.

Vom Endt, D., Kijne, J.W. and Memelink, J. (2002) Transcription factors controlling plant secondary metabolism: what regulates the regulators? Phytochemistry 61, 107–114.

Walker, A.R., Davison, P.A., Bolognesi-Winfield, A.C., James, C.M., Srinivasan, N., Blundell, T.L., Esch, J.J., Marks, M.D. and Gray, J.C. (1999) The *TRANSPARENT TESTA GLABRA1* locus, which regulates trichome differentiation and anthocyanin biosynthesis in Arabidopsis, encodes a WD40 repeat protein. Plant Cell 11, 1337–1350.

Weiss, M.R. (1995) Floral colour change: a widespread functional convergence. Am. J. Bot. 82, 167–185.

Weiss, D. (2000) Regulation of flower pigmentation and growth: multiple signaling pathways control anthocyanin synthesis in expanding petals. Physiol. Plant. 110, 152–157.

Wellmann, F., Griesser, M., Schwab, W., Martens, S., Eisenreich, W., Matern, U. and Lukacin, R. (2006) Anthocyanidin synthase from *Gerbera hybrida* catalyzes the conversion of (+)-catechin to cyanidin and a novel procyanidin. FEBS Lett. 580, 1642–1648.

Wilmouth, R.C., Turnbull, J.J., Welford, R.W., Clifton, I.J., Prescott, A.G. and Schofield, C.J. (2002) Structure and mechanism of anthocyanidin synthase from *Arabidopsis thaliana*. Structure 10, 93.

Winefield, C. (2002) The final steps in anthocyanin formation: a story of modification and sequestration. Adv. Bot. Res. 37, 55–74.

Winefield, C.S., Lewis, D.H., Swinny, E.E., Zhang, H., Arathoon, H.S., Fischer, T.C., Halbwirth, H., Stich, K., Gosch, C., Forkmann, G. and Davies, K.M. (2005) Investigation of 3-deoxyanthocyanin biosynthesis in *Sinningia cardinalis*. Physiol. Plant. 124, 419–430.

Xie, D.-Y., Sharma, S.B., Pavia, N.L., Ferreira, D. and Dixon, R.A. (2003) Role of anthocyanidin reductase, encoded by *BANYULS* in plant flavonoid biosynthesis. Science 299, 396–399.

Yamaguchi, T., Fukada-Tanaka, S., Inagaki, Y., Saito, N., Yonekura-Sakakibara, K., Tanaka, Y., Kusumi, T. and Iida, S. (2001) Genes encoding the vacuolar Na^+/H^+ exchanger and flower coloration. Plant Cell Physiol. 42, 451–461.

Yamazaki, M., Gong, Z., Fukuchi-Mizutani, M., Fukui, Y., Tanaka, Y., Kusumi, T. and Saito, K. (1999) Molecular cloning and biochemical characterization of a novel anthocyanin 5-*O*-glucosyltransferase by mRNA differential display for plant forms regarding anthocyanin. J. Biol. Chem. 274, 7405–7511.

Yamazaki, M., Yamagishi, E., Gong, Z., Fukuchi-Mizutani, M., Fukui, Y., Tanaka, Y., Kusumi, T., Yamaguchi, T. and Saito, K. (2002) Two flavonoid glucosyltransferases from *Petunia hybrida*: molecular cloning, biochemical properties and developmentally regulated expression. Plant Mol. Biol. 48, 401–411.

Zimmermann, I.M., Heim, M.A., Weisshaar, B. and Uhrig, J.F. (2004) Comprehensive identification of *Arabidopsis thaliana* MYB transcription factors interacting with R/B-like BHLH proteins. Plant J. 40, 22–34.

Zuker, A., Tzfira, T., Ben-Meir, H., Ovadis, M., Schklarman, E., Itzhaki, H., Forkmann, G., Martens, S., Neta-Sharir, I., Weiss, D. and Vainstein, A. (2002) Modification of flower colour and fragrance by antisense suppression of the flavanone 3-hydroxylase gene. Mol. Breed. 9, 33–41.

Zrÿd, J.-P. and Christinet, L. (2004) Betalains. In: Davies, K. (ed.) *Plant Pigments and their Manipulation*. Blackwell Publishing, Oxford, UK, pp. 185–213.

4

Prevalence and Functions of Anthocyanins in Fruits

W.J. Steyn

University of Stellenbosch, Department of Horticultural Science, wsteyn@sun.ac.za

Abstract. This chapter reviews possible visual, nutritional and physiological functions of anthocyanins in fruits. Merits of the various functions are considered and discussed with reference to the prevalence of different fruit colours and the contribution of anthocyanins thereto as well as anthocyanin accumulation in response to environmental factors, seed disperser visual systems and fruit quality parameters. Blue, purple, black and most red fruits derive their colour from anthocyanin accumulating during ripening. Red and black are the most common colours of small bird-consumed fruits whereas larger "mammalian" fruits are more typically orange, brown or green in colour. Red fruits are conspicuous to birds with their tetrachromatic vision, but cryptic to unintended dichromatic mammalian frugivores and opportunistic insect seed predators. Blue fruits are scarce, probably because they are easy to detect by disadvantageous frugivores. Black fruits, with their very high anthocyanin levels, are fairly common despite being inconspicuous to dispersers. The prevalence of black fruits could relate to the powerful antioxidant ability of anthocyanins. On the other hand, blackness also correlates with fruit maturity and, thus, quality. Finally, the presence of anthocyanins in immature fruits and its regulation by environmental factors could relate to the photoprotective ability of anthocyanins. Since anthocyanins are able to fulfil a range of functions in plants, their adaptive value in fruits should be interpreted against a background of interaction with dispersers, genotype and environment.

4.1 Introduction

Colourful pigmentation is an appealing feature of ripe fruits. Various colours are derived from only four pigment groups; the chlorophylls, carotenoids, betalains and anthocyanins. Of these pigments, anthocyanins are the most prominent, imparting red, blue and black hues to the fruits in which they accumulate (Macheix et al. 1990). In ten of the twelve families of the order Caryophyllales (including various cacti fruit such as the prickly pear) they are replaced by the chemically unrelated betalains that impart a similar range of colours (Stafford 1994). Green fruit owe their colour to chlorophylls, while yellow fruits and some red fruits (e.g. tomatoes and peppers) derive their colour from various carotenoids (Macheix et al. 1990). Oxidation and polymerization products of phenolic compounds give rise to brown fruit colours (Macheix et al. 1990).

The primary concern of this chapter is the functions of anthocyanins in fruit. Most eminent of these functions is the ability of anthocyanins to impart colour and so attract seed dispersers (Willson and Whelan 1990). Fruit pigmentation may also

K. Gould et al. (eds.), *Anthocyanins*, DOI: 10.1007/978-0-387-77335-3_4,
© Springer Science+Business Media, LLC 2009

signal fruit quality and afford protection against seed predators (Willson and Whelan 1990). Apart from fruits and flowers, anthocyanins also accumulate in vegetative tissues where they are considered to confer protection against various biotic and abiotic stresses (Gould and Lister 2006 for a recent review). Anthocyanins are able to protect tissues from photoinhibition caused by high levels of visible light (Smillie and Hetherington 1999) and from oxidative damage (Neill and Gould 2003). It has also been proposed that foliar anthocyanins may reduce herbivory by signalling defensive strength (Schaefer and Rolshausen 2005). Anthocyanins could also potentially fulfil these functions in fruits (Stintzing and Carle 2004).

Before exploring the functions of anthocyanins, we need to consider broad patterns in the occurrence and distribution of anthocyanins in fruits. We also need to assess the regulation of anthocyanin synthesis by endogenous and environmental factors. This is to ensure that supportive evidence for the various functions concur with observed patterns in anthocyanin accumulation. As there are very few studies on the physiology of colour development in wild fruits, this review will refer to numerous such studies in cultivated fruits to provide more insight into the environmental regulation of colour development.

4.2 Prevalence of Fruit Colours

It is difficult to assess the frequencies of different colours and the contribution of anthocyanins thereto. This is due to differences in the subjective colour classes into which fruits are grouped. For example, some authors group orange and red together (Traveset et al. 2004) while dull dark red fruit may be classified as brown (Wheelwright and Janson 1985). Authors may distinguish between blue (including violet and purple) and black fruits (Wheelwright and Janson 1985) or may have an additional category for blue-black fruits (Traveset et al. 2004).

Despite these limitations, a distinct relationship between fruit colour and dispersal by either birds or mammals can be discerned. Fruits adapted for dispersal by birds (small, soft pulp and no protective covering) generally tend to be blue, white, red or black when ripe (Janson 1983). Red and black are the most common of these colours (24–64% and 26–50% of fruits, respectively) (Janson 1983; Traveset et al. 2004; Wheelwright and Janson 1985; Willson and Thompson 1982). Green, yellow and brown are uncommon bird-fruit colours, while orange only appears to be of importance in the tropics (Janson 1983; Wheelwright and Janson 1985). In contrast, fruits adapted for dispersal by mammals (large, fibrous, and covered by a hard husk) tend to be green, brown, yellow or orange in colour although primates also include red and purple "bird" fruits in their diets (Regan et al. 2001; Voigt et al. 2004). This dichotomy in colour preference persists even where the diet of monkeys overlaps with that of large birds such as the hornbills that are potentially able to utilize relatively large fruits (Poulsen et al. 2002). Hence, regional differences in the frequency of fruit colours generally seem to relate to differences in dispersal assemblages (Voigt et al. 2004) and are not consistently associated with any geographical or environmental factor (Wheelwright and Janson 1985). Red and black, the principal colours of bird-fruits, are derived mainly from anthocyanins. Considering this, anthocyanins appear to be of more importance in bird-dispersed fruits.

Many wild and cultivated plant species display stable polymorphisms in fruit colour (Traveset et al. 2001, 2004; Whitney and Lister 2004; Willson 1986). This means that plants of the same species carry fruits that differ in colour. Comparison between species does not indicate a consistent preference of dispersers and seed predators for any particular colour. Even intraspecific preferences, when present, are generally not very strong. New Zealand mistletoe provides an exception. Bach and Kelly (2004) found that birds strongly preferred red morphs over orange morphs, but this preference extended also to green unripe fruits of red morphs. Therefore, this preference probably relates to other correlated aspects of fruit quality and not to red colour *per se*.

Differences in fruit colour between morphs may also correlate to differences in various other aspects of plant morphology and physiology. This is because genes involved in regulation of anthocyanins may also be involved in the regulation of various other pathways. These pleiotropic effects may manifest as differences in, for example, germination ability, germination rate, growth requirements and vigour (Traveset et al. 2001; Traveset and Willson 1998; Whitney 2005; Whitney and Lister 2004; Whitney and Stanton 2004; Willson and O'Dowd 1989). Hence, maintenance of stable colour polymorphisms seems to result from disperser and seed predator preferences for a particular morph interacting with pleiotropic effects of colour genes.

Ripe fruits often co-occur with unripe fruits and/or accessory structures (i.e. persisting calyces and sepals, peduncles, bracts, capsules and arils) with contrasting colours (Fig. 4.1a–c). Red with black is the most common combination in multicolour displays and occurs together in about 18% of tropical plant species (Wheelwright and Janson 1985). In almost all of these red and black displays, ripe fruits are black while unripe fruits or accessory structures are red.

Fig. 4.1 Fruit and flowers with contrasting colours. (A) Black ripe fruits and red calyxes of *Ochna serrulata*. (B) Red unripe and black ripe fruits of *Colpoon compressum*. (C) Black ripe fruits of *Halleria lucida* are displayed with red flowers. Fruits turn from green to black without any intermediary colours. Photos by E. van Jaarsveld. See Plate 5 for colour version of these photographs

4.3 Developmental Patterns

While immature fruits are generally green in colour, very few fruits remain green until maturity. This is due to the general pattern of loss of chlorophyll and/or accumulation of carotenoids, betalains and anthocyanins that characterize the ripening phase in most fruits (Macheix et al. 1990). Anthocyanin-accumulating fruits often display a range of intermediary colours progressing from green to white or pink, then red or blue and finally purple to black with increasing anthocyanin and decreasing chlorophyll levels (Wheelwright and Janson 1985; Willson and Thompson 1982). Rapid direct colour changes from green to black also occur (Willson and Thompson 1982; Fig. 4.1c). The anthocyanin profile of many fruits (e.g. cherry) does not change during fruit development; rather, changes in colour arise from increasing anthocyanin concentrations. Some other fruits display qualitative changes in their anthocyanin profile during fruit development resulting in fruit colour changing from red in immature fruits to blue and purple in ripe fruits (e.g. blueberry) (Macheix et al. 1990).

Anthocyanins do not always follow the general pattern of accumulation towards maturity. For example, fruit of some red pear (Steyn et al. 2004a; Fig. 4.2a) and pineapple cultivars (Gortner 1965) attain their highest anthocyanin levels while immature and red colour gradually fades towards harvest. Many apple cultivars, including varieties that are typically yellow or green at maturity, display a peak in anthocyanin levels shortly after fruit set (Lancaster et al. 2000). However, these intermediary colours are seldom reported and may often go unnoticed (Wheelwright and Janson 1985; Willson and Thompson 1982).

Fig. 4.2 Anthocyanins in fruit. (A) Immature and mature "Forelle" pears. Pears attain their highest anthocyanin concentrations during early fruit development whereafter red colour gradually fades towards harvest. Photo by W.J. Steyn. (B) Anthocyanin accumulation surrounding sunburn lesion in apple. Photo by J. Gindaba. (C) Grape berries displaying a bloom. Photo by M. Huysamer. See Plate 6 for colour version of these photographs

4.4 Distribution of Anthocyanins in Fruit

Most anthocyanic fruits contain two aglycones, with cyanidin being the most prevalent (present in 90% of 40 common fruits and 82% of fruits of 44 angiosperm species) (Macheix et al. 1990). On a quantitative basis, anthocyanin contents of fruits vary considerably with concentrations ranging from 0.25 mg per 100 g FW in pear peel to >200 mg per 100 g FW for black fruits rich in anthocyanins (Macheix et al. 1990). However, much higher anthocyanin contents have been reported; anthocyanins contribute 10.7% to the dry weight of fruits of the Mediterranean shrub, *Coriaria myrtifolia* (Escribano-Bailón et al. 2002). The effect of increasing anthocyanin concentration on fruit colour is well illustrated by the example of cherries. Cultivars with bright red peel were found to contain 30 mg anthocyanins per 100 g peel whereas dark red to black cultivars where anthocyanins are located throughout the fruit contained 350–450 mg per 100 g FW (Macheix et al. 1990). At 4 mg per 100 g FW, anthocyanins levels in pink cultivars were about 10 and 100-fold lower compared to red and black fruits, respectively.

Anthocyanins are most prevalent in the epidermal and hypodermal layers of the fruit skin, but may also occur throughout the fruit (many berries), in tissues surrounding seeds or pip (pomegranate and peach, respectively), or may be confined to the flesh (blood orange) (Gross 1987; Harborne 1976). Not all cells in red tissues necessarily contain anthocyanins; a large proportion of colourless cells are randomly dispersed throughout red anthocyanin-rich apple peel (Li et al. 2004). Some apple and pear fruit exhibit longitudinal stripes of more intense colour on their red sun-exposed sides, whereas stripes have not been reported in bird-fruits (Wheelwright and Janson 1985). Fruit of some cultivated crops (e.g. apple, mango and peach) may even be bicoloured due to a light requirement for anthocyanin synthesis (effects of light are discussed in the following section). This light requirement results in fruits having a red sun-exposed side and a green shaded side that may turn yellow when ripe (Awad et al. 2000; Hetherington 1997). Shaded fruit from the inside of the tree canopy do not accumulate anthocyanin and remain green or ripen to yellow (Awad et al. 2000; Erez and Flore 1986). Such bi-colouration may also occur in wild fruits, but is not reported.

4.5 Environmental Regulation of Colour Development

Unlike in vegetative tissues (reviewed by Steyn et al. 2002), very little is known about the environmental regulation of anthocyanin synthesis and other characteristics of wild fruits (Hampe 2003). The same applies to the majority of cultivated fruits, with apple and grape being the most notable exceptions. In these fruits, financial incentives have seen to numerous detailed studies on the physiology of fruit colouration. Ecologists, however, seldom refer to pigmentation patterns in cultivated fruits. Their concern is that years of domestication and selection may have separated fruit characteristics from their original ecological and physiological intent (Regan et al. 2001). However, dependency on environmental factors is certainly not selected for in the domestication of fruit crops. Hence, it seems reasonable to speculate that similar environmental regulation of anthocyanin synthesis also occurs in wild fruits.

4.5.1 Light

Some temperate Rosaceous fruits, i.e., apple, pear, peaches, nectarines and apricots, are considered to have an absolute light requirement for anthocyanin synthesis in the skin (Allen 1932). Other fruits including strawberries, blackberries, grapes, cherries and plums are considered able to develop colour, albeit to a lesser extent, in the absence of light (Allen 1932). An absolute light requirement has also been documented for ripening figs (Puech et al. 1976). It can also be inferred for at least some mango and pomegranate cultivars from the absence of anthocyanins from shaded skin (Gil et al. 1995; Hetherington 1997). Anthocyanin levels in cranberry also increase with light exposure (Zhou and Singh 2002). The effect of light levels on anthocyanin levels in red-fleshed fruits (e.g. strawberry, blood orange and some apple, pear, plum and peach genotypes) has generally not been recorded. The exception is pomegranate where, contrary to the effect on skin anthocyanins, seed coats of fruit from the outer canopy were less pigmented than in fruit from the inner canopy (Gil et al. 1995).

Cultivars differ in their responsiveness to light. Lighter red cultivars are generally more responsive compared to dark red, purple and black fruits (Kliewer 1970; Steyn et al. 2004b). This could relate to the observation that much greater changes in anthocyanin concentration are required to induce a comparable change in colour in highly pigmented fruit skin compared to skin containing little anthocyanin (Steyn et al. 2004b). Fruit colour becomes darker with addition of pigment due to a reduction in reflection of incident light. When anthocyanins reach a level where nearly all light is absorbed, addition of pigment does not result in further colour change. Fruit will appear black.

The extent of light-responsive red colour development in apple is dose dependent. Red colour increases with increasing light exposure at the level of the individual fruit, position within the infrutescence and position within the canopy of the tree (Awad et al. 2000; Erez and Flore 1986). In addition, the ability of apple peel to synthesize anthocyanins after harvest was found to decrease in proportion to light exposure during fruit development (Lancaster et al. 2000). Furthermore, fruit are apparently able to actively degrade anthocyanins during warm conditions or in the absence of light. For example, shading of previously exposed pear fruit resulted in the rapid loss of anthocyanin and red colour (Steyn et al. 2004b). This stringent regulation of anthocyanin levels by light suggests a strong link between other light-regulated processes (e.g. photosynthesis) and anthocyanin accumulation.

4.5.2 Temperature

Second to light, temperature is the most important environmental factor that influences anthocyanin synthesis. Low temperatures either before harvest and/or during storage generally favour anthocyanin synthesis (reported for apple, pear, grape, blackberry and cranberry) (Curry 1997; Hall and Stark 1972; Kliewer 1970; Naumann and Wittenburg 1980; Steyn et al. 2004a). In contrast, high temperatures

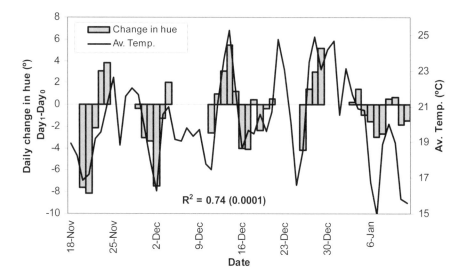

Fig. 4.3 Daily changes in the hue of "Rosemarie" pear and average daily temperatures. Hue angles reported fluctuate between 0° (red-purple) and 90° (yellow). (Adapted from Steyn et al., 2004b)

are associated with anthocyanin degradation and the preharvest loss of red colour in pears (Steyn et al. 2004b). Consequently, the colour of some pear cultivars has been found to fluctuate between red and green in response to the passing of cold fronts and intermittent warmer conditions (Steyn et al. 2004b; Fig. 4.3). In addition to inductive low temperatures at night, anthocyanin synthesis in mature apples requires mild day temperatures between 20 and 25°C (Curry 1997). However, anthocyanin synthesis in immature apple fruit requires lower temperatures (Faragher 1983). Differential cultivar responsiveness to low temperatures has been recorded in grape (Kliewer and Schultz 1973) and pear (Steyn et al. 2004a).

4.5.3 Other Factors

Nitrogen deficiency generally favours anthocyanin pigmentation in vegetative tissues (Steyn et al. 2002) and the same apparently applies to fruit. Nitrogen application reduced anthocyanin concentrations in grape and apple (Kliewer 1977; Reay et al. 1998), but may also reduce the intensity of red colour by increasing chlorophyll levels (Marsh et al. 1996; Reay et al. 1998). The stimulation of anthocyanin synthesis in nitrogen-deficient leaves apparently relates to the accumulation of carbohydrate (Steyn et al. 2002). Anthocyanin synthesis is strongly inducible by sugars (Larronde et al. 1998; Solfanelli et al. 2006).

Wounding and postharvest application of methyl jasmonate induced anthocyanin synthesis in apple skin in the presence of light (Faragher and Chalmers 1977; Rudell et al. 2002). Apples sometimes display a red anthocyanin halo in affected tissues surrounding necrotic lesions caused by thermal injury to the skin (Fig. 4.2b).

4.6 Anthocyanins in Attraction

Biotic dispersal of fleshy seeds or fruits has been present since the Permian period (300 MYA) and has become widespread during the Tertiary period (65 MYA) (Tiffney 2004). At present, up to 56% of angiosperm and 85% of gymnosperm species make use of primarily birds and mammals, but also ants, reptiles and fish, to disperse their seeds (Tiffney 2004). Mack (2000) postulated that fleshy fruits may have initially evolved to defend seeds against invertebrate seed predation and fungal attack. Subsequent interaction with frugivores led to the adaptation of fruits for the new task of attracting and rewarding their consumers. The development of attractive and conspicuous colours at maturity is thought to be one of these adaptations to alert frugivores to the presence of ripe fruits (Willson and Whelan 1990). As previously mentioned, fruits dispersed by birds and mammals differ in colour, but primates also include fruits with typical "bird" colours in their diets. These differences in fruit colour between the different disperser groups may relate to differences in the visual abilities of the different disperser groups. However, the visual systems of seed predators should also be considered.

4.6.1 Visual Systems

The ability of animals to discriminate colour depends on the presence of distinct retinal photoreceptors. These photoreceptors have peak light absorbance at different wavelengths and neural mechanisms to compare their input (Mollon 1989). Colour discrimination is only possible for wavelengths where absorbance of the different photoreceptors overlaps. Most mammals are dichromats possessing only two photoreceptors; one sensitive to short wavelengths (λ-max at about 420 nm) and another sensitive to longer wavelengths (λ-max 510–570 nm) (Jacobs 1993). Absorbance of these two photoreceptors overlaps over the green and blue parts, but not over the orange to red part of the light spectrum. Hence, mammalian dichromats can discriminate between blue and green, but not between green and orange/red. In addition to the short wavelength photoreceptor, trichromatic Old World apes and monkeys and New World howler monkeys have two photoreceptors at the longer wavelength end of the spectrum (Jacobs 1993; Mollon 1989). These so-called middle (λ-max ~530 nm) and long (λ-max ~560 nm) wavelength photoreceptors overlap over the green and red parts of the light spectrum, thus allowing colour discrimination between green and red (Jacobs 1993). Humans form part of this trichromatic group. In other New World monkeys, a single polymorphic gene encoding for various green and red alleles resides on the X chromosome. The result is that heterozygous females are trichromats and have colour vision over a continuous light spectrum from blue to red (Regan et al. 1998). On the other hand, all males as well as homozygous females are dichromats and unable to distinguish between green and red. The persistence of dichromatic genotypes among New World monkeys may relate to their apparent greater ability to penetrate colour camouflage of predators and prey (Morgan et al. 1992; Surridge et al. 2003).

Frugivorous bird species have tetrachromatic colour vision with a fourth retinal photoreceptor (λ-max 355-380 or 403-426 nm). This is potentially able to extend their colour perceptive ability well into the ultraviolet (UV) range (Ödeen and Håstad 2003). Frugivourous rodents and reptiles also have UV vision (Honkavaara et al. 2002). Many insects studied have UV, blue and green light receptors (λ-max at ~350, ~440 and ~530 nm, respectively) (Briscoe and Chittka 2001). Red receptors (λ-max >565 nm) are a more recent acquisition in some species in the Odonata, Hymenoptera, Lepidoptera and Coleoptera. To summarize, birds and trichromatic primates are able to see orange to red colours whereas dichromatic mammals cannot. In addition to trichromatic colour vision, birds can also see UV. Insects can see UV, but generally not red.

4.6.2 Red Fruits

The minimal variation in background hue of mature leaves allows for significant colour contrast with non-green fruits (Sumner and Mollon 2000). Considering differences in visual abilities, red fruits would be visible to birds and trichromatic primates. However, red fruits would be cryptic to insect seed predators and undesirable mammalian frugivores that may be poor dispersers. Laboratory experiments as well as field studies on fruit removal rates from different colour morphs of polymorphic species indicate that birds generally do not have an innate preference for red or black fruits despite their prevalence (Schmidt et al. 2004; Traveset et al. 2001; Willson et al. 1990; Willson and O'Dowd 1989). Birds do, however, show a strong preference for displays where fruit colour contrasts with background colour (Schmidt et al. 2004). Red fruits are highly conspicuous to birds against a green background (Burns and Dalen 2002; Schmidt et al. 2004). They are also easily seen from a distance (Schaefer et al. 2006).

Several recent studies indicate a distinct advantage for trichromatic primates compared to dichromatic primates when foraging for red/yellow foods against a green background (Caine and Mundy 2000; Osorio et al. 2004; Osorio and Vorobyev 1996; Regan et al. 1998). The difficulties experienced by red-green colour blind people in assessing the ripeness of fruits (~40% of 37 dichromats surveyed) and in detecting red flowers and fruits against green foliage (57% of dichromats) further highlights this advantage (Steward and Cole 1989). Yet, since trichromatic vision also enables primates to detect young leaves or conspecifics against the background of green leaves, there is no direct evidence to support the notion that trichromatic colour vision in primates evolved as a means of finding fruit (Regan et al. 2001).

Due to the inability of most insects to see red directly, it has been suggested that red pigmentation could potentially reduce seed predation (Schaefer and Schmidt 2004). Of course, frugivorous insects also make use of olfactory signals to locate fruits (Howell 1991; White and Elson-Harris 1992) and only some insects are frugivores and potential seed predators. Specialist frugivores such as tephritid fruit flies (about 4000 species of which about 35% attack fruits; White and Elson-Harris 1992) are attracted by ripe fruit colours, including red, black, blue and UV-reflecting, of their host fruit species (Drew et al. 2003; Katsoyannos and Kouloussis 2001 and references therein). For example, female olive fruit flies were found to prefer red and black plastic spheres over other colours (Katsoyannos and Kouloussis 2001). Anyhow, judging from their range of host species, high levels of anthocyanins are not a safeguard against fruit fly attack (White and Elson-Harris 1992). Immature fruits generally rely on cryptic green colouration, unpalatability and/or toxicity to reduce insect predation (Willson and Whelan 1990). Anthocyanins do not appear to have bioactivity against herbivores, but have been proposed to signal defensive strength of vegetative tissues due to their correlation with various defensive compounds (Schaefer and Rolshausen 2005; see Chapter 2). The merit of this hypothesis and whether it pertains to anthocyanins in immature fruits still remains to be confirmed.

In summary, it appears that the prevalence of red fruits may stem from their conspicuousness against a background of green leaves to a specific set of preferred dispersers (Schmidt et al. 2004). An additional possibility is that red fruits may be cryptic to inadvertent consumers and opportunistic insect frugivores. Anthocyanins also impart colour to blue and black fruits. The visibility of these colours and whether it relates to their prevalence will be considered next.

4.6.3 Blue and Black Fruits

Considering the visual systems of different disperser groups, blue fruits, like red fruits, should be conspicuous to both birds and trichromatic primates. However, unlike red fruits, blue fruits should also be conspicuous to undesirable frugivores such as insects and dichromatic mammals. This may explain the relative scarcity of blue fruits (Wheelwright and Janson 1985). Regan et al. (2001) suggested that the scarcity of blue fruits could be due to phylogenic limitations in the ability to produce blue anthocyanins. In contradiction to this suggestion, but in accordance with the visual explanation, blue is quite a common colour in flowers where it serves to attract insect pollinators (Harborne and Grayer 1994).

In contrast to red and blue fruits, dispersers cannot rely on colour vision for the detection of black fruits. This is because black fruits only present an achromatic luminance signal due to their nearly complete absorbance of incident visible light. Birds have been shown to detect the achromatic signal of black fruits against a background of higher, but uniform brightness (Schaefer et al. 2006). The same should apply to mammals. However, the achromatic signal may be lost against the huge variation in background luminance that is typical of the tree canopy and forest understorey (Regan et al. 2001; Sumner and Mollon 2000). Consequently, black fruits may be relatively inconspicuous to dispersers under most viewing conditions in the natural environment (Regan et al. 2001).

Plants may use different methods to increase the conspicuousness of black fruits. One of these entails displaying ripe black fruit together with unripe red fruits and/or red accessory structures (Wheelwright and Janson 1985; Figs. 4.1a–c). Of course, the effectiveness of the red part of the display in attraction precludes such a role for the black fruits. Plants can also increase the conspicuousness of black fruits by means of ultraviolet (UV) reflection (Honkavaara et al. 2002; Schaefer et al. 2006; Siitari et al. 1999). In a Panamanian survey, Altshuler (2001) found about 60% of fruits to reflect UV. UV reflection was attributed to absorbent properties of epidermal compounds. Furthermore, an UV-reflecting epicuticular wax layer was found to accumulate on ripe fruits of about half the surveyed temperate plant species

(Burkhardt 1982; Willson and Whelan 1989; Fig. 4.2c). The wax layer also increases reflectance within the human-visible spectrum (Willson and Whelan 1989). Since black fruits with a wax covering appear blue to humans, it is possible that many fruits that were previously broadly classified as black (Wheelwright and Janson 1985) may fall into the blue-black UV-reflecting group.

Several recent studies have studied the conspicuousness of UV-reflecting fruits. Rubbing off the wax layer of blue fruits abolished UV reflectance; fruits subsequently appeared black in colour and were much less conspicuous to birds (Schaefer et al. 2006; Siitari et al. 1999). Altshuler (2001) found that *in situ* placement of UV filters above plants, markedly reduced removal of UV-reflecting fruit compared to plants under ambient light or UV-transmitting filters. Although these studies focused on UV reflection as pertaining to the visibility of black fruits, UV-reflection is not confined to black fruits and not all black fruits reflect UV (Burkhardt 1982). Some of the UV-reflecting fruits listed by Burkhardt (1982) were orange, white and green in colour. Black fruits with their high levels of UV-absorbing anthocyanins most probably also do not fall into the class of wax-less fruits that reflect UV. Therefore, UV-reflection does not appear to be a strategy aimed specifically at increasing the conspicuousness of black fruits.

It is also possible that the attraction of birds could simply be a fortuitous and unintentional side-effect of the epicuticular wax layer. Many fruits that are not dispersed by birds, e.g., the apple, also develop an epicuticular wax layer. This is in agreement with the supposed primary function of epicuticular waxes in protecting fruit against water loss, UV light and pathogens (Shepherd and Griffiths 2006; Veraverbeke et al. 2001).

In summary, black fruits appear to be relatively inconspicuous to dispersers and various strategies are required to make them more visible. With regard to attraction of frugivores, the accumulation of very high levels of anthocyanins in black fruits seems counterproductive when fruits would be highly conspicuous to dispersers at relatively low anthocyanins levels. However, it is possible that high pigment levels may be a reward in itself (Wheelwright and Janson 1985). Apparently birds prefer diets that are rich in anthocyanins irrespective of colour (Schaeffer, unpublished data). The next section will consider the possibility of communicating fruit quality with colour.

4.7 Fruit Quality and Composition

4.7.1 Health Benefits

Unfortunately, data on the health benefits of ingesting anthocyanins are almost exclusive to humans. Anthocyanins are potent radical scavengers and antioxidants with proven health benefits and activity against a range of chronic human diseases (Lila 2004). Various studies have shown highly pigmented berry fruits to possess considerable antioxidant capacity, at least partially due to their high anthocyanin concentrations (Deighton et al. 2000; Kähkönen et al. 2003; Moyer et al. 2002; Reyes-Carmona et al. 2005). In 107 blueberry, blackberry and blackcurrant genotypes, total anthocyanins were found to correlate significantly ($r = 0.57–0.93$) to total phenolics and to average 34% of total phenolics (Moyer et al. 2002). However, antioxidant capacity correlated better to total phenolics than to anthocyanin levels. This is why fruits containing low levels of anthocyanins, but high levels of total phenolics may still display high antioxidant capacity (Deighton et al. 2000). The beneficial effects of anthocyanins may be modified by complex interactions with various other fruit constituents (Lila 2004). There is also still much uncertainty regarding the extent of uptake of anthocyanins from the human intestine (Manach 2004).

4.7.2 Nutritional Content and Defensive Strength

It is possible that peel colour can provide information regarding the nutritional content and defensive strength of fruits. Schaefer and Schmidt (2004) found that surface reflection from yellow, orange and green fruits provide information regarding their protein, tannin and carbohydrate contents. In contrast, reflection from red and black fruits yielded no information regarding their nutritional value. None of the colours indicated levels of phenolics, which Schaefer et al. (2003b) found to deter consumption of tropical fruits. Anthocyanins are, of course, the most visible fruit phenolic, occurring in large quantities in black "bird" fruits and contributing significantly to total phenolics in these fruits. This paradox may be explained if the deterrence by phenolics resides with another of its components (e.g. phenolic acids) and is based on taste.

Yellow, orange and red fruits are conspicuous to both birds and trichromatic primates. Their placement in different disperser syndromes suggests that colour-linked fruit characteristics may increase their relative attractiveness to different disperser groups. Voigt et al. (2004), however, found that fruit colouration gave a better indication than morphological and chemical fruit traits of the frugivore assemblages of a specific region. In addition, chemical constituents in fruits show considerable intraspecific variation. This suggests that birds do not exert strong selective pressure on fruit composition (Hampe 2003; Izhaki et al. 2002). Intraspecific differences in the chemical composition of fruits may rather arise from constraints imposed by the abiotic environment (Izhaki et al. 2002). However, Cipollini et al. (2004) found that the nutritional content and defensive attributes of *Solanum* sp. fruits are maintained despite large differences in nutrition and water availability. This contrasts with the effect of excess nitrogen on fruit colour in several woody perennials. Maybe the difference is due to the greater plasticity of growth in herbaceous plants compared to woody perennials.

In contrast to wild fruits, cultivated fruits have been subject to numerous studies that have looked into the effects of cultural and environmental factors on their chemical composition. These studies often report a positive correlation between anthocyanin levels and sugar concentration of fruits (Crisosto et al. 2002). At the basis of the correlation is the positive relationship between light exposure and both anthocyanin and total soluble solids content (Erez and Flore 1986; Kliewer 1977). Anthocyanins are also strongly sugar-inducible (Larronde et al. 1998; Solfanelli et al. 2006). In grape berries, anthocyanin accumulation commences with the onset of sugar accumulation (Boss et al. 1996). Solfanelli et al. (2006) speculated that this correlation may be causal. Sugar levels of wild fruits decrease with shading (Levey 1987), but it is not known how light exposure influences anthocyanin levels in these fruits. Both mammals and birds have a predilection for sweet foods (Harborne 1988). With regard to birds, Schaefer et al. (2003a) found that four tanager species were able to detect 1% differences in sugar levels and preferred the diet highest in sugar. Such a difference in sugar levels is well within the range found between fruits of the same species and between fruits on the same plant at different maturities. In humans, the colour of food products influences the perception of sweetness and flavour. To illustrate, a consumer panel perceived an increase in the sweetness of "fruit of the forest" yoghurt with addition of red, purple and violet, but not yellow colorant (Calvo et al. 2001). Addition of pigment also increased the perception of flavour of orange, red, purple and violet yoghurts. All yoghurts had the same sugar concentration. It seems compelling to suggest that dispersers could potentially judge the sweetness and carbohydrate content of fruits from the extent and intensity of their red colour.

In support, Riba-Hernández et al. (2005) found a positive correlation ($r^2 = 0.493$, $p = 0.02$) between time spent by trichromatic spider monkeys foraging on a particular fruit species and the average glucose concentration therein. In contrast, dichromatic spider monkeys did not allocate more time towards consumption of glucose-rich fruits. Since the red-green, but not the blue-yellow, colour vision signal of fruits correlated with their glucose concentration, the authors inferred that trichromatic spider monkeys can assess the glucose level of fruits. Other data indicate that the colour signal of fruit may only be of value within a species. For example, Li et al. (2004) found comparable levels of non-reducing and reducing sugars in red "Stark-rimson" and yellow "Golden Delicious" apples grown in the same region. The soluble solid content of blackberry cultivars also did not correlate with anthocyanin, total phenolic or antioxidant potential (Reyes-Carmona et al. 2005). Since black fruits absorb almost all incident light, they are limited in their ability to signal fruit composition using colour (Schaefer and Schmidt 2004). The signal value of dark fruit colours may, however, reside in their association with fruit maturity.

4.7.3 Maturity

Whereas red fruits can be ripe or unripe, black is a ripe fruit colour only. Thus black fruits can be considered ripe and can be eaten without any consideration or acquired knowledge of the ripening patterns of the particular fruit species (Wheelwright and Janson 1985).

Humans also widely make use of changes in skin colour as a simple indicator of whether fruit are ready to be picked (Crisoto et al. 2002). This can be attributed to a strong relationship between skin colour and quality attributes such as soluble solids content (Crisoto et al. 2002). While fruit colour changes may enable consumers to visually assess ripeness and fruit internal quality, high anthocyanin surface coverage in bi-coloured fruits can also make it harder to accurately assess ripeness. In "Royal Gala" apple, the extent of the red blush on the sun exposed sides of fruit and the green to yellow colour change on the non-blushed shaded sides of fruit were the preferred independent cues used by consumers to assess ripeness (Richardson-Harman et al. 1998). Consumers considered fruit with a yellow background colour and greater extent of blush coverage as being riper. However, whereas background colour was a reliable indicator of fruit maturity, the extent of blush coverage was unrelated to maturity. Fruits with 66–100% blush coverage were considered to be overripe even if other unrelated maturity parameters indicated them to be at optimal maturity. Hence, high blush coverage can decrease the accuracy of ripeness assessment.

4.8 Anthocyanin and Fruit Size

Whatever the signal value of black fruits, very high anthocyanin levels, and therefore dark fruit colours, are scarce amongst large cultivated and wild fruits. This probably relates to colour preferences of mammalian dispersers (Janson 1983). Evidently the green-yellow colour change displayed by many large fruits functions adequately in attracting mammalian frugivores and signalling maturity. The scarcity of large black fruits could also relate to a cost factor in the production of anthocyanins (Wheelwright and Janson 1985; Willson and Whelan 1990). In fruits species where anthocyanins accumulate in the skin only, anthocyanin concentrations generally increase with decreasing fruit size due to the change in surface area to volume ratio (Moyer et al. 2002). High levels of anthocyanins also accumulate in the flesh of some wild berry fruits, but these fruits are usually relatively small in size and the fruit skin may be thin (Moyer et al. 2002). If high anthocyanin levels were to constitute a reward to mammalian dispersers, large concentrations of anthocyanins would have to accumulate in the pulp of large fruits. A cost factor in the production of anthocyanins would be expected to result in a compensatory reduction in growth, defence or reproduction in plants with highly pigmented fruits (Willson and Whelan 1990). No such compensation could be found even between different colour morphs of polymorphic species (Willson and Whelan 1990). However, various dark red pear cultivars selected from spontaneous bud mutations do show reduced vigour and lower crop loads compared to their wild type green-fruited parents (Martin et al. 1997). Physiological comparison of these red cultivars with their parents suggests that changes in phenology could be due to pleiotropic effects of the mutated gene(s).

Anthocyanins in large fruits (e.g. the apple) are usually confined to the sun-exposed sides of outer-canopy fruits (Awad et al. 2000). Even then, not all cells are pigmented (Li et al. 2004), maybe utilizing the fact that not all cells need to be red in order to obtain a continuous red surface (cf. photographic images). This apple example may represent a compromise between attraction and the cost of anthocyanin accumulation. Alternatively, the restriction of anthocyanins to exposed peel relates to the photoprotective ability of anthocyanins.

4.9 Photoprotection

Anthocyanins in chlorophyllous tissues provide protection against photoinhibition and photobleaching of chlorophyll by acting as a selective screen that reduces the incidence of blue-green light on chloroplasts (Neill and Gould 2003; Smillie and

Hetherington 1999). Immature fruit have been shown to display comparable rates of photosynthesis on a unit chlorophyll basis as mature leaves (Aschan and Pfanz 2003). Fruit photosynthesis can contribute significantly to its own carbon budget and as such may increase the viability of seeds or prevent abscission during stress and at sensitive stage during fruit development. The green colour of immature fruits also makes them cryptic to the eyes of seed predators. Whatever the reason for their presence, fruit chloroplasts makes fruit skin susceptible to damage when light in excess of what can be used in photosynthesis is absorbed. Due to much lower chloroplast densities (Aschan and Pfanz 2003), fruit skin is likely to experience greater light stress at a particular light level compared to leaves. As in leaves (Steyn et al. 2002), the potential for photoinhibition would be greatest during early fruit development and during ripening concomitant with the decrease in chlorophyll levels. As mentioned previously, these are the stages during fruit development when anthocyanins are likely to accumulate. Stresses such as low temperatures and nitrogen deficiency would also render fruit more sensitive to light stress (Smillie and Hetherington 1999). Anthocyanin accumulation in immature fruits and in response to low temperatures would be in agreement with a photoprotective function, but not with a visual function.

Recent studies clearly demonstrated the ability of anthocyanins to afford photoprotection. At very high light levels anthocyanins were found to decrease the bleaching of chlorophyll in red compared to green apples (Merzlyak and Chivkunova 2000). Peel of purple mango cultivars was also more tolerant of light stress than peel of green cultivars (Hetherington 1997). Pods of a purple *Bauhinia variegata* cultivar were more tolerant of blue-green and white light than pods of a green cultivar (Smillie and Hetherington 1999). However, purple and green pods showed comparable susceptibility to red light not absorbed by the anthocyanins.

Unfortunately, it is much more difficult to establish whether the photoprotection provided by anthocyanins is deliberate. By virtue of their light-absorbent properties, anthocyanins, whenever present in fruit skin, will provide photoprotection to an extent relating to their concentration (Pietrini and Massacci 1998). The protection provided can of course be unintentional and a fortuitous side-effect of the presence of the anthocyanins. While fruit may turn red at maturity to attract dispersers, anthocyanins that accumulate will also afford photoprotection. This being said, the presence of anthocyanins in immature fruits and the regulation of anthocyanin synthesis by environmental factors such as light and temperature strongly suggest that anthocyanins in fruits may have a primary photoprotection function, at least during early fruit development (Hetherington 1997; Smillie and Hetherington 1999).

4.10 Perspectives

The diversity of fruit colours is generally believed to be indicative of increased specialization to facilitate biotic dispersal (Willson and Thompson 1982; Willson and Whelan 1990). Distinct relationships may develop over time between a specific small set of dispersers in which case fruits could be any in a range of colours – birds do not seem to have a preference for any particular colour. At the other end of the spectrum, plants may use abiotic dispersal in addition to biotic dispersal and may cater for a range of mammalian, avian and insect dispersers. However, it is possible to form at least some general conclusions regarding the functions of anthocyanins in fruits. First of all, red fruits and red accessory structures are highly conspicuous against background foliage to tetrachromatic birds and trichromatic primates, but not to non-specialist mammalian and opportunistic insect frugivores. Red colour is, therefore, effective in attracting a specific set of dispersers while being cryptic to potential harmful organisms. Blue fruits are also conspicuous to dispersers, but without the benefit of being cryptic to potential insect pests and inadvertent fruit consumers. The adaptive value of black fruits with their high anthocyanin content is more obscure, but their prevalence may relate to the association between dark colours and fruit maturity. High anthocyanin levels may also protect consumers from oxidative stress.

Briscoe and Chittka (2001) caution that the great diversity in insect visual systems increases the danger of finding erroneous support for at least one in a multitude of ecological hypotheses that explain these patterns. Similar care should be taken when assigning adaptive functions to the abundant fruit colouration patterns. Anthocyanins are multifunctional compounds that, in addition to their obvious visual functions, are potentially able to confer protection against various abiotic and biotic stresses (Gould and Lister 2006 for a recent review). Their synthesis is also closely integrated with plant stress-signalling networks and phenolic metabolism. This manifests in the pleiotropic effects of genes involved in anthocyanin regulation. In many instances environmental factors (e.g. light, temperature and nutrition) have similar effects on fruit and foliar anthocyanins suggesting that anthocyanins originally fulfilled and may still be able to perform similar functions in fruit skin and vegetative tissues (Stintzing and Carle 2004). One of these functions could possibly be photoprotection, which would explain the presence of anthocyanins in immature fruits and its regulation by environmental factors such as light and temperature. Evidently the adaptive value of fruit anthocyanins should be seen and interpreted against a background of interaction with dispersers, genotype and environment.

References

Allen, F.W. (1932) Physical and chemical changes in the ripening of deciduous fruits. Hilgardia 6, 381–441.

Altshuler, D.L. (2001) Ultraviolet reflectance in fruits, ambient light composition and fruit removal in a tropical forest. Evol. Ecol. Res. 3, 767–778.

Aschan, G. and Pfanz, H. (2003) Non-foliar photosynthesis – a strategy of additional carbon acquisition. Flora 198, 81–97.

Awad, M.A., De Jager, A. and Van Westing, L.M. (2000) Flavonoid and chlorogenic acid levels in apple fruit: characterization of variation. Sci. Hort. 83, 249–263.

Bach, C.E. and Kelly, D. (2004) Effects of forest edges, fruit display size, and fruit colour on bird seed dispersal in a New Zealand mistletoe, Alepsis flavida. New Zeal. J. Ecol. 28, 93–103.

Boss, P.K., Davies, C. and Robinson, S.P. (1996) Analysis of the expression of anthocyanin pathway genes in developing Vitis vinifera L. cv Shiraz grape berries and the implications for pathway regulation. Plant Physiol. 111, 1059–1066.

Briscoe, A.D. and Chittka, L. (2001) The evolution of color vision in insects. Annu. Rev. Entomol. 46, 471–510.

Burkhardt, D. (1982) Birds, berries and UV. Naturwissenschaften 69, 153–157.

Burns, K.C. and Dalen, J.L. (2002) Foliage color contrasts and adaptive fruit color variation in a bird-dispersed plant community. Oikos 96, 463–469.

Caine, N.G. and Mundy, N.I. (2000) Demonstrating of a foraging advantage for trichromatic marmosets (*Callithrix geoffroyi*) dependent on food colour. Proc. R. Soc. Lond. B 267, 439–444.

Calvo, C., Salvador, A. and Fiszman, S.M. (2001) Influence of colour intensity on the perception of colour and sweetness in various fruit-flavoured yoghurts. Eur. Food Res. Technol. 213, 99–103.

Cipollini, M.L., Paulk, E., Mink, K., Vaughn, K. and Fischer, T. (2004) Defense tradeoffs in fleshy fruits: effects of resource variation on growth, reproduction, and fruit secondary chemistry in *Solanum carolinense*. J. Chem. Ecol. 30, 1–17.

Crisosto, C.H., Crisosto, G.M. and Ritenour, M. (2002) Testing the reliability of skin colour as an indicator of quality for early season "Brooks" (*Prunus avium* L.) cherry. Postharvest Biol. Tech. 24, 147–154.

Curry, E.A. (1997) Temperatures for optimal anthocyanin accumulation in apple tissue. J. Hort. Sci. 72, 723–729.

Deighton, N., Brennan, R., Finn, C. and Davies, H.V. (2000) Antioxidant properties of domesticated and wild *Rubus* species. J. Sci. Food Agric. 80, 1307–1313.

Drew, R.A.I., Prokopy, R.J. and Romig, M.C. (2003) Attraction of fruit flies of the genus *Bactrocera* to colored mimics of host fruit. Entomol. Exp. Appl. 107, 39–45.

Erez, A. and Flore, J.A. (1986) The quantitative effect of solar radiation on "Redhaven" peach fruit skin color. HortScience 21,1424–1426.

Escribano-Bailón, M.T., Santos-Buelga, C., Alonso, G.L. and Salinas, M.R. (2002) Anthocyanin composition of the fruit of *Coriaria myrtifolia* L. Phytochem. Anal. 13, 354–357.

Faragher, J.D. (1983) Temperature regulation of anthocyanin accumulation in apple skin. J. Exp. Bot. 34, 1291–1298.

Faragher, J.D. and Chalmers, D.J. (1977) Regulation of anthocyanin synthesis in apple skin. III Involvement of phenylalanine ammonialyase. Austr. J. Plant Physiol. 4, 133–141.

Gil, M.A., García-Viguera, C., Artés, F. and Tomás-Barberán, F.A. (1995) Changes in pomegranate juice pigmentation during ripening. J. Sci. Food Agric. 68, 77–81.

Gortner, W.A. (1965) Chemical and physical development of the pineapple fruit. IV. Plant pigment constituents. J. Food Sci. 30, 30–32.

Gould, K.S. and Lister, C. (2006) Flavonoid functions in plants. In: Ø. M. Andersen and K.R. Markham (Eds), *Flavonoids. Chemistry, Biochemistry and Applications*. CRC Press, Boca Raton, pp. 397–442.

Gross, J. (1987) *Pigments in Fruits*. Academic Press, London.

Hall, I.V. and Stark, R. (1972) Anthocyanin production in cranberry leaves and fruit, related to cool temperatures at a low light intensity. Hort. Res. 12, 183–186.

Hampe, A. (2003) Large-scale geographical trends in fruit traits of vertebrate-dispersed temperate plants. J. Biogeogr. 30, 487–496.

Harborne, J.B. (1976) Functions of flavonoids in plants. In: T.W. Goodwin (Ed), *Chemistry and Biochemistry of Plant Pigments*. Academic Press, London, pp. 736–778.

Harborne, J.B. (1988) *Introduction to Ecological Biochemistry*. Academic Press, London.

Harborne, J.B. and Grayer, R.J. (1994) Flavonoids and insects. In: J.B. Harborne (Ed), *The Flavonoids. Advances in Research Since 1986*. Chapman & Hall/CRC, Boca Raton, pp. 589–618.

Hetherington, S.E. (1997) Profiling photosynthetic competence in mango fruit. J. Hort. Sci. 72, 755–763.

Honkavaara, J., Koivula, M., Korpimäki, E., Siitari, H. and Viitala, J. (2002) Ultraviolet vision and foraging in terrestrial vertebrates. Oikos 98, 505–511.

Howell, J.F. (1991) Reproductive Biology. In: L.P.S. van der Geest and H.H. Evenhuis (Eds), Tortricid Pests. Elsevier Science Publishers, Amsterdam, pp. 157–174.

Izhaki, I., Tsahar, E., Paluy, I. and Friedman, J. (2002) Within population variation and inter-relationships between morphology, nutritional content, and secondary compounds of Rhamnus alaternus fruits. New Phytol. 156, 217–223.

Jacobs, G.H. (1993) The distribution and nature of colour vision among the mammals. Biol. Rev. 68, 413–471.

Janson, C.H. (1983) Adaptation of fruit morphology to dispersal agents in a Neotropical forest. Science 219, 187–189.

Kähkönen, M.P., Heinämäki, J., Ollilainen, V. and Heinonen, M. (2003) Berry anthocyanins: isolation, identification and antioxidant activities. J. Sci. Food Agric. 83, 1403–1411.

Katsoyannos, B.I. and Kouloussis, N.A. (2001) Captures of the olive fruit fly Bactrocera oleae on spheres of different colours. Entomol. Exp. Appl. 100, 165–172.

Kliewer, W.M. (1970) Effect of day temperature and light intensity on coloration of Vitis vinifera L. grapes. J. Amer. Soc. Hort. Sci. 95, 693–697.

Kliewer, W.M. (1977) Influence of temperature, solar radiation and nitrogen on coloration and composition of Emperor grapes. Am. J. Enol. Viticult. 28, 96–103.

Kliewer, W.M. and Schultz, H.B. (1973) Effect of sprinkler cooling of grapevines on fruit growth and composition. Amer. J. Enol. Viticult. 24, 17–26.

Lancaster, J.E., Reay, P.F., Norris, J. and Butler, R.C. (2000) Induction of flavonoids and phenolic acids in apple by UV-B and temperature. J. Hortic. Sci. Biotech. 75, 142–148.

Larronde, F., Krisa, S., Decendit, A., Chéze, C., Deffieux, G. and Mérillon, J.M. (1998) Regulation of polyphenol production in Vitis vinifera cell suspension cultures by sugars. Plant Cell Rep. 17, 946–950.

Levey, D.J. (1987) Sugar-tasting ability and fruit selection in tropical fruit-eating birds. Auk 104, 173–179.

Li, X-J., Hou, J-H., Zhang, G-L., Liu, R-S., Yang, Y-G., Hu, Y-X. and Lin, J-X. (2004) Comparison of anthocyanin accumulation and morpho-anatomical features in apple skin during color formation at two habitats. Sci. Hort. 99, 41–53.

Lila, M.A. (2004) Plant pigments and human health. In: K.M. Davies (Ed), Plant Pigments and Their Manipulation. Annual Plant Reviews. Blackwell Publishing/ CRC Press, Boca Raton, pp. 248–274.

Macheix, J-J., Fleuriet, A. and Billot, J. (1990) Fruit Phenolics. CRC Press, Boca Raton.

Mack, A.L. (2000) Did fleshy fruit pulp evolve as a defence against seed loss rather than as a dispersal mechanism? J. Biosci. 25, 93–97.

Manach, C. (2004) Polyphenols: food sources and bioavailability. Am. J. Clin. Nutr. 79, 727–747.

Marsh, K.B., Volz, R.K. and Reay, P. (1996) Fruit colour, leaf nitrogen level, and tree vigour in "Fuji" apples. New Zeal. J. Crop Hort. 24, 393–399.

Martin, M.M., Larsen, F.E., Higgins, S.S., Ku, M.S.B. and Andrews, P.K. (1997) Comparative growth and physiology of selected one-year-old red- and green-fruited European pear cultivars. Sci. Hort. 71, 213–226.

Merzlyak, M.N. and Chivkunova, O.B. (2000) Light-stress-induced pigment changes and evidence for anthocyanin photoprotection in apples. J. Photochem. Photobiol. B: Biol. 55, 155–163.

Mollon, J.D. (1989) "Tho' she kneel'd in that place where they grew..." The uses and origins of primate colour vision. J. Exp. Biol. 146, 21–38.

Morgan, M.J., Adam, A. and Mollon, J.D. (1992) Dichromats detect colour-camouflaged objects that are not detected by trichromats. Proc. R. Soc. Lond. B 248, 291–295.

Moyer, R.A., Hummer, K.E., Finn, C.E., Frei, B. and Wrolstad, R.E. (2002). Anthocyanins, phenolics, and antioxidant capacity in diverse small fruits: *Vaccinium*, *Rubus*, and *Ribes*. J. Agric. Food. Chem. 50, 519–525.

Naumann, W.D. and Wittenburg, U. (1980) Anthocyanins, soluble solids, and titratable acidity in blackberries as influenced by preharvest temperatures. Acta Hort. 112, 183–190.

Neill, S. and Gould, K.S. (2003) Anthocyanins in leaves: light attenuators or antioxidants? Funct. Plant Biol. 30, 865–873.

Ödeen, A. and Håstad, O. (2003) Complex distribution of avian color vision systems revealed by sequencing the SWS1 opsin from total DNA. Mol. Biol. Evol. 20, 855–861.

Osorio, D. and Vorobyev, M. (1996) Colour vision as an adaptation to frugivory in primates. Proc. R. Soc. Lond. B 263, 593–599.

Osorio, D., Smith, A.C., Vorobyev, M. and Buchanan-Smith, H.M. (2004) Detection of fruit and the selection of primate visual pigments for color vision. Am. Nat. 164, 696–708.

Pietrini. F., and Massacci, A. (1998) Leaf anthocyanin content changes in *Zea mays* L. grown at low temperature: significance for the relationship between quantum yield of PS II and the apparent quantum yield of CO_2 assimilation. Photosynth. Res. 58, 213–219.

Poulsen, J.R., Clark, C.J., Connor, E.F. and Smith, T.B. (2002). Differential resource use by primates and hornbills: implications for seed dispersal. Ecology 83, 228–240.

Puech, A.A., Rebeiz, C.A. and Crane, J.C. (1976) Pigment changes associated with the application of Ethephon ((2-chloroethyl)phosphonic acid) to fig (*Ficus carica* L.) fruits. Plant Physiol. 57, 504–509.

Reay, P.F., Fletcher, R.H. and Thomas, V.J. (1998) Chlorophyll, carotenoids and anthocyanin concentrations in the skin of "Gala" apples during maturation and the influence of foliar applications of nitrogen and magnesium. J. Sci. Food Agric. 76, 63–71.

Regan, B.C., Julliot, C., Simmen, B., Viénot, F., Charles-Dominique, P. and Mollon, J.D. (1998) Frugivory and colour vision in *Alouatta seniculus*, a trichromatic platyrhine monkey. Vision Res. 38, 3321–3327.

Regan, B.C., Julliot, C., Simmen, B., Viénot, F., Charles-Dominique, P. and Mollon, J.D. (2001) Fruits, foliage and the evolution of primate colour vision. Phil. Trans. R. Soc. Lond. B 38, 3321–3327.

Reyes-Carmona, J., Yousef, G.G., Martínez-Peniche, R.A. and Lila, M.A. (2005) Antioxidant capacity of fruit extracts of blackberry (*Rubus* sp.) produced in different climatic regions. J. Food Sci. 70, S497-S503.

Riba-Hernández, P., Stoner, K.E. and Lucas, P.W. (2005) Sugar concentration of fruits and their detection via color in the Central American spider monkey (*Ateles geoffroyi*). Am. J. Primatol. 67, 411–423.

Richardson-Harman, N., Phelps, T., McDermott, S. and Gunson, A. (1998) Use of tactile and visual cues in consumer judgments of apple ripeness. J. Sens. Stud. 9, 121–132.

Rudell, D.R., Mattheis, J.P., Fan, X. and Fellman, J.K. (2002) Methyl jasmonate enhances anthocyanin accumulation and modifies production of phenolics and pigments in "Fuji" apples. J. Amer. Soc. Hort. Sci. 127, 435–441.

Schaefer, H.M., Levey, D.J., Schaefer, V. and Avery, M.L. (2006) The role of chromatic and achromatic signals for fruit detection in birds. Behav. Ecol. 17, 784–789.

Schaefer, H.M. and Rolshausen, G. (2005) Plants on red alert: do insects pay attention? BioEssays 28, 65–71.

Schaefer, H.M. and Schmidt, V. (2004) Detectability and content as opposing signal characteristics in fruits. Proc. R. Soc. Lond. B 271, (Suppl.), S370–S373.

Schaefer, H.M., Schmidt, V. and Bairlein, F. (2003a) Discrimination abilities for nutrients: which difference matters for choosy birds and why? Anim. Behav. 65, 531–541.

Schaefer, H.M., Schmidt, V. and Winkler, H. (2003b) Testing the defence trade-off hypothesis: how contents of nutrients and secondary compounds affect fruit removal. Oikos 102, 318–328.

Schmidt, V., Schaefer, H.M. and Winkler, H. (2004) Conspicuousness, not colour as foraging cue in plant-animal signalling. Oikos 106, 551–557.

Shepherd, T. and Griffiths, D.W. (2006) The effects of stress on plant cuticular waxes. New Phytol. 171, 469–499.

Siitari, H., Honkavaara, J. and Viitala, J. (1999) Ultraviolet reflection of berries attracts foraging birds. A laboratory study with redwings (*Turdus iliacus*) and bilberries (*Vaccinium myrtillus*). Proc. R. Soc. Lond. B 266, 2125–2129.

Smillie, R.M. and Hetherington, S.E. (1999) Photoabatement by anthocyanin shields photosynthetic systems from light stress. Photosynthetica 36, 451–463.

Solfanelli, C., Poggi, A., Loreti, E., Alpi, A. and Perata, P. (2006) Sucrose-specific induction of the anthocyanin biosynthetic pathway in Arabidopsis. Plant Physiol. 140, 637–646.

Stafford, H.A. (1994) Atnhocyanins and betalains: evolution of the mutually exclusive pathways. Plant Sci. 101, 91–98.

Steward, J.M. and Cole, B.L. (1989) What do color vision defectives say about everyday tasks? Optometry Vision Sci. 66, 288–295.

Steyn, W.J., Holcroft, D.M., Wand, S.J.E. and Jacobs, G. (2004a) Regulation of pear color development in relation to activity of flavonoid enzymes. J. Amer. Soc. Hort. Sci. 129, 6–12.

Steyn, W.J., Holcroft, D.M., Wand, S.J.E. and Jacobs, G. (2004b) Anthocyanin degradation in detached pome fruit with reference to preharvest red color loss and pigmentation patterns of blushed and fully red pears. J. Amer. Soc. Hort. Sci. 129, 13–19.

Steyn, W.J., Wand, S.J.E., Holcroft, D.M. and Jacobs, G. (2002) Anthocyanins in vegetative tissues: a proposed unified function in photoprotection. New Phytol. 155, 349–361.

Stintzing, F.C. and Carle, R. (2004) Functional properties of anthocyanins and betalains in plants, food, and in human nutrition. Trends Food Sci. Tech. 15, 19–38.

Sumner, P. and Mollon, J.D. (2000) Chromaticity as a signal of ripeness in fruits taken by primates. J. Exp. Biol. 203, 1987–2000.

Surridge, A.K., Osorio, D. and Mundy, N.I. (2003) Evolution and selection of trichromatic vision in primates. Trends Ecol. Evol. 18, 198–205.

Tiffney, B.H. (2004) Vertebrate dispersal of seed plants through time. Annu. Rev. Ecol. Evol. Syst. 35, 1–29.

Traveset, A., Riera, N. and Mas, R.E. (2001). Ecology of fruit-colour polymorphism in *Myrtus communis* and differential effects of birds and mammals on seed germination and seedling growth. J. Ecol. 89, 749–760.

Traveset, A. and Willson, M.F. (1998) Ecology of fruit-colour polymorphism in *Rubus spectabilis*. Evol. Ecol. 12, 331–345.

Traveset, A., Willson, M.F. and Verdú, M. (2004) Characteristics of fleshy fruits in southeast Alaska: phylogenetic comparison with fruits from Illinois. Ecography 27, 41–48.

Veraverbeke, E.A., Van Bruaene, N., Van Oostveldt, P. and Nicolaï, B.M. (2001) Non destructive analysis of the wax layer of apple (*Malus domestica* Borkh.) by means of confocal laser scanning microscopy. Planta 213, 525–533.

Voigt, F.A., Bleher, B., Fietz, J., Ganzhorn, J.U., Schwab, D. and Böhning-Gaese, K. (2004) A comparison of morphological and chemical fruit traits between two sites with different frugivore assemblages. Oecologia 141, 94–104.

Wheelwright, N.T. and Janson, C.H. (1985) Colors of fruit displays of bird-dispersed plants in two tropical forests. Am. Nat. 126, 777–799.

White, I.M. and Elson-Harris, M.M. (1992) *Fruit Flies of Economic Significance: Their Identification and Bionomics*. CAB International, Wallingford.

Whitney, K.D. (2005) Linking frugivores to the dynamics of a fruit color polymorphism. Am. J. Bot. 92, 859–867.

Whitney, K.D. and Stanton, M.L. (2004). Insect seed predators as novel agents of selection on fruit color. Ecology 85, 2153–2160.

Whitney, K.D. and Lister, C.E. (2004) Fruit colour polymorphism in *Acacia ligulata*: seed and seedling performance, clinal patterns, and chemical variation. Evol. Ecol. 18, 165–186.

Willson, M.F. (1986) Avian frugivory and seed dispersal in eastern North America. Curr. Ornithol. 3, 223–279.

Willson, M.F., Graff, D.A. and Whelan, C.J. (1990) Color preferences of frugivorous birds in relation to the colors of fleshy fruits. Condor 92, 545–555.

Willson, M.F. and O'Dowd, D.J. (1989) Fruit color polymorphism in a bird-dispersed shrub (*Rhagodia parabolica*) in Australia. Evol. Ecol. 3, 40–50.

Willson, M.F. and Thompson, J.N. (1982) Phenology and ecology of color in bird-dispersed fruits, or why some fruits are red when they are "green". Can. J. Bot. 60, 701–713.

Willson, M.F. and Whelan, C.J. (1989) Ultraviolet reflectance of fruits of vertebrate-dispersed plants. Oikos 55, 341–348.

Willson, M.F. and Whelan, C.J. (1990) The evolution of fruit color in fleshy-fruited plants. Am. Nat. 136, 790–809.

Zhou, Y. and Singh, B.R. (2002) Red light stimulates flowering and anthocyanin biosynthesis in American cranberry. Plant Growth Regul. 38, 165–171.

5

Anthocyanin Biosynthesis in Plant Cell Cultures: A Potential Source of Natural Colourants

Simon Deroles

New Zealand Institute for Crop & Food Research Ltd, Private Bag 11600, Palmerston North, New Zealand, deroless@crop.cri.nz

Abstract. Increasing concern over the use of artificial food colourants has resulted in a steady increase in demand for anthocyanins as a natural alternative. Anthocyanins offer a range of colours from red to blue as well as orange. In addition they offer the added benefit of therapeutic and medicinal properties such as: antioxidant, anti-inflammatory, anti-convulsant, and chemoprotectant activities. Anthocyanins have been implicated lowering the risk of cardiovascular disease and certain cancers. Anthocyanins are present in may plant tissues and there is much research on finding new sources for the production of natural colourants. Currently most anthocyanin colourants are isolated from grape skins, as well as red cabbage, black carrots and sweet potato. Over the last 30 years researchers have also explored the idea of generating plant cell cultures for the production of natural colourants. This offers the advantages of consistent supply and quality as well as the opportunity to control the type of pigment produced. In spite of the amount of interest in this field no commercially viable system has been developed. This review describes current progress on the development of cell cultures within individual plant species for the production of anthocyanins. It details the processes that enhance anthocyanin production as well as factors that place limits on final yield. Finally it also offers ideas on techniques to overcome the production barriers leading to commercial viability.

5.1 Introduction

5.1.1 The Anthocyanins

Anthocyanins are part of the plant-derived flavonoid compounds and are responsible for colours ranging from pale pink to red to purple and deep blue. They are present in a wide range of plant tissues, principally flowers and fruit, but also storage organs, roots, tubers and stems. Within the plant kingdom they are almost universally present

K. Gould et al. (eds.), *Anthocyanins*, DOI: 10.1007/978-0-387-77335-3_5,
© Springer Science+Business Media, LLC 2009

in higher plants, with the exception of the betalain producers, and are also present in lower plants such as algae, liverworts, mosses and ferns. The anthocyanins are water-soluble pigments and are mainly stored in the vacuole. Over 500 anthocyanin pigments have been described in the literature (Andersen and Jordhein 2006). However, most anthocyanin types are based around three primary structures, pelargonidin, cyanidin and delphinidin, with each type being characterised by the number of hydroxyl groups on the B-ring. A single hydroxyl group (pelargonidin type) produces a more red pigment, with increasing B-ring hydroxylation (2 = cyanidin, 3 = delphinidin) causing a colour shift into the blue spectrum. Further modifications to the basic structure include glycosylation, acylation and methylation. These cause smaller colour shifts, but dramatically improve the stability of the final anthocyanin and also its ability to be transported into the vacuole.

Anthocyanin biosynthesis is probably the most studied plant secondary metabolite pathway. Genes have been isolated for almost every biosynthetic step and a considerable body of knowledge is available on the mechanisms that regulate their expression within the plant cell (Schwinn and Davies 2004; Andersen and Jordhein 2006; Davies and Schwinn 2006). The biosynthetic pathway can be divided into two sections, the early and late biosynthetic genes (Fig. 3.2). The early biosynthetic genes lead to the formation of the dihydroflavonols, which are the first committed compounds in the formation of anthocyanins and comprise: phenylalanine ammonia lyase (PAL), cinnimate 4-hydroxylase (C4H), 4-coumarate:CoA ligase (4CL), chalcone synthase (CHS), chalcone isomerase (CHI), and flavanone 3-hydroxylase (F3H). The late biosynthetic genes lead to the formation of the anthocyanin molecule: dihydroflavonol reductase (DFR), anthocyanidin synthase (AS), UDPGlucose:flavonoid 3-O-glucosyltransferase (F3GT), and the enzymes that control the shift between the major anthocyanin groups (pelargonidin, cyanidin and delphinidin), the flavonoid 3'- and 3'5'-hydroxylases (F3'H and F3'5'H). The latter biosynthetic group also includes the anthocyanin modification enzymes such as glycosyltransferases, methyltransferases and acyltransferases.

Anthocyanins perform a number of roles in the plant cell. Perhaps the most obvious is as an attractant for pollinators via flower colour and for seed dispersal agents via brightly coloured fruit. Moreover, one of the primary determinants in consumer purchasing of flowers and fruit is their bright and vibrant colour, much of which is due to anthocyanin pigments. In addition to these roles, anthocyanins also play important roles in photo-protection, as an antioxidant, biological defence and also in symbiotic functions between microbes and plant cells (Gould and Lister 2006).

People have used anthocyanins since the dawn of civilisation, using crude plant extracts for art and self-decoration. The earliest use of anthocyanins as a food colourant is thought to be about 1500 B.C. when the Egyptians used wine to enhance the colour of candies (Downham and Collins 2000). Early industrial use was for fabric dyes. More recently, with the advent of prepared and processed foods they have become increasingly popular as natural food colourants (Francis 1989; Bridle and Timberlake 1996; Downham and Collins 2000). However, in addition to their ability to provide vibrant colours they are now widely acknowledged as having significant health-giving properties, primarily centred around their role as an antioxidant but also through their anti-inflammatory, anti-ulcer and wound healing properties (Lila

2004). Research has also shown that many artificial pigments are actually detrimental to our health. With an increasing consumer preference for healthy foods there is now considerable demand for the use of anthocyanins as natural colourants, due to their natural pedigree and healthful properties. As a direct result, the production of anthocyanin pigments is one of the fastest growing segments of the food colourant industry (Delgado-Vargas et al. 2000; Downham and Collins 2000).

The only commercial sources for anthocyanin pigments are from whole plant extracts, with the most common source being grape skins from the wine industry. Other plant sources include red cabbage leaves, sweet potato, carrot and potato (Downham and Collins 2000). In 2002 the sales of anthocyanins isolated from grape skin alone had an estimated value of US$200 million Worldwide. The demand for natural colourants is estimated to grow by 5 to 15% per year.

However, there are several limitations to the current supply of anthocyanin pigments. The number of different plant sources is very limited. As a result, food manufacturers have a narrow colour range from which to choose. In addition, the supply can be subject to long cultivation times, seasonal effects, climate variations, pest/disease attack and the increasing cost of intensive agriculture coupled with the decreasing availability of low-cost arable land (Ramachandra Rao and Ravishankar 2002). All of these factors affect quality and quantity of supply. As the demand for more prepared and processed food increases and the expectation of high health products also rises, food manufactures are requiring a wider range of colourants and antioxidant additives. One alternative source for the production of anthocyanins is through the use of plant cell cultures (Delgado-Vargas et al. 2000; Ramachandra Rao and Ravishankar 2002; Lila 2004).

5.1.2 Plant Cell Cultures

For the purposes of this review, a plant cell culture can be defined as a collection of undifferentiated plant cells, either as cellular suspensions in liquid or as disorganised solid masses (callus). Plant cell cultures have been established from many plant species. Within a cell culture there are usually several cell types, ranging from completely undifferentiated cells, which are usually large and of random shape, to semi-organised cellular masses that are often the precursor structures in the initiation of embryogenesis, or in some cases organogenesis. These cells are often smaller in size with increasing cytoplasmic density and a more regular spherical shape. Manipulation of nutrient and phytohormone levels can alter the differentiation state of the cultures, with increasing auxin content usually resulting in an increase in undifferentiated cells. Liquid suspension cultures usually consist of small multicellular aggregates rather than a true single cell suspension. The size of these aggregates can vary widely from as small as 100 μm to as large as several millimetres. In general, liquid cell suspensions grow faster than callus, probably because of improved nutrient supply.

Plant cell cultures are very useful as a research tool as they represent a simplified plant cell system when compared with a whole plant, due to their limited number of cell types and relatively undifferentiated state. In addition, they have a rapid growth rate and higher rate of metabolism than whole plants, making the study of biosynthetic pathways considerably easier (Lila 2004).

   The high growth rate and relative uniformity of cell types makes plant cell culture an attractive option for the production of metabolic compounds. Plant cell cultures are mostly totipotent and hence are able to differentiate into specific cell types when exposed to the correct phytohormone and nutrient conditions. This means the researcher is able to activate defined metabolic pathways by controlling the differentiation state of the cells. Plant cells are able to produce an enormous range of metabolites, with approximately 4000 new compounds discovered in the past few years and over 100,000 currently known (Verpoorte et al. 1999). Many of these compounds have potential use as pharmaceuticals, nutraceuticals, insecticides, medicinal compounds, and food additives. As result there is constant research on the most effective means of producing such compounds.

   It is widely acknowledged that secondary metabolism in plant cells is tightly linked to its differentiation state (Endress 1994). Thus, when a cell is completely dedifferentiated, secondary metabolite pathways are often completely shut off. This feature is the primary limitation of using plant cell cultures as a source of secondary metabolite production. In order to maximise secondary metabolite production, the cell culture must be optimised for maximum growth whilst maintaining a differentiation state that enhances secondary metabolism. In many cases, maximum growth rate is achieved by lowering the differentiation state, which of course is in complete conflict with the requirements for secondary metabolism. This means that cell cultures used for secondary metabolite production must be carefully maintained to achieve a differentiation state that is an effective balance between growth rate and secondary metabolism. This can be a costly process and is one of the principal reasons why cell cultures have not been more successful as biofactories for secondary metabolites. Examples of successful industrial production of secondary metabolites from plant cell cultures include shikonin from *Lithospermum erythrorhizon*, berberine from *Coptis japonica* and sanguinarine from *Papaver somniferum* (Endress 1994; Ramachandra Rao and Ravishankar 2002).

   Plant cell cultures have been used extensively in the study of the regulation of anthocyanin biosynthesis. There has been considerable and ongoing interest in the production of anthocyanins from plant cell cultures for use as natural colourants and, more recently, as nutraceuticals, due to their potent antioxidant properties. Despite this level of interest there are no commercially viable anthocyanin production systems for the manufacture of anthocyanins. The three most common species referred to in the literature on anthocyanin production from cell cultures are carrot (*Daucus carota*), grape (*Vitis vinifera*) and strawberry (*Fragaria ananassa*). To date anthocyanin production in cell cultures has been established from approximately 33 different plant species.

   This review is about the generation of anthocyanins in plant cell cultures and is grouped into sections that provide an overview by plant species. Each section will include different aspects of anthocyanin production, from their use as a research tool in the study of anthocyanin and secondary metabolism, to work on their development as biofactories for natural colour and antioxidant production, including optimisation of production, different cell culture techniques and the use of modern biotechnology methods.

5.2 *Daucus carota* (Carrot)

Anthocyanin production in carrot cell cultures has been studied since the late 1970s. Carrot cell cultures have been used extensively as a model system in the study of anthocyanin biosynthesis. In addition, carrot cell cultures have also been considered as a possible source for the commercial production of anthocyanins.

5.2.1 Types of Anthocyanins

Early reports on the anthocyanin content of carrot cell cultures described the dominant pigments in carrot cell cultures as non-acylated, mono- and di-*O*-glycosides of cyanidin (Hemingson and Collins 1982). Cell cultures derived from a black Afghan wild carrot show the presence of six anthocyanin types, all containing cyanidin glycosylated with xylosylgalactoside or with a branched xylosylglucosylgalactoside and with some showing acylation (Harborne et al. 1983; Hopp and Seitz 1987; Gläßgen et al. 1992a, 1992b). The pattern of anthocyanin types closely resembled those found in the roots and flowers of the intact plants, the only difference being minor variations in relative quantity. This demonstrates that carrot cell cultures could be used to synthesise anthocyanin preparations characteristic of the pigments from intact plants. In a recent publication, Narayan and Venkataraman (2000) looked at the pigments in a cell culture derived from a local pale orange variety. The two major pigments (mono- and di-*O*-glycosides of cyanidin) were non-acylated, in contrast to the pigments from the black Afghan carrot derived cultures.

Dougall (1989) and Hopp and Seitz (1987) showed that in their cell lines, acylation with sinapic acid is essential for anthocyanin accumulation. However, it is still possible to isolate cell lines that accumulate non-acylated anthocyanins (Baker et al. 1994; Dougall et al. 1998; Narayan and Venkataraman 2000). Baker et al. (1994) and Dougall et al. (1998) have also shown that feeding different cinnamic and benzoic acid derivatives can alter the acylation pattern of the anthocyanin. This shows that the anthocyanin-specific acyltransferases in carrot can accommodate a range of carboxylic acids. In most cases mono-acylation of the anthocyanins occurred at the C-6 of the glucose moiety. Careful selection of particular cell lines can yield different anthocyanins depending on the identity of the original type. This raises the possibility of using cell cultures to produce specific anthocyanin species not available from the parent plant material that are highly acylated and thus have a more stable colour at neutral pH.

5.2.2 Glucosyltransferases from Carrot Cell Cultures

As mentioned above the dominant pigments in carrot cell cultures are acylated cyanidin-triglycosides. The cyanidin aglycone has a galactose moiety attached at the C3 position, which in turn has xylose and glucose molecules attached. Acylation of this complex is on the glucose molecule. Analysis of intermediate complexes present in the cell cultures shows only the presence of cyanidin di-*O*-glycosides containing galactose and xylose. This suggests that the galactose molecule is attached first then followed by the xylose moiety. Rose et al. (1996) have isolated the enzymes responsible for both of these steps, the anthocyanidin 3-O-galactosyltransferase and the anthocyanin 3-O-galactoside 2″-O-xylosyltransferase (CGXT). An earlier report describes the isolation of a cyanidin 3-O-glucosyltransferase with the production of cyanidin-3-O-glucoside (Petersen and Seitz 1986). However, this molecule is not present in carrot cell cultures. The isolated enzyme is also able to accept the flavonols quercetin and kaempferol as substrates. Rose et al. (1996) suggest that this enzyme is not involved in anthocyanin biosynthesis but instead plays a role in the formation of the flavonol glycosides that are present in carrot cell cultures.

5.2.3 Phytohormones

Plant cell cultures are in general a heterogeneous collection of cell types based on their level of differentiation, which in turn is controlled by the phytohormone regime in which they are grown. Anthocyanin-producing carrot cell cultures show quite a variable response to the presence of phytohormones. In some cell lines anthocyanin production is enhanced by the addition of auxins (e.g. 2,4-D) and reduced by the addition of cytokinins (e.g. kinetin) (Kinnersley and Dougall 1980; Gleitz and Seitz 1989). This effect was related to aggregate size, with anthocyanin yield being highest in small cell aggregates, which are promoted by a lack of cytokinin in the media. Other cell lines respond to a mix of auxin and low levels of cytokinin, with phytohormone preferences differing between solid and liquid phase cultures (Narayan et al. 2005). In contrast, Ozeki and Komamine (1981) have developed a cell line capable of rapid growth in the presence of 2,4-D but unable to produce anthocyanins. In that cell line anthocyanin biosynthesis was induced after removal of 2,4-D and addition of zeatin. This treatment also slowed cell growth rate and induced embryogenesis. This result indicates a close linkage between embryogenesis/differentiation and anthocyanin biosynthesis (Ozeki and Komamine 1981, 1986). Further studies using a DNA synthesis inhibitor showed that 2,4-D was able to regulate the production of anthocyanins irrespective of cell division (Ozeka and Komamine 1982). The only point of similarity between the cultures of Kinnersley and Dougall (1980) and Ozeki and Komamine (1981) is that both cell lines required the selection of small cell aggregates for anthocyanin production. Such differences in response to phytohormone regimes could be due to the origins of the different cell lines and their endogenous levels of naturally synthesised phytohormones.

Ozeki and co workers have used their inducible cell line to study the regulation of the early biosynthetic enzymes, PAL and CHS, in relation to anthocyanin biosynthesis. PAL and CHS activity and gene expression are up-regulated in conjunction with the induction of anthocyanin biosynthesis when 2,4-D is removed from the media (Ozeki and Komamine 1985; Ozeki et al. 1987, 1990). In addition to the induction associated with anthocyanin biosynthesis, PAL activity is also induced immediately after transfer of the cell line to any new media. This rapid and transient response is independent of the presence of 2,4-D and is not related to the production of anthocyanins (Ozeki et al. 1990). Analysis of early and late PAL transcripts showed that the PAL transcript up-regulated in conjunction with anthocyanin biosynthesis (ANTPAL), is different from the transcript activated as part of the stress induced transfer effect (TRNPAL) (Ozeki and Takeda 1994). The differential expression of these genes is due to their respective promoter regions, as shown by fusion experiments to the luciferase reporter gene (Ozeki and Takeda 1994). The promoter region of the ANTPAL gene contains two putative binding sites for MYB transcription factors, and these sites are required for gene activity during anthocyanin biosynthesis (Ozeki et al. 2000). In addition, a second region was isolated (box A and P') that is responsible for regulation by 2,4-D. Two myb genes have been isolated (*Dcmyb8* and *Dcmyb10*) that closely follow the activity of ANTPAL, indicating that one or both of them may be responsible for the regulation of PAL activity during anthocyanin biosynthesis (Ozeki et al. 2000). An alternative mechanism for the regulation of PAL expression is via the alteration of chromatin structure. The promoter region of ANTPAL contains two miniature inverted-repeat transposable elements (MITES) that bind to nuclear matrices and possibly act as a topological switch for gene expression. It is interesting to note that another PAL gene, identical to ANTPAL (*gDcPAL4*) but lacking the MITE regions, is silent in all anthocyanin-producing cells. Ozeki and co-workers have also isolated the promoter of the CHS gene upregulated during anthocyanin biosynthesis and have shown that it is suppressed in the presence of 2,4-D and in the absence of light, confirming its role in the regulation of anthocyanin biosynthesis in carrot cell cultures (Ozeki et al. 1993). The results from Ozeki and co-workers show that in their inducible system, anthocyanin biosynthesis is primarily controlled through the differential expression of specific PAL and CHS genes (Ozeki 1996).

In cell lines that produce anthocyanins, supplementation with intermediate metabolites, such as naringenin and dihydroquercitin, increased anthocyanin production (Vogelien et al. 1990). This indicates a rate-limiting step in the early part of the pathway and matches the observations of Ozeki and co-workers on the influence of PAL and CHS activity on anthocyanin production (Vogelien et al. 1990). However, in clonally related lines that could not make anthocyanin, no *de novo* synthesis was seen after supplementation with the precursors. These colourless subclones accumulate dihydroquercitin and catechin, indicating that synthesis or accumulation is halted beyond leucocyanidin. Vogelien et al. (1990) suggest that the colourless subclones may not be able to carry out the acylation step necessary for stable accumulation of anthocyanins. It is interesting to note that loss of anthocyanin production of carrot cell cultures can be the result both of interruptions in early (CHS and PAL) and of late (acylation) steps of the pathway.

5.2.4 GA₃

The influence of GA_3 on anthocyanin biosynthesis in carrot cell cultures has been well documented. Addition of GA_3 rapidly shuts down anthocyanin biosynthesis at the PAL and CHS steps (Heinzmann and Seitz 1977; Noe and Seitz 1982; Hinderer et al. 1984; Ozeki and Komamine 1986; Ilan and Dougall 1994; Ilan et al. 1994). However, unlike the addition of 2,4-D, GA_3 does not promote the degradation of existing anthocyanins, resulting in a much slower loss of colour (Ozeki and Komamine 1986). Feeding of precursors such as dihydroquercitin overcame inhibition of anthocyanin biosynthesis by GA_3 (Hinderer et al. 1984). In addition, GA_3 is responsible for the shut-down of a single isoform of CHS that is responsible for anthocyanin biosynthesis (Ilan and Dougall 1994). Endogenous levels of gibberellins can also regulate anthocyanin biosynthesis. When carrot cell cultures were exposed to growth retardants that inhibit gibberellin biosynthesis, anthocyanin production was improved (Ilan and Dougall 1992). The exact mechanism of GA_3 regulation of anthocyanin biosynthesis in cell cultures is still unknown.

5.2.5 Nutrients (Carbon, Nitrogen, Phosphate)

Different carbon and nitrogen sources can dramatically influence the growth rate and production of anthocyanins. In carrot liquid cultures, sucrose is rapidly broken up into glucose and fructose, with glucose being preferentially used as the carbon source (Kanabus et al. 1986). Nagarajan et al. (1989) showed that a mixed carbon source of galactose and sucrose produced the most efficient combination of growth rate and anthocyanin production. In addition, they noted that the inoculum density also influenced the behaviour of cultures on different carbon sources. Zwayyed et al. (1991) found that glucose gives a high growth rate and fructose gives higher anthocyanin content. However, a combination of both sources reduced both factors. No explanation is currently available for the difference in behaviour of the two carbon sources. Elevated carbon sources can enhance anthocyanin production through increased energy supply and/or increased stress via elevated osmotic pressure. Rajendran et al. (1992) subjected cultures to high concentrations of both sucrose and mannitol and showed that the resultant increase in anthocyanin production was the result of increased osmotic pressure rather than an enhanced energy supply. Both Nagarajan et al. (1989) and Zwayyed et al. (1991) observed that anthocyanin accumulation in carrot cells occurred during the growth phase and stopped when cell division ceased, indicating that anthocyanin production was growth associated.

The two common nitrogen sources for plant cell cultures are ammonia and nitrogen. In carrot cells grown on ammonia as the sole nitrogen, fructose is the preferred carbon source rather than sucrose (Dougall and Frazier 1989). In addition, they showed that anthocyanin accumulation is inhibited by ammonia levels higher than 3–5 mM and a pH greater than 4.5. Reduced nitrogen conditions suppress growth rate and enhance anthocyanin production (Rajendran et al. 1992). Narayan and Venkataraman (2002) achieved maximum anthocyanin production using sucrose and a combination of ammonia and nitrogen. The absolute level of nitrogen, the ratio of the two nitrogen sources, and the carbon source significantly influenced anthocyanin accumulation. These studies clearly illustrate the complexity of interaction between carbon and nitrogen sources and their influence on both the growth rate and production of anthocyanins. Significant improvements in anthocyanin production can only be gained by precise control of these factors during the entire cell culture growth cycle.

Low concentrations of phosphate also limit growth rate and enhance anthocyanin production (Rajendran et al. 1992). Cultures under stress, either from nutrient starvation or osmotic pressure, produce more anthocyanins. It is possible that this is due to the increased availability of precursors for secondary metabolism, arising from reduced primary metabolic activity (Rajendran et al. 1992), although environmental stress can also induce anthocyanin biosynthesis in intact plants of many species.

5.2.6 Elicitation

Carrot cultures also respond to exposure to fungal elicitors. Dark-grown carrot cultures with no anthocyanin production show a rapid onset of PAL transcription when exposed to elicitors from *Pythium aphanidermatum*, resulting in the accumulation of the phytoalexin, 4-hydroxybenzoic acid. No induction of any other anthocyanin biosynthetic enzymes occurred and no anthocyanin was produced. When UV-A induced cell cultures (that are producing anthocyanins) are exposed to the same elicitors, anthocyanin biosynthesis is shut down and production of 4-hydroxybenzoic acid is induced. PAL activity remained high in these cultures but the later anthocyanin biosynthetic enzymes were rapidly shut down (Gleitz et al. 1991; Seitz et al. 1994; Gläßgen et al. 1998). These results show that there is a co-ordinated regulation of anthocyanin and phytoalexin biosynthesis with phytoalexins taking precedence when inducers of both pathways are present.

In contrast to the results using *P. aphanidermatum* elicitors, other elicitor sources can significantly enhance anthocyanin production. Suvarnalatha et al. (1994) tested culture filtrates and cell extracts from a range of bacteria and yeast species and showed that in all cases anthocyanin production was induced, with the best candidate (a cell extract from *Candida albicans*) resulting in a 1.8-fold increase in anthocyanin production. Rajendran et al. (1994) showed that a range of fungal elicitors were able to enhance anthocyanin biosynthesis, with the best candidate being mycelial extracts of *Aspergillus flavus*. Ramachandra Rao et al. (1996) also showed that the blue pigment phycocyanin from the alga *Spirulina platensis* was able to induce a 2-fold increase in anthocyanin production.

Elicitation with extracts of *Aspergillus niger* results in a significant increase in anthocyanin production by day 3 (Sudha and Ravishankar 2003a). Sudha and Ravishankar also showed that elicitation elevates cytoplasmic Ca^{2+} and that removal of Ca^{2+} halts the stimulation of anthocyanin production (Sudha and Ravishankar 2003a). Exposure of the cells to an ionophore that increases calcium channel activity (increasing cytoplasmic Ca^{2+}) results in a dramatic increase in anthocyanin content, and an increase in growth rate. Addition of a channel blocker reversed the ionophore effect and resulted in reduced anthocyanin accumulation (Sudha and Ravishankar 2003a). The addition of putrescine, salicylic acid (SA) and methyljasmonate (MeJA) also enhanced anthocyanin production, most likely via an increase in cytoplasmic Ca^{2+} (Sudha and Ravishankar 2003b, 2003c). These results show that calcium flux across the plasma membrane is an essential part of the signal transduction pathway leading to the induction of anthocyanin biosynthesis. In addition, elicitation by SA and MeJA had opposite effects on the induction of ethylene, indicating that ethylene does not play a role in the induction of anthocyanin biosynthesis in this carrot cell line (Sudha and Ravishankar 2003c).

Enhancement of anthocyanin production by elicitors is a useful tool in the analysis of the regulation of anthocyanin biosynthesis in carrot cell systems. In addition, it could be a useful tool in maximising production at the commercial scale. Fine-tuning of these treatments is required to avoid toxic effects adversely affecting cell growth rate and survival. As more of the signal transduction pathway becomes known, it may be possible to add specific inducers of anthocyanin biosynthesis rather than the crude elicitor, thus avoiding any adverse toxic effects.

5.2.7 Light

Takeda, Ozeki and co workers have also shown that the induction of anthocyanin biosynthesis by light is closely associated with the induction by removal of 2,4-D. Takeda (1988) reported that the induction of anthocyanins in carrot cell cultures was possible by irradiation with white fluorescent light. However, the light-mediated induction was dependent on the state of the cells as controlled by the hormone regime. Cells grown on 2,4-D were unable to make anthocyanins either in the light or dark, whereas cells grown on zeatin were light-inducible. As a result, Takeda (1988) suggested that 2,4-D controls the transition between a light-sensitive and light-insensitive state. Analysis of enzyme activity showed that light-induced anthocyanin biosynthesis was preceded by rises in the activity of several anthocyanin biosynthetic enzymes (PAL, 4CL, C4H, CHS and CHI) (Takeda 1990). However, PAL activity was induced after irradiation in both light-sensitive and insensitive cells, indicating that its role after irradiation is more than just a precursor for anthocyanin biosynthesis. Analysis of PAL activity and gene expression in light-induced cultures showed two peaks of PAL activity, the first as early as 2 h after irradiation and a second at the same time as the other anthocyanin biosynthetic enzymes (Takeda 1990; Takeda et al. 1993). The two peaks of PAL activity correspond to the expression of the two different PAL genes as reported in the previous section, namely TRNPAL and ANTPAL (Ozeki et al. 1990). Anthocyanin biosynthesis under continuous irradiation is primarily regulated by the light-induced expression of the ANTPAL and CHS genes and is dependent on a light-sensitive cell state induced by the absence of 2,4-D (Takeda et al. 1993).

Anthocyanin production in response to pulses of UV-B light shows a different path of regulation, using the early TRNPAL gene and CHS (Takeda et al. 1994). Surprisingly, anthocyanin accumulation and the corresponding ANTPAL expression at 23 h were negligible. This indicates that the short UV-B pulse was not able to activate this pathway. In addition, the response to UV-B was significantly enhanced by exposure to red light either before or after the UV-B treatment (Takeda and Abe 1992). This indicates the involvement of phytochrome in the regulation of anthocyanin biosynthesis. Exposure to red light alone had little effect.

UV-A irradiation is also able to induce anthocyanin biosynthesis in carrot cells. Gleitz and Seitz (1989) have studied a carrot cell line that is able to accumulate anthocyanins in the dark and in the presence of 2,4-D. From this line they have isolated a colourless line that does not accumulate anthocyanins in the dark. After irradiation with UV-A, the coloured cell line shows significantly enhanced accumulation of anthocyanins whereas the colourless line failed to respond (Gleitz and Seitz 1989; Seitz et al. 1994; Suzuki 1995). Analysis of CHS enzyme activity showed that two forms of the enzyme were induced in the coloured line and only one form in the colourless line, showing that one of the isoforms is specific for anthocyanin biosynthesis. The function of the latter form, present in both cell lines, remains unknown since neither cell line forms other flavonoid compounds upon UV-A irradiation.

Induction of the coloured cell line by UV-A light also results in the induction of other anthocyanin biosynthetic enzymes, including the late enzymes that cause glycosylation and acylation (Gleitz et al. 1991; Seitz et al. 1994; Gläßgen et al. 1998).

All the enzymes in the pathway showed successive delays in their induction profiles according to their position in the pathway. No mechanism has yet been found to explain this co-ordinated induction pattern. Transcription of *Dc*CHS1, *Dc*F3H, *Dc*DFR1 and *Dc*ANS are up-regulated in the coloured cell line after UV-A irradiation (Hirner et al. 2001). However, the only gene specifically expressed in the uncoloured cell line in response to UV-A irradiation is the *Dc*CHS2 gene. The increase in level of anthocyanins acts as a UV blocker, protecting the cells from oxidative damage. Evidence of this can be seen in the expression of the α-tubulin gene that shows greater loss of expression under UV-A irradiation in the uncoloured cell line. In addition, the growth rate of the uncoloured cell line is slowed after UV-A irradiation when compared to the coloured cell line (Hirner et al. 2001).

5.2.8 Aggregate Size

Another factor that can influence anthocyanin production is the aggregate size within the culture. As mentioned above, elevated colour density in dark-grown cultures has been associated with reduced cell aggregate size and is related to the differentiation state of the cells (Kinnersley and Dougall 1980; Gleitz and Seitz 1989). Madhusudhan and Ravishankar (1996) have shown that an increase in aggregate size beyond 850 μm causes a loss of anthocyanin production in the centre of the aggregate. However, their cell lines were grown in the presence of light and they suggest that lack of the light signal in the centre of the large aggregates causes the loss of anthocyanin production. This emphasizes the importance of the light signal in anthocyanin production and also the advantage of using cell lines for commercial pigment production that can synthesise anthocyanins in the dark.

5.2.9 Future Strategies for Enhanced Production of Anthocyanins

Carrot cell cultures have been used extensively as a model system to study the activities of the anthocyanin biosynthetic pathway. Considerable knowledge has been gained on how the pathway is activated through exogenous stimulants and through the physiology of the cells themselves. Some effort has also gone into the optimisation of anthocyanin accumulation with a view to commercial production. Despite the efforts in this area, large-scale production remains uneconomic, principally due to the association between cellular differentiation and secondary metabolism. It is clear that in carrot cell cultures the early anthocyanin biosynthetic genes are activated with a change in differentiation state, and that this alteration also causes a reduction of growth rate. External stimulus by elicitors or irradiation can enhance anthocyanin production but not by enough to make it commercially economical. An alternative method to enhance production is via genetic modification to release the anthocyanin biosynthetic pathway from regulation via cellular differentiation. Much work has been carried out on the genetic control of anthocyanin biosynthesis (Davies 2004). Studies of differential gene expression between pigmented and non-pigmented carrot cells showed that a significant number of genes are differentially regulated between the different cell populations. However, to date no genes have been isolated that relate directly to anthocyanin biosynthesis (Marshall et al. 2002). The first step in developing transgenic cell cultures for anthocyanin production is to establish a gene transfer system for the target lines. Deroles et al. (2002) have established transformation methods for two carrot cell lines derived from those used by Prof Don Dougall and co-workers. Other methods that will improve our ability to up regulate anthocyanins production are developments in image analysis to better understand secondary metabolite production. Current studies of factors that influence anthocyanin biosynthesis and also studies on gene expression rely on the collection of a number of cells for analysis. Such a collection means that the observed response is only an average response across all the cell types. Ceoldo et al. (2005) have developed image analysis tools that enable the analysis and tracking of individual cells to better understand the inter-relationship between anthocyanin production and cell type. These new technologies will be useful in the elucidation of the regulatory mechanisms controlling anthocyanin biosynthesis in carrot cell systems and in the future development of high-producing carrot cell lines.

5.3 *Vitis vinifera* (Grape)

One of the most common sources of natural colourants for food use is anthocyanin extracts from grape skins, a by-product of the wine industry. As a result grape cell cultures have been extensively studied in an attempt to provide an alternative supply of natural colourants.

Most of the research on grape cell cultures has used cell lines from two sources. The most popular cell line was established by Dr J.C. Pech (Toulouse, France) in 1978, using the pulp of young fruits from the cultivar Gamay Freaux var. Teinturier (described in Cormier et al. 1990). In most instances this culture produces anthocyanins under continuous light. However, a derivative has been established that can also produce anthocyanins in the dark (Pepin et al. 1995). The other popular cell line was established from material of the cultivar Bailey Alicant A (Yamakawa et al. 1982, 1983a, 1983b). Cell lines were originally selected for anthocyanin production in the dark, but the cell line can also be grown under low light for improved yield.

In both cell lines most cells remain colourless on maintenance medium, resulting in a low level of colour production. Anthocyanin production is dramatically enhanced upon induction via metabolic or osmotic stress, but a high portion of the cells still remain colourless (see below). Anthocyanin production in cultures derived from the Gamay and Bailey Alicant A lines are unstable, with loss of colour over a period of subcultures (Yamakawa et al. 1982, 1983b; Hirasuna et al. 1991; Qu et al. 2005). This is because the higher growth rate of uncoloured cells enables them to out-grow the coloured population, leading to a steady reduction in the proportion of coloured cells (Hirasuna et al. 1991; Qu et al. 2005). In addition, the level of anthocyanins in individual culture vessels varied widely. Enhancement of anthocyanin production through elicitation, precursor feeding and light exposure showed that lighter coloured cell lines had a greater response than the darker lines. These results clearly illustrate the heterogeneous nature of these cell culture systems.

5.3.1 Types of Anthocyanins

The anthocyanin types present in cell lines derived from the Gamay cell cul^{..}re have been reported in numerous papers, with the dominant pigment being peonidin 3-*O*-glucoside (e.g. Cormier et al. 1994; Do et al. 1995; Conn et al. 2003; Curtin et al. 2003; Aumont et al. 2004). This pigment is also dominant in the fruit pulp of the source tissue. Other pigments present are cyanidin-, petunidin-, malvidin- and delphinidin-3-*O*-glucosides. In addition, several of these compounds are present in acylated form, usually with *p*-coumaric acid (e.g. peonidin 3-*O*-*p*-coumaroylglucoside). Cell lines derived from Bailey Alicant A have a similar anthocyanin profile (cyanidin-3-*O*-glucoside, peonidin-3-*O*-glucoside and peonidin-3-*O*-5-*O*-diglucoside) (Yamakawa et al. 1982, 1983b) Some acylation of these pigments was also detected.

Cell lines derived from a third source (Muscat Bailey A) yielded two major peaks corresponding to cyanidin-3-*O*-glucopyranoside and peonidin-3-*O*-glucopyranoside (Tamura et al. 1989). These pigments are also present in the intact plant. Feeding of organic acids such as cinnamic and *p*-coumaric acids altered the ratio between the two anthocyanin types, showing that the anthocyanin type produced from grape cell cultures can be influenced by culture conditions in a similar manner to carrot cell lines (Baker et al. 1994; Dougall et al. 1998). Krisa et al. (1999a) have also shown that it is possible to change the anthocyanin composition of grape cell cultures through careful selection and purification of clonal cell lines. By selecting highly coloured multicell aggregates at each subculture these researchers generated a range of cell lines, some of which showed a significant shift in their anthocyanin composition resulting in a 5-fold increase in the portion of malvidin-based anthocyanins.

5.3.2 Modification of Anthocyanins in Grape Cell Cultures

An enzyme catalysing the glycosylation of anthocyanins at the 3 position has been isolated from the Gamay cell line (Do et al. 1995). The enzyme had the highest affinity for cyanidin, followed by delphinidin and to a lesser extent their methylated derivatives, peonidin and malvidin. Analysis of substrate specificity for the grape methyltransferase showed that its preferred substrate was cyanidin- or delphinidin-3-*O*-glucoside rather than the aglycones (Bailly et al. 1997). In addition, when the cultures are placed on media optimised for anthocyanin production there is an increase in methylation to produce peonidin-based pigments. Bailly and co-workers also showed that the isolated methyltransferase was unable to use the acylated cyanidin 3-*O*-glucoside, indicating that methylation can only occur after glycosylation and before acylation of the anthocyanin (Bailly et al. 1997).

5.3.3 Phytohormones

Because of the specificity of culture media required to maintain anthocyanin producing grape cell cultures there are few publications on the effect of different phytohormones on anthocyanin production. Kokubo et al. (2001) showed that elevated levels of 2,4-D (10 mg/L) enhanced the glycosylation of the flavonoid quercetin and was associated with an increase in F3GT activity. Hirasuna et al. (1991) showed that low levels of 2,4-D (0.5 mg/L) were most effective in producing anthocyanins compared with other auxins. Cytokinins reduced anthocyanin production.

Krisa et al. (1999a) observed that auxin was required for cell growth, with NAA performing better than 2,4-D for growth rather than anthocyanin production. Higher levels of auxin resulted in a reduction of anthocyanin yield. Addition of cytokinins did not alter cell growth. Kinetin did not alter anthocyanin production significantly, but addition of more than 1 µMol BA reduced anthocyanin yield. Based on these results Krisa et al. (1999a) used 1 µm kinetin and 0.5 µm NAA in their anthocyanin induction medium.

5.3.4 Nutrients (Carbon, Nitrogen, Phosphate)

In intact plants, water stress enhances secondary metabolism, in particular flavonoid and anthocyanin biosynthesis, as part of the stress response. Several studies have shown that in grape cell cultures a similar response exists. Increasing the osmotic potential, thus placing the cells under water stress, slows down cell division and enhances anthocyanin production. Gamay derived cell cultures subjected to increasing levels of both metabolisable sugars (sucrose) and non-metabolisable sugars (mannitol) showed an increase in anthocyanin production and a concomitant reduction in cell division (Cormier et al. 1990; Do and Cormier 1990, 1991a). This result shows that the enhanced anthocyanin production is due to osmotic stress rather than an increased availability of the carbon source (e.g. sucrose). Osmotic stress increased the portion of coloured cells (indicating a change in differentiation state), and the increase in anthocyanin levels was almost solely due to the increase in peonidin-3-*O*-glucoside rather than the cyanidin pigment. This indicates that high sucrose levels also increase the activity of anthocyanin methyltransferases. Once osmotic pressure was reduced, colour was lost and the number of coloured cells returned to basal levels. This result was also observed in a dark-grown anther-derived cell line, where the elevated anthocyanin content was due to both an increase in the anthocyanin content in individual cells and an increase in the portion of coloured cells in the culture (Yamakawa et al. 1983a; Suzuki 1995).

Other workers have also reported on the effect of high sugar levels on anthocyanin production, but with differing results. When using a Gamay-derived cell line Larronde et al. (1998) showed that elevated levels of sucrose and glucose are required for enhanced anthocyanin biosynthesis and that non-metabolisable sugars cannot duplicate this result. This is in direct contrast to the results of Do and Cormier (1990, 1991a). Larronde et al. (1998) suggest that the difference in results is due to the physical state of the cultures. In their system, an increase in sugar level did not result in a reduction of cell division as in the experiments of Do and Cormier. This result suggests that enhanced anthocyanin biosynthesis is due to an increased carbon source rather than increased osmotic potential. However, Larronde et al. (1998) also report that the addition of mannose can enhance anthocyanin biosynthesis. Mannose can be phosphorylated but not metabolised via glycolysis. From this result they suggest that anthocyanin biosynthesis is stimulated by the presence of high levels of hexose sugars, mediated through the activity of hexokinase as a sugar sensor and not through the role of these sugars as a general carbon source (Jang et al. 1997). It is possible that this mechanism of anthocyanin enhancement may be regulated through the sugar-dependent expression of CHS as reported by Tsukaya et al. (1991). Continuation of this work is reported in Vitrac et al. (2000) who confirm the role of the hexokinase sugar sensor through the use of a hexokinase inhibitor (mannoheptulose) and glucose analogs not phosphorylated by hexokinase. In addition, they show that elevated levels of calcium ions (Ca^{2+}), calmodulin and protein kinases/phosphatases are involved in the signal transduction pathway, leading to the probable up-regulation of CHS and subsequent anthocyanin accumulation.

Another important component of the culture media that significantly influences the level of anthocyanin accumulation is the level and form of nitrogen. Several reports study this variable in conjunction with the manipulation of sugar levels in an effort to maximise anthocyanin production. Do and Cormier (1991b) showed that in addition to elevated sugar levels, a reduction in the availability of inorganic nitrogen (nitrate) also reduced growth rate and improved anthocyanin biosynthesis. Their explanation of this phenomenon is in line with their result for elevated sugars in that both treatments act by reducing cell division, which in turn makes nutrient resources more available for secondary metabolism, thus leading to enhanced anthocyanin biosynthesis. Maximum anthocyanin production occured during periods of reduced cell division. Similar results for reduced nitrogen were also obtained by Yamakawa et al. (1983a), Park et al. (1989), Hirasuna et al. (1991) and Decendit and Mérillon (1996). However, Decendit and Mérillon (1996) show that maximum anthocyanin accumulation occurred during the exponential growth phase, i.e. during cell division. This is in agreement with their results for elevated sugar levels and in direct contrast with the results from the labs of Cormier et al.

Hirasuna et al. (1991) have demonstrated that elevated sucrose and reduced nitrogen could induce anthocyanin production in a colourless cell line derived from the cultivar Vignoles (Ravat 51). These researchers observed that enhancement of anthocyanin production occurred at a specific level of inorganic nitrogen rather than from a gradual decrease in concentration. They propose that this is a result of the removal of nitrogen-mediated inhibition of a tonoplast ATPase that may control anthocyanin transport into the vacuole, thus enhancing anthocyanin production.

Elevated levels of ammonia increase cell division and reduce anthocyanin accumulation (Do and Cormier 1991c). The reduction in anthocyanins may be the result of increased cell division and also alteration of the media pH. In addition, elevated ammonia also increased acylation activity, causing a rise in the relative level of peonidin 3-O-p-coumaroylglucoside.

Since there is a conflict between growth rate and high pigment production, maximum production may be obtained through a two-phase system in which the cultures are grown on a medium for maximum growth rate then transferred to pigment induction medium containing high sucrose and low nitrogen (Park et al. 1989).

Phosphate starvation of light- and dark-grown grape cell lines leads to an early onset of anthocyanin production associated with reduced cell division (Yamakawa et al. 1983a; Dedaldechamp et al. 1995; Decendit and Mérillon 1996; Dedaldechamp and Uhel 1999). This is very similar to the response to nitrogen starvation and may also be due to the redirection of nutrients from primary to secondary metabolism. In addition, phosphate starvation increases DFR enzyme activity. The importance of the increase in DFR activity was illustrated by the increase in anthocyanin production after feeding the cultures with dihydroquercitin. Analysis of the relative increase in DFR activity and final anthocyanin yield indicated that other steps in the biosynthetic pathway are also regulated.

The effects of high sugars, and low nitrogen and phosphate show that there is a close correlation between cell division status and anthocyanin production. In general anthocyanin production is enhanced in the absence of cell division and represents a redirection of resources from primary to secondary metabolism. Further evidence for this model comes from Hirose et al. (1990) who show that inhibition of DNA synthesis also leads to enhancement of anthocyanin production in Bailey Alicante A cell cultures grown in the dark. They observed that the onset of anthocyanin accumulation began only at the stationary phase. During the growth phase (active cell division) the level of phenylalanine was low but increased significantly at the beginning of the stationary phase (Sakuta et al. 1994). At the same time PAL and CHS gene expression also increased, shortly followed by anthocyanin accumulation. Based on these results, Sakuta and co-workers propose that the onset on anthocyanin biosynthesis is regulated at the PAL and CHS steps and is controlled by the level of the substrate (phenylalanine) pool. In actively dividing cells, phenylalanine is preferentially used for primary metabolism and remains low. As cell division ceases, the substrate pool increases. This results in induction of PAL and CHS gene expression and anthocyanin accumulation. Further evidence for this model comes from the addition of phosphate just after the onset of the stationary phase. This returns the culture to active cell division, the substrate pool falls, PAL and CHS gene expression are turned off and anthocyanin accumulation stops (Kakegawa et al. 1995). Likewise, addition of phenylalanine during the mid-log phase activated PAL and CHS gene expression and anthocyanin accumulation. As in other plant systems, two PAL mRNAs (PAL1 and PAL2) were found, and PAL2 gene expression was closely correlated with anthocyanin production.

5.3.5 pH, Conditioned Media and Feeder Layers

Anthocyanin production is affected by the pH of the media, with low pH (4.5) being best (Suzuki 1995). Increasing the pH results in loss of both anthocyanin production and cell growth rate. Limitations on the minimum plating density of grape cells on solid media can be alleviated both by the addition of conditioned medium and the use of feeder layers. Conditioned medium and feeder layers isolated from either grape or tobacco cell lines are effective (Yamakawa et al. 1985). In suspension culture inoculation density affects the lag time and can also affect the culture's response to different nutrient conditions (Decendit and Mérillon 1996).

5.3.6 Elicitation and Light

In addition to alteration of nutrient and osmotic conditions, the use of elicitors and light can also significantly enhance anthocyanin production. Zhang et al. (2002a) showed that addition of the elicitor jasmonic acid (JA) to dark grown cell cultures derived from the Gamay cell line significantly enhanced anthocyanin production as did exposure to continuous light. In addition, they showed that a combination of the two treatments led to a synergistic effect, resulting in enhancement greater than the sum of the two separate treatments. Since both these treatments up-regulate CHS and DFR gene expression in other plant systems, it is probable that both stimulants are acting on the same parts of the anthocyanin biosynthetic pathway in grape cell cultures. The major pigments present in their cell lines were cyanidin, peonidin and malvidin 3-O-glucosides and their acylated (coumaroyl) derivatives. Induction by JA resulted in a high degree of methylation, forming a majority of peonidin and malvidin pigments. Induction by light increased the amounts of all the anthocyanins but did not alter their relative ratios compared with the control. Combination of the two treatments showed the same distribution of anthocyanins as JA alone but the amounts were augmented (Curtin et al. 2003). Curtin and co-workers have also developed a derivative cell line selected for high anthocyanin production in the dark. That cell line shows enhanced levels of malvidin 3-O-glucoside and malvidin 3-O-p-coumaroylglucoside. Elicitation of that cell line using JA did not increase the proportion of methylated anthocyanins and the level of induction was not as high, perhaps indicating an upper level of anthocyanin content for grape cells.

In contrast to that work, Saigne-Soulard et al. (2006) studied the effect of MeJA on the same cell line grown in the light. In their system MeJA reduced anthocyanin accumulation and left the rate of cell division unaffected. The researchers used ^{13}C labelled phenylalanine to study the induction of anthocyanin biosynthesis. They showed that stilbene biosynthesis was up-regulated rather than anthocyanin biosynthesis by MeJA. Exposure to high levels of sucrose resulted in an increase in anthocyanin production and no change in stilbenes. The pattern of labelled intermediates in the stilbene and anthocyanin pathways after exposure to MeJA suggests that differential regulation of stilbene and anthocyanin biosynthesis is at the STS/CHS step. Saigne-Soulard and co-workers provide no explanation for the contrast between their results and those of Curtin et al. (2003) on the effect of MeJA /JA.

In contrast to the enhancing effect of the elicitor JA, the compound eutypine from the fungus *Eutypa lata* inhibits anthocyanin accumulation in Gamay cell cultures at concentrations over 200 μMol/l (Afifi et al. 2003). However, the rate of cell proliferation was unaffected. Analysis of gene expression showed that the toxin specifically acted on suppression of the F3GT gene, thus preventing accumulation of anthocyanins in the plant cell.

5.3.7 Localisation of Anthocyanins in the Plant Cell

Plant cells use active transport mechanisms to transfer anthocyanins into the vacuole, and glycosylation of anthocyanins is a prerequisite for this process. Cormier et al. (1990) and Calderon et al. (1993) observed the formation of anthocyanin-containing bodies called anthocyanoplasts in cell cultures producing high amounts of anthocyanins. Anthocyanoplasts were present within the cytoplasm in cells with both coloured and uncoloured vacuoles. Larger anthocyanoplasts are perhaps the result of fusions between smaller bodies. It was suggested (Calderon et al. 1993) that anthocyanins are formed within a multi-enzyme complex bound to the endoplasmic reticulum, and collected in the anthocyanoplasts, which in turn discharge them into the vacuole.

In a recent publication on this subject Conn et al. (2003) report the presence of anthocyanic vacuolar inclusions (AVIs) in grape cells from a Gamay derived cell line and suggest that they are similar to the anthocyanoplasts described by Calderon et al. (1993). AVIs are insoluble matrices containing complexed anthocyanins that are present in the vacuoles of many plant species (Markham et al. 2000; Zhang et al. 2006). It has been suggested that the role of AVIs is to remove anthocyanins from solution, thus enabling the plant cell to store higher concentrations of the compound. Markham et al. (2000) and Conn et al. (2003) have also shown that the AVIs in lisianthus petals and grape cells preferentially bind acylated anthocyanins.

The presence of AVIs in grape cell cultures offers a possible mechanism for enhancing anthocyanin accumulation with a view to using these cultures as an industrial source of anthocyanins. In order to increase the level of AVI formation a greater understanding of their structure and genetic control of formation is needed.

Anthocyanins are now widely recognised as a highly beneficial part of our diet, mainly due to their antioxidant activity (Frankel et al. 1993, 1995). It has been suggested that the anthocyanins in red wine contribute to the low level of heart disease in France, known as the French paradox (Renaud and De Lorgeril 1992; Gronbaek et al. 1995). Because of the health-giving properties of anthocyanins there is much interest in alternative supplies to satisfy an ever-increasing demand for anthocyanins as natural colourants and nutraceuticals in our food. Grape cell cultures are an ideal system for the production of labelled anthocyanins for analysis of *in vivo* and *in vitro* absorption and metabolism in animal systems. Because their anthocyanin profile is similar to that of red wine they are also a useful system for the study of the anticancer and antioxidative properties of red wine. Krisa et al. (1999b) have developed a grape cell culture system based on the Gamay cell line to produce anthocyanins labelled with ^{13}C in order to investigate their absorption and *in vivo* metabolism in the gut. In their hands the Gamay cell line shows maximal production of anthocyanins at the end of the exponential growth phase. In addition, their cell line shows high growth rates whilst producing anthocyanins. As mentioned above, this finding is direct contrast to other workers who have found that maximum anthocyanin production in this cell line is associated with reduced growth rate. Phenylalanine supplementation increased the amount of anthocyanin produced but also reduced cell growth rate. Using supplementation with ^{13}C labelled phenylalanine Krisa et al. (1999b) also succeeded in producing labelled cyanidin, peonidin and malvidin-3-*O*-glucosides labelled at the C4 position. This system has been scaled up into a 2l bioreactor and has produced the additional anthocyanins petunidin and delphinidin 3-*O*-ß-glucosides (Aumont et al. 2004). The rate of enrichment for ^{13}C ranged from 40 to 66%. Vitrac et al. (2002) have also used the same system to produce ^{14}C-labelled anthocyanins for use in *in vitro* metabolic studies.

Anthocyanins from grape cell cultures have been compared with other natural anthocyanin colourants such as berry liquid, grape skin, elderberry, red cabbage and purple corn powder (Cormier et al. 1997). Extracts from grape cell cultures are free of brown oxidised phenolics, thus giving superior clarity over other colour sources. In addition, the hue is closer to primary red than current products such as red cabbage extracts. The major disadvantage of grape cell extracts is the lack of colour stability at elevated pH and temperatures. This is most likely due to the presence of more highly acylated anthocyanins in other sources such as red cabbage.

5.3.8 Physical Parameters

There has been much research on the regulation of anthocyanin biosynthesis and its relationship to primary metabolism in order to maximise pigment production. An essential part of this research is the study of the physical culture conditions and the ability to scale-up cell cultures to commercially viable volumes. Grape cell cultures in shake flasks have two growth phases, cell division followed by biomass accumulation (Pepin et al. 1995). In the second phase cell division stops but biomass is still increased, mainly due to nutrient and water absorption. This has important implications in systems where the desired production is dependent on the number of cells rather than total biomass. The shift between the two states may be induced by osmotic changes in the medium and the availability of ammonium and phosphate.

An alternative to liquid suspension culture is to grow plant cells immobilised in an inert medium. When grape cells are grown on immobilised polyurethane foam their growth rate is slower than that of freely suspended cells (Iborra et al. 1994). However, their rate of anthocyanin accumulation is higher. This inverse relationship between growth rate and secondary metabolism is similar to the behaviour of grape cells cultured in growth-limited medium. The relationship between growth rate and secondary metabolism may be influenced by cell-to-cell contact as a result of multicell aggregation, resulting in enhanced communication between cells activating secondary metabolism (Yeoman et al. 1982). Alternatively, there is competition for substrate between primary and secondary metabolism, and secondary metabolism can only proceed apace once primary metabolism has slowed as a result of reduced cell division. Analysis of the immobilised cells did not show an enhanced degree of cell aggregation or the presence of multicellular clumps when compared to the free floating cells, indicating that the enhanced anthocyanin production may be due to lack of competition with primary metabolism, highlighting the importance of nutrient supply to both primary and secondary metabolism. Guardiola et al. (1995) have developed a model linking cell growth and secondary metabolism that provides a good representation of growth patterns in both batch and semi-continuous cultures. This model can be used to predict plant cell behaviour under different conditions, thereby enabling greater optimisation for secondary metabolite production.

One of the crucial factors in scaling up a cell line for commercial production is the cell line's ability to withstand the increased shear forces present in a stirred system rather than the more gentle environment of the shake flask that is most commonly used in the laboratory. Decendit et al. (1996) transferred a Gamay derived cell line from shake flasks to a 20 L stirred bioreactor and showed that this cell line can maintain its growth rate in a stirred environment and can also make anthocyanins at levels similar to shake flask cultures.

Shear force in a cell culture system can also influence the size distribution of multicell aggregates. Nagamori et al. (2001) studied the growth and anthocyanin production of Bailey Alicant derived cell cultures grown in shake flasks in a viscous medium. They found that increasing the viscosity slowed growth rate but dramatically increased anthocyanin production. In addition, they found that anthocyanin production was constant after multiple subcultures. This is in striking contrast to cells grown in the control medium, which completely lost all colour production after several subcultures. When these colourless cells were transferred to a viscous medium, anthocyanin production was re-established. Analysis of the cultures showed that in the viscous medium the proportion of large cell masses (>0.6 mm) was much larger and all of these masses were highly coloured. Cell masses smaller than 0.2 mm were prevalent in the control medium and over 50% of them were colourless. Nagamori and co-workers concluded that the increase in viscosity reduced shear forces, thus enabling the formation of larger cell masses, and these cell masses were able to stably produce large amounts of anthocyanin. This report highlights the influence that shear stress can have on the yield of secondary metabolites from cell cultures. These results were repeated when the cultures were tested with the same media in a 500 mL airlift bioreactor (Honda et al. 2002). The average cell aggregate size was 1.6 times larger than the control media, turbulence was 0.4 times lower in the viscous medium and anthocyanin content increased 2-fold.

Another method to enhance anthocyanin yield from plant cell cultures is via release of the pigment from the vacuole into the surrounding media. Cormier et al. (1992) enabled the release of 70–80% of the anthocyanins through chemical permeabilisation and the use of a solid absorbant in a two-phase system. However, the release was accompanied by a loss in cell viability.

5.3.9 Conclusions

Grape cell cultures offer much promise as a future source for natural colourants, in particular because their anthocyanin profile is already in use as a natural colourant within the processed food industry. There remain some ongoing challenges to increase yields to commercial levels through greater consistency and stability of anthocyanin production and increased robustness to withstand high shear forces. Grape cell cultures will also continue to be a valuable model system for the study of interactions between cell growth, environmental stimulus and secondary metabolism.

5.4 *Fragaria ananassa* (Strawberry)

5.4.1 Types of Anthocyanins

Light-dependent anthocyanin-producing callus and suspension cultures have been established from a variety of strawberry explants such as immature strawberry fruit, the apical meristem, leaf and petiole tissue (Hong et al. 1989; Mori et al. 1993; Asano et al. 2002). There was a significant difference in the level of anthocyanin production from lines derived from the different explants, with leaf tissue generating the most productive cell line.

Analysis of the anthocyanins from all the cell lines showed that their composition was different from that found in strawberry fruit. Two major anthocyanins are pre-

Plate 1. Thorns and their mimics. (A) Anthocyanic thorns on a rose stem. (B) Red mucron at the apex of a leaf of *Limonium angustifolium*. (C) Red fruit of *Erodium laciniatum*. (D) Variegated red Autumn leaf of Acer may undermine the crypsis of herbivorous insects. Imperfect defence, and the exceptions. (E) Aphids feeding from the midrib of the red underside of a rose leaf. (F) Anthocyanic leaves of the New Zealand sundew *Drosera spathulata* (photo: Dean O'Connell)

Plate 2. Top row: Flowers of lisianthus cultivars pigmented predominately by pelargonidin- (left), cyanidin- (centre) or delphinidin-based (right) anthocyanins. Centre row left: Solutions of, from left to right, pelargonidin, cyanidin, delphinidin, apigeninidin (3-deoxypelargonidin) and luteolinidin (3-deoxycyanidin). Centre row right: 3-Deoxyanthocyanin pigmented flowers of *Sinningia cardinalis*. Bottom row left: Floral organs of the Black Bat Flower. Bottom row centre and right: The floral organs of calla lily Treasure and a close up of the epidermis of the spathe of the same cultivar

Plate 3. Top row: Anthocyanic vacuolar inclusions in petals of carnation (left) and lisianthus (centre and right). Centre row and bottom row right: Examples of floral pigmentation patterning in painted tongue (*Salpiglossis sinuate*, centre left), *Paphiopedilum* orchid (centre right), and the snapdragon cultivar Venosa (bottom right). Bottom row left: 35S:LC transgenic petunia plants with purple-leaf phenotypes growing in field trials

Plate 4. Top row left: Flowers of a non-transgenic line of torenia (left) and two lines contain-ing a transgene for inhibition of ANS gene activity. Top row right: A flower of a transgenic rose cultivar engineered to accumulate delphinidin-derived anthocyanins. (Top row photo-graphs courtesy of Suntory Ltd, http://www.suntory.co.jp). Centre row: Examples of trans-genic carnation cultivars engineered to accumulate delphinidin-derived anthocyanins (photo-graphs courtesy of Suntory Ltd and Florigene Pty Ltd, http://www.florigene.com). Bottom row left: Flowers of a non-transgenic line of lisianthus (top) and a line containing a transgene for inhibition of F3′,5′H gene activity. Bottom row right: Flowers of a line of lisianthus containing a transgene for inhibition of CHS gene activity, with all flowers shown being on a single plant

Plate 5. Fruit and flowers with contrasting colours. (A) Black ripe fruits and red calyxes of *Ochna serrulata*. (B) Red unripe and black ripe fruits of *Colpoon compressum*. (C) Black ripe fruits of *Halleria lucida* are displayed with red flowers. Fruits turn from green to black without any intermediary colours. Photos by E. van Jaarsveld

Plate 6. Anthocyanins in fruit. (A) Immature and mature "Forelle" pears. Pears attain their highest anthocyanin concentrations during early fruit development whereafter red colour gradually fades towards harvest. Photo by W.J. Steyn. (B) Anthocyanin accumulation surrounding sunburn lesion in apple. Photo by J. Gindaba. (C) Grape berries displaying a bloom. Photo by M. Huysamer

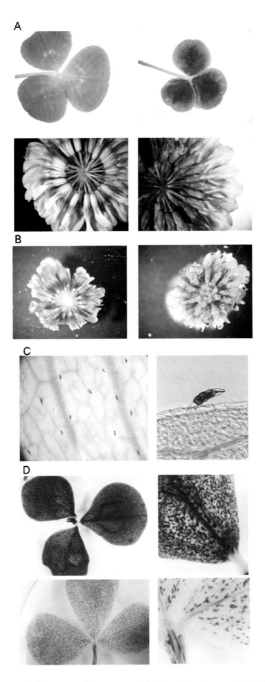

Plate 7. A - Leaves and flowers of common (*left*) and color variant (*right*) *T. repens*;
B - DMACA staining of PA's in common (*left*) and color variant (*right*) *T. repens* flowers;
C - DMACA staining of PA's in *T. repens* leaf trichomes; D - DMACA staining of PA's in *L.
pedunculatus* (*top*) and *L. corniculatus* (*bottom*) leaves

sent in strawberry fruit, pelargonidin 3-*O*-glucoside followed by cyanidin 3-*O*-glucoside. All the cell lines produced peonidin 3-*O*-glucoside rather than pelargonidin 3-*O*-glucoside, indicating the upregulation of both F3′H and methyltransferase activities (Hong et al. 1989; Mori et al. 1993; Asano et al. 2002). This is unusual in that most cell lines show a loss of pigment complexity rather than a gain. Explant source did not influence the anthocyanin types.

5.4.2 Phytohormones

2,4-D is the most effective auxin for a combination of good growth rate and anthocyanin production, particularly when used in combination with BA (Hong et al. 1989; Mori et al. 1993, 1994a). Variation in BA levels had little effect on anthocyanin production compared with 2,4-D (Mori et al. 1994a). At an early stage of culture, high anthocyanin production was achieved with low 2,4-D concentrations, probably due to stress in the plant cells from growth limitation. However, in the latter growth stages anthocyanin production was stimulated by high 2,4-D concentrations. Mori et al. (1994a) point out that 2,4-D is not desirable in a commercial production system and suggest its replacement with other phytohormones.

The Nyoho cell line established by Asano et al. (2002) used NAA rather than 2,4-D and was progressively habituated to reducing concentrations of phytohormones until the line was able to grow rapidly on hormone-free medium and still produce large amounts of anthocyanin. It is rare for habituated cell lines to maintain their secondary metabolite ability.

Strawberry cell cultures behave in a similar manner to grape and carrot cells in that limitation of growth (e.g. by low phytohormone levels) stimulates anthocyanin production. In addition, it also stimulates methyltransferase activity leading to increased levels of peonidin 3-*O*-glucoside at the expence of its cyanidin precursor (Mori et al. 1994a; Nakamura et al. 1998).

5.4.3 Nutrients (Carbon, Nitrogen, Phosphate)

For carbon supply, strawberry cell cultures grew best and had optimal anthocyanin yields on sucrose, or its monomeric units glucose and fructose, compared with other carbon sources such as xylose, mannose, rhamnose, arabinose and galactose (Mori and Sakurai 1994; Miyanaga et al. 2000a). High sugar concentrations improved anthocyanin yield, in particular when coupled with low ammonium levels. However, they also reduced cell growth. This is very similar to the behaviour of grape and carrot cells. Addition of mannitol enhanced anthocyanin production and gave high levels of sucrose, indicating that all or part of the high sucrose response is due to a stress-related increase in osmotic pressure (Sato et al. 1996). The increase in anthocyanin content under these conditions was due to an increase in the proportion of pigmented cells in the culture (Sato et al. 1996; Miyanaga et al. 2000a).

Anthocyanin production and cell growth were maximised by low ammonium and low total nitrogen (Mori and Sakurai 1994; Sato et al. 1996; Miyanaga et al. 2000a). The increase in anthocyanin content under these conditions was due to an increase in the proportion of pigmented cells in the culture (Sato et al. 1996). In addition, the proportion of cyanidin-3-*O*-glucoside compared to peonidin-3-*O*-glucoside also increased with low ammonium and nitrogen levels, indicating a shut-down of methyltransferase activity (Mori and Sakurai 1994).

Strawberry cell suspensions grown on LS and B5 medium show quite different behaviour, mainly due to the difference in nitrogen supply (Zhang et al. 2001). Cells on LS medium grow faster but produce more anthocyanin per g fresh weight (FW) on B5 medium. For commercial production the largest possible amount of anthocyanins is needed in the shortest culture time. It is possible that alternating a cell culture between the two media may maximise yield. In addition, the performance of a repeated batch culture system is influenced by the state of the inoculating cells, which come from the end of the previous batch culture. As a result the period of the batch culture will heavily influence the overall productivity. Zhang et al. (2001) tested two batch periods of 9 and 14 days and found that a 9-day period (end of exponential growth in strawberry cells) performed best. However, they found that alternating between the two different media did not improve production compared with a single media formulation.

Phosphate starvation dramatically improves anthocyanin yield below levels limiting for cell growth (Sato et al. 1996; Miyanaga et al. 2000a). The increase in anthocyanin production is due to an increase in the portion of pigmented cells (Sato et al. 1996). Careful balancing between the degree of cell growth and level of anthocyanin production through phosphate starvation is needed to maximise total anthocyanin yield.

Growth limitation through low phosphate and nitrogen, osmotic pressure and elicitation (see below) enhances anthocyanin production through an increase in the number of cells capable of producing anthocyanins as well as the total amount of anthocyanins produced per cell. This is in keeping with the results from carrot and grape and may also represent a shift in metabolism from primary to secondary. It is possible that the mechanism for this action is the accumulation of phenylalanine through reduction in the activity of primary metabolism. In support of this theory, Edahiro et al. (2005a) showed that supplementation of cell cultures with phenylalanine resulted in a significant increase in total anthocyanin production. In addition, maximum accumulation occurred earlier, thus shortening batch culture times. Cellular phenylalanine concentration was maintained by repetitive feeding and its accumulation was shortly followed by increased anthocyanin biosynthesis. However, repetitive feeding was only able to maintain anthocyanin production during the first phase of cell growth (days 8–10). This finding indicates that other processes overcome this effect in the later stages of culture.

5.4.4 Conditioned Medium

The growth rate of strawberry suspension cultures is heavily influenced by the amount of inoculation. Low inoculation levels (less than 2 g/100 mL) lead to failure of the culture to thrive and little anthocyanin production (Mori and Sakurai 1994). Optimum anthocyanin production was achieved using an inoculation volume of 2 g/100 mL. Volumes higher than this caused an early peak in anthocyanin production, resulting in a reduced anthocyanin yield. In addition, the stationary phase was reached progressively earlier, resulting in a steady decrease in the final cell volume. It is possible that the influence of inoculation density is related to the release of a conditioning factor into the medium that encourages cell division (Yamakawa et al. 1985). If this is the case then a medium purified from a growing cell line (called a conditioned medium) that may contain such a factor should alleviate the limitation of inoculation density. Mori et al. (1994b) and Sakurai et al. (1996) showed that the addition of conditioned media reduced inoculation density from 2 to 1 g/100 mL. Boiling of the conditioned medium had no effect on its stimulating potential, suggesting that the critical factor is non-volatile and thermostable (Mori et al. 1994b). Analysis of the conditioned medium revealed several changes in nutrients known to influence cell growth such as nitrogen and sucrose level. Attempts to mimic the effect of the conditioned medium by duplicating the altered nutrient levels have failed to reproduce the improvements in cell growth and anthocyanin biosynthesis (Sakurai and Mori 1996). This indicates that the plant cells specifically release the relevant factor(s) in the conditioned medium. The use of conditioned medium also resulted in an enrichment of cyanidin in early culture (down-regulation of methyltransferase) that was dependent on the concentration of conditioned medium (Mori et al. 1994b; Sakurai et al. 1996, 1997b).

Addition of the strawberry conditioned medium to uncoloured rose cell cultures resulted in the *de novo* production of anthocyanins (Sakurai et al. 1997a), whereas conditioned media prepared from red and white grape cells promoted strawberry cell growth but not rose (Mori and Sakurai 1998). These results show that the factors in conditioned medium can operate across species barriers and that anthocyanin production in the cell line from which the conditioned media is isolated is not a requirement. The factor(s) responsible for early stimulation of anthocyanin biosynthesis are present in both sub- and super-10,000 Da dialysed fractions. Stimulation at day 15 was due to factor(s) in the sub-10,000 Da fraction. Alkali treatment removed the effect of the factor(s) greater than 10,000 Da (Mori and Sakura 1999). Both PAL and CHS activity were significantly greater in cell lines grown in conditioned media (Mori et al. 2001). The exact mechanism behind stimulation by conditioned medium is not yet known.

5.4.5 Elicitation

Addition of MeJA and riboflavin to low yielding cell lines was able to increase anthocyanin production, but the growth rate was reduced via riboflavin-mediated degradation of auxins. When used in conjunction with high sucrose a further increase in anthocyanin production was achieved (Mori and Sakurai 1995; Miyanaga et al. 2000a). Addition of MeJA and riboflavin to a high anthocyanin producing line did not result in an increase in anthocyanin production as the increase in anthocyanin yield was offset by the reduced growth rate (Mori and Sakurai 1995). This illustrates the influence that the host explant tissue can have on the performance of a cell line. Analysis of the timing of anthocyanin production showed that riboflavin specifically enhanced anthocyanin biosynthesis in the early stages of the growth cycle. In addition, riboflavin was able to enhance anthocyanin production in nitrogen regimes more suitable for cell growth than anthocyanin yield (Mori and Sakurai 1996). Riboflavin also showed some influence on the distribution of anthocyanin types, with an increase in the proportion of peonidin-3-O-glucoside, indicating up-regulation of methyltransferase activity (Mori and Sakurai 1996).

5.4.6 Physical Parameters

To scale up strawberry cell cultures for commercial production, transfer from small laboratory shake flasks to large-scale bioreactors is necessary and often results in increased hydrodynamic stress. In order to better understand the cells' response to such stress Takeda et al. (2003) isolated a gene fragment (*tuf*) specifically expressed under hydrodynamic stress in strawberry cell cultures. The *tuf* fragment was homologous to nuclear-binding-site domains in plant disease resistance genes. In addition, they showed the up-regulation of a calcium-dependent protein kinase and also that Ca^{2+} ion channel blockers suppressed the expression of *tuf*.

One of the biggest drawbacks to using strawberry cell cultures as a commercial source of anthocyanins is their need for light, as continuous exposure to light in large-scale bioreactors is hard to achieve. Kurata et al. (2000) have shown that for strawberry cell suspensions, intermittent pulses (20 s) contributing to a total of 75% total light were able to maintain anthocyanin production. In light-dependent strawberry cell cultures, alteration of light intensity and duration improves anthocyanin production through increased intracellular pigment density rather than raising the population of pigmented cells (Sato et al. 1996). Enzymes early in the anthocyanin biosynthetic pathway such as PAL and CHS increase in activity under high light conditions. In addition, one of the isozymes (DS-Mn) of the enzyme DAHP synthase is also up-regulated. This enzyme is part of the shikimate pathway leading to the synthesis of phenylalanine (Mori et al. 2000).

Nakamura et al. (1999) have isolated a cell line derivative that can produce anthocyanins in the dark. The line is stable and contains more than 90% coloured cells compared with 26–36% for its parent line. It is possible that this cell line carries a mutation in the regulatory mechanism linking anthocyanin production to light irradiation. Interestingly, when the new cell line was exposed to light in solid culture, cell growth rate and anthocyanin production fall, indicating that light has a detrimental effect on the viability of the cell line and its ability to synthesise anthocyanins.

A common aspect of cell cultures producing secondary metabolites is the relationship between cell aggregate size and metabolic activity. In the dark-grown cell line developed by Nakamura et al. (1999) the majority of pigment production is associated with large cell aggregates, which can be disadvantageous in large-scale culture due to their susceptibility to shear stress (Edahiro and Seki 2006). The formation of large-cell aggregates in this cell line is directly influenced by the level of phenylpropanoid metabolism. When PAL activity is inhibited, anthocyanin production is shut down and the average aggregate size falls. It is probable that phenylpropanoid products are increasing cellular cohesiveness. Both lignins and tannins are known to perform this function in intact plant tissues (Edahiro and Seki 2006).

Strawberry cells in suspension have a strong pH buffering capacity and rapidly return the media to a pH of 4.5–5.0 within the first three days of culture (Zhang and Furusaki 1997). The initial lag phase in cell growth after subculturing corresponded to the time required to return the pH to ideal conditions. However, total cell growth over a 15-day period is unaffected by altering initial pH levels. In contrast, initial pH levels do affect anthocyanin production, with the highest yield coming from a starting pH of 8.7, significantly higher than the common media pH of 5.7–5.8. The increase in anthocyanin production from an elevated initial pH was from increased anthocyanin content in the pigmented cells. There was no increase in the overall proportion of pigmented cells indicating that alteration of pH is not capable of inducing anthocyanin biosynthesis in unpigmented cells (Zhang and Furusaki 1997).

Like most other pigmented cell lines, strawberry cell cultures consist of a mix of coloured and uncoloured cells. Over the course of a growth cycle the anthocyanin content of individual cells increases but the rate of accumulation is not related to any specific position in the cell cycle and is independent of cell division (Miyanaga et al. 2000b). The cells used in that study were not subject to growth limitation. It would be interesting to observe the relationship between cell division and anthocyanin production under such conditions. In order to enrich for anthocyanin production, cell cultures have to be constantly selected for a high proportion of coloured cells. This is a time-consuming process and is one of the reasons why cell cultures are not yet considered as a commercial source for anthocyanin production. Edahiro et al. (2005b) report on a rapid method for the selection of coloured cells that goes some way to alleviate this problem. An aqueous two-phase partition system (ATPS) containing a PEG/Dextran Sulphate mix is able to sort coloured strawberry cells based on their cell surface properties. The method is simple and quick (2.5 min to partition) and would considerably reduce the time necessary for maintenance of coloured cells lines containing mixed populations.

Commercial production of anthocyanins from cell cultures would be greatly improved if there were a mechanism to release the anthocyanins into the media without killing the cells. Takeda et al. (2003) released anthocyanins into the media after heat treatment at 45°C. However, the resulting subculture of these cells was slow-growing and colourless, showing that heat treatment is not a viable mechanism for the harvest of anthocyanins from strawberry cell cultures (Takeda et al. 2003).

Cultivation temperature affects both cell growth and anthocyanin production (Zhang et al. 1997). Maximal growth rate of strawberry cell cultures was achieved at 30°C and lag phase was progressively reduced with increasing temperature. However, anthocyanin content decreased with rising temperature. When this is balanced with the growth rate response, maximum anthocyanin production on a per litre of culture basis was achieved by culturing at 20°C. In order to maximise the benefits of high growth rate and anthocyanin production, a two-phase culture was tested in which the cells were grown at 30°C for 3 days then reduced to 20°C. Such a two-phase system resulted in further gains in total anthocyanin production. The response to temperature variations shows again the tight linkage between cell growth rate and the ability to manufacture secondary products such as anthocyanins. The growth characteristics of strawberry cell cultures and their response to differing conditions has been modelled to optimise the process for commercial production of anthocyanins (Zhang et al. 1998).

5.4.7 Conclusions

Strawberry cell cultures can produce anthocyanins under both light and dark regimes. Their response to outside stimuli and media supply are similar to both grape and carrot. The anthocyanin profile is relatively simple and may provide a novel natural colourant compared with grape skins. The common conflict remains between growth rate and secondary metabolism that limits the yield of pigments from these cultures. In addition, aggregate size will limit the ability to grow the culture on a commercial scale.

5.5 *Ajuga* Species

Ajuga species are used as medicinal plants, in salads, as a cotton dye and also as an ornamental ground cover (Callebaut et al. 1993). Cell cultures have been established from two species. A homogeneous fast-growing blue callus line that is capable of colour production in the dark was established from young flowers of *Ajuga reptens* (Callebaut et al. 1988, 1990a, 1993). A light-grown callus and suspension cell line was established from leaf pieces of *A. pyramidalis*, a purple leafed ornamental groundcover plant (Madhavi et al. 1996).

5.5.1 Types of Anthocyanin

Most of the pigments in the *A. reptens* cell line are acylated derivatives of cyanidin and delphinidin-3,5-diglucosides (Callebaut et al. 1993, 1997; Terahara et al. 1996, 2001). The main anthocyanin is cyanidin 3-*O*-(2-*O*-(6-*O*-(E)-*p*-coumaryl-ß-D-glucopyranosyl)-(6-*O*-(E))-*p*-coumaryl)-ß-D-glucopryanosyl)-5-*O*-(6-*O*-malonyl-ß-D-glucipyranoside) (CyGGPPGm), and the most common minor pigment is the delphinidin equivalent (DpGGPPGm). The other minor components are non-malonylated and acylated derivatives (CyGGPPG, DpGGPPG, CyGG). The pattern of anthocyanin types is different from the parental flower tissue, which is predominantly delphinidin-based rather than cyanidin-based. Subculturing over several years has resulted in a further drop in the delphinidin-based pigments with a concomitant rise in cyanidin-based ones. This indicates a steady loss of F3′5′H activity over time within these cell lines (Callebaut et al. 1990a, 1997). When solid cultures were transferred to liquid an even more dramatic decrease in delphinidin-based pigments occurred, as well as a decrease in the percentage of acylated pigments. As a result it seems that transfer from solid to liquid culture results in the further loss of F3′5′H activity as well as the loss of acyltransferases. Two acyltransferases have been isolated from the *A. reptens* cell line, a hydroxycinnamoyl transferase and a malonyl transferase (Callebaut et al. 1996). Substrate specificity for both enzymes was broader than the substrate range available in the cell cultures. The highly acylated anthocyanin species present in these cultures show a significant amount of antioxidant activity coupled with good stability, making them good candidates as a novel source of natural food colourants (Terahara et al. 2001).

A. pyramidalis cell cultures are able to produce pigments under light, the major pigment being a cyanidin with three glucose substitutions, plus two substitutions of ferulic acid and one of malonic acid. The anthocyanin profile of the cultures closely resembled the profile from the parental leaf tissue. Analysis of anthocyanin stability under light irradiation showed that the pigments from the cell cultures were significantly more stable than the same pigment from the leaf. It is possible that the cell culture produces co-pigments such as flavonols at a higher level than the parental tissue and that the presence of these compounds prolongs the lifetime of the pigment (Madhavi et al. 1996).

5.5.2 Phytohomones

Modification of 2,4-D concentrations in the *A. repens* culture medium affected both growth and anthocyanin production, with low concentrations leading to increased aggregate formation and enhanced anthocyanins (Callebaut et al. 1988). Replacement of 2,4-D with NAA reduced growth rate and hence total anthocyanin production. BAP gave a higher anthocyanin yield than kinetin. Supplementation with gibberellic acid also reduced growth rate and pigment production (Callebaut et al. 1993). For the *A. pyramidalis* cell lines a combination of 2,4-D and kinetin was best for cell growth, and IAA/NAA with zeatin produced the most anthocyanin (Madhavi et al. 1996).

5.5.3 Nutrients (Carbon, Nitrogen, Phosphate)

In an *A. reptens* cell line, carbon source affected both growth and anthocyanin production with sucrose providing the best result under light and glucose in the dark (Callebaut et al. 1990a, 1993). Fructose inhibited anthocyanin production and cell growth. Very high levels of sucrose slightly enhanced the yield of anthocyanins per gram dry weight (DW) but did not affect overall anthocyanin production due to a concomitant reduction in growth rate (Callebaut et al. 1990a, 1993). Alteration of osmotic potential through the use of mannitol showed that the primary effect of high sucrose in increasing anthocyanin biosynthesis was via increasing osmotic pressure (Callebaut et al. 1993). The differences in anthocyanin production under different carbon sources reflected changes in cell growth rate rather than alterations in pigment production. In an attempt to reduce the cost of nutrient media the *A. reptens* cell line was also grown on a medium containing milk whey as the only carbon source (Callebaut et al. 1990b). After several rounds of selection for strong colour and high growth rate on milk whey, the authors recovered a line with production characteristics identical to the parent cell line. However, unlike the parental line, the new cell line could not grow under light.

For an *Aspergillus pyramidalis* cell line fructose as the sole carbon source gave the best growth rate and was significantly higher than sucrose (Madhavi et al. 1996). This is different from cell cultures of *A. repens* in which fructose inhibited growth (Callebaut et al. 1990a). Galactose produced the highest yield of anthocyanins, with a growth rate only slightly below that of fructose. The cell lines were unable to use lactose, a finding similar to the results for *A. repens*. However, as noted above,

Callebaut et al. (1990b) were able to select a derivative cell line that was habituated to lactose as the carbon source.

The *A. reptens* cell line responded to low nitrogen and phosphate concentrations by increasing anthocyanin levels. Anthocyanin production occurs more rapidly in the latter half of the cell cycle and is delayed by rapid cell division (Callebaut et al. 1993). This is a common characteristic of cell cultures and may be caused by elicitation via loss of nutrients such as nitrogen. Transfer of the *A. reptens* cell line from shake flasks to a stirred bioreactor (2 L) resulted in a reduction in pigment yield which the researchers suggested may be due to limitations in aeration of the culture compared with shake flasks (Callebaut et al. 1993)

5.5.4 Physical Parameters

The dark-grown *A. repens* callus line responds to the presence of light with an increase in anthocyanin biosynthesis and cell growth rate. However, liquid cultures of this line were not able to produce anthocyanins without the aid of light and were also less stable in their anthocyanin yield over repeated subcultures than were callus cultures (Callebaut et al. 1990a, 1993). Analysis of different light/dark regimes on anthocyanin production from another *A. repens* cell line showed that pigment formation was proportional to the duration of light exposure (How and Smith 2003).

A. pyramidalis cell cultures have been used to establish a non-destructive image analysis system for the estimation of cell growth and pigment production (Smith et al. 1995). Callus mass (FW) can be accurately determined from top and side images of the callus without removal from the sterile environment. In cell suspension cultures, image analysis using hue, saturation and intensity parameters was able to estimate pigment content and cell growth. Such a non-destructive system will be very useful in the set-up and maintenance of commercial-scale cell culture facilities based on cell level phenomena.

5.5.5 Conclusions

Ajuga cell cultures are able to make pigment in the absence of light. In addition, a cell line has been isolated that can utilise a waste stream (milk whey) as the carbon source, thus lowering the cost of production. In addition, *Ajuga* cell lines make highly acylated pigments with good stability and have high antioxidant capacity, thus making them good candidates for the commercial production of natural pigments.

5.6 *Ipomea batatas* (Sweet Potato)

Highly coloured sweet potato tubers of the cultivar Ayamurasaki are used to make a juice that is popular in Japan and show high colour stability and antioxidative activity (Suda et al. 1997; Konczak-Islam et al. 2000). Konczak-Islam et al. (2000) have established a cell line (PL) from this cultivar that can produce large amounts of anthocyanin in the dark. Pigment accumulation was associated with reduced growth rate and the maximum pigment concentration was similar to the parental tuber tissue (Konczak-Islam et al. 2000).

5.6.1 Types of Anthocyanin

In tuber tissue the main anthocyanins are peonidin and cyanidin 3-O-sophoroside-5-O-glucosides acylated with ρ-hydroxybenzoic, caffeic or ferulic acids, with peonidin-based pigments in the majority. In the PL cell line the dominant pigment is cyanidin 3-O-sophoroside-5-O-glucoside followed by peonidin 3-O-sophoroside-5-O-glucoside (Konczak-Islam et al. 2000, 2003a; Konczak et al. 2005). This indicates that when grown on a basal medium, the cell line has reduced methyltransferase and acyltransferase activities compared with the parental tissue. The PL cell line has been maintained for over seven years during which the level of anthocyanin production has been steadily increased by selection for highly pigmented cells. The portion of non-acylated to acylated anthocyanins has also risen steadily in conjunction with an increase in growth rate, probably due to habituation of the cell lines (Konczak et al. 2005).

Acylation of anthocyanins in PL cell cultures uses p-coumaric acid as a substrate. When PL cell lines were fed p-coumaric acid, almost all of the anthocyanins were converted to their acylated derivatives (Plata et al. 2003). This indicates that the presence of non-acylated anthocyanins in this culture is mainly due to a limitation of substrate. Feeding of hydroxycinnamic acid also had a similar effect. As well as altering anthocyanin composition, p-coumaric acid also increased the total amount of anthocyanin produced and growth rate was unaffected.

Sweet potato cell lines have also been established from sweet potato storage roots (Nozue and Yasuda 1985). Two lines were established, a coloured line with over 95% of the cells capable of producing pigment, and an uncoloured derivative. The coloured line showed a high stability of anthocyanin production over multiple sub-cultures, with the major pigments the same as the PL line. Unlike the PL line, however, anthocyanin biosynthesis required exposure to continuous light.

5.6.2 Phytohormones

The basal medium for the PL line includes 2,4-D (2 mg/L). An increase in 2,4-D results in an increase in acylated pigments and a decrease in growth rate. Lower 2,4-D concentrations result in an increase in total pigment accumulation (Konczak et al. 2005). For the Nozue and Yasuda (1985) line optimum pigment production was obtained using lower levels of 2,4-D. Higher amounts of 2,4-D, or the use of other auxins and cytokinins, reduced the pigment content (Nozue et al. 1987).

5.6.3 Nutrients (Carbon, Nitrogen, Phosphate)

Elevated sucrose concentrations stimulated anthocyanin production and, up to 5% sucrose, did not affect the cell growth rate. No alteration of anthocyanin types was observed (Konczak-Islam et al. 2001).

A reduction of ammonium nitrate to below 7.5 mM causes a dramatic shift in anthocyanin type, with a reduction in cyanidin 3-O-sophoroside-5-O-glucoside and a corresponding increase in its acylated derivatives. Further reduction in nitrate levels did not alter acylation patterns, leading to the conclusion that excess ammonium ions inhibit acylation activity (Konczak-Islam et al. 2001). Reduced nitrogen availability also increased the total amount of anthocyanin produced without affecting cell growth. Other cell lines, such as grape, strawberry and carrot, often show significantly reduced cell growth when anthocyanins are induced under low nitrogen conditions (see relevant sections in this chapter).

Elevated temperature (30°C) also had an effect on anthocyanin composition, with reduced levels of acylation and methylation (Konczak-Islam et al. 2001). Cell growth was also reduced. The optimum temperature for cell growth and anthocyanin production was 25°C. Light was tested for its influence on anthocyanin production in the PL cell line and no effect was observed. This is probably due to the original selection of a light-independent culture (Konczak et al. 2005). Taking all these factors in account an anthocyanin induction medium was developed comprising low nitrogen, high sucrose, and no growth regulators (Konczak-Islam et al. 2003a). Analysis of the anthocyanin profile of the culture grown in this media showed that the anthocyanins gradually shifted from a predominance of non-acylated pigment to increasing levels of acylation (Konczak-Islam et al. 2003a).

5.6.4 Elicitation

Fungal elicitors are often used to enhance secondary metabolite production. When the PL line was exposed to MeJA no change in anthocyanin content or growth rate was observed. However, the proportion of acylated anthocyanins was significantly increased in a manner similar to the cells' response to low ammonium ion concentrations (Plata et al. 2003). It is possible that the lack of anthocyanin stimulation is due to the lack of light in the culture system. Zhang et al. (2002a) report a synergistic effect between MeJA and light for grape cells. However, the impact of MeJA can be to increase the proportion of coloured cells in a mixed cell culture (such as strawberry). Thus, since MeJA principally affects the uncoloured cells, and the PL line is already uniformly coloured, it is possible that MeJA treatment may have no benefit in this case.

The PL cell line is a good candidate for the commercial production of anthocyanin pigments as natural colourants (Konczak-Islam et al. 2003a). Highly acylated pigments are more stable and hence more useful as food additives. In addition, the pigments and other secondary metabolites produced by the PL cell line show high levels of antioxidant activity (Konczak-Islam et al. 2003b; Konczak et al. 2004; Terahara et al. 2004).

Much research has been done using the Nozue et al. cell line to understand the mechanism of anthocyanin storage in the vacuole. Anthocyanin-containing bodies, called anthocyanoplasts, are formed in the cytoplasm of the line at the same time as the onset of anthocyanin biosynthesis, and are then transported into the vacuole (Nozue and Yasuda 1985; Nozue et al. 1993). A possible mechanism for the formation of anthocyanoplasts is via interaction with the protein VP24, which is found in the anthocyanoplasts when anthocyanins are synthesised (Nozue et al. 1995, 1997). Analysis of the VP24 DNA sequence suggests that it is a novel vacuolar localised amino peptidase and plays a role in the concentration of anthocyanins into anthocyanoplasts to enhance the cells' capacity for storage of large amounts of anthocyanins (Nozue et al. 2003). A greater understanding of the mechanism and genetic control of the formation of anthocyanoplasts or AVIs (as mentioned in the section on grape cell cultures) is needed to improve the anthocyanin storage capacity of cell cultures and hence their use as an alternative means of anthocyanin production.

5.6.5 Conclusions

Sweet potato cell cultures offer both a model system for the analysis of anthocyanin storage and a potential source of natural colourants. Their pigment profile and associated biological activity are highly desirable. The principal drawback with current cell lines is the inability to grow in large volumes due to limitations in aggregate size. This is similar to the problems facing strawberry cell lines.

5.7 *Perilla frutescens*

Perilla frutescens is a medicinal herb that is used in Chinese medicine and in Japan as a food garnish and natural colourant. Highly coloured cell lines have been produced from *P. frutescens* by selective cloning of coloured cell aggregates (Zhong et al. 1991; Wang et al. 2004). Optimum anthocyanin production and cell growth was achieved using a medium containing 2,4-D and BA and the use of either continuous light or 16-hour days (Zhong et al. 1991; Wang et al. 2004).

5.7.1 Carbon Source and Elicitation

High levels of sucrose (5%) result in a significant increase in both cell growth rate and anthocyanin production. A unique feature of this cell line is its ability to release anthocyanins into the medium at high sucrose concentrations, probably due to the increased permeability of the cell membrane (Zhong et al. 1994a). This feature could be very useful in the large-scale production of anthocyanins, since extraction from the medium is easier than from the cell mass. Supplemental feeding of sucrose during the culture time improved anthocyanin biosynthesis. However, once an upper limit of sucrose concentration was reached, anthocyanin biosynthesis decreased rapidly but cell growth was unaffected (Zhong and Yoshida 1995). An increase in inoculum size also increases the anthocyanin content per cell and hence total anthocyanin production. When combined with elevated sucrose levels, Zhong and Yoshida (1995) showed an additive effect of the two treatments to obtain an anthocyanin content of almost 6 g/L.

P. frutescens cell cultures are susceptible to yeast elicitors, resulting in an increase in anthocyanin production of up to 10% DW (Wang et al. 2004).

5.7.2 Physical Parameters

Increasing the light intensity increased anthocyanin production but had no effect on cell growth in *P. frutescens* cell cultures (Zhong et al. 1991). Analysis of the specific light requirement showed that in flask cultures, illumination for the first 7 days of the

14-day culture period was enough to maximise anthocyanin production. This cell line has been successfully transferred to a stirred bioreactor where the cell culture was able to maintain the same growth rate and anthocyanin production characteristics as in shake flasks. However, a further increase of light intensity in the bioreactor system caused a collapse in cell growth and loss of anthocyanin. Maximum anthocyanin production occurred at the transition between exponential growth and the stationary phase (Zhong et al. 1991). This is similar to other cell culture systems.

Oxygen supply to the cells in culture is critical for optimum anthocyanin production. Improving the oxygen transfer rate into solution through the use of a fine air sparger more than doubled the production of anthocyanins in a stirred bioreactor (Zhong et al. 1993a). A further increase in dissolved oxygen can be achieved by increasing the airflow rate. However, the authors found that beyond a threshold flow rate the cell growth rate slowed slightly and anthocyanin production reduced significantly. They suggest that the loss of performance is due to frothing from excessive turbulence in the medium caused by the higher gas flow rate. When optimum light and oxygen parameters are combined in one treatment there is a surprising loss in anthocyanin production but no significant change in cell growth rate (Zhong et al. 1993a). This indicates that in *P. frutescens* cell culture systems there is an unknown complex interaction between oxygen availability and light intensity that regulates anthocyanin biosynthesis. It is also interesting to note that high oxygen levels were able to produce anthocyanins at the same rate as optimum irradiation levels, raising the possibility that these cultures may not need light supplementation in large-scale bioreactors.

Anthocyanin production in *P. frutescens* cell lines is very sensitive to temperature fluctuations (Zhong and Yoshida 1993). An increase in temperature from 25°C to 28°C resulted in a slight increase in specific growth rate but an irreversible loss of anthocyanin production. Anthocyanin production at 22°C was reduced, mainly due to a loss in cell growth rate.

One of biggest challenges in transferring cell cultures from shake flasks to a stirred bioreactor is the effect of the shear force from agitation by the rotating blade and aeration rate. As shown for other cell lines, cell viability of *P. frutescens* cells falls with increasing shear force (Zhong et al. 1994b). However, the authors also noted that even at low shear forces the viability of the cells still decreased over time and there did not appear to be a threshold where shear force damage ceased. Optimum conditions in a stirred bioreactor were achieved with an impeller speed of 120–170 rpm. At this speed cell growth and pigment production per cell was optimised. Higher impeller speeds resulted in a reduction in cell expansion, possibly due to the destruction of larger cells. Anthocyanin production on a per cell basis was also reduced (Zhong and Yoshida 1994; Zhong et al. 1994b). A significant increase in both growth rate and anthocyanin yield was obtained by the addition of the surfactant Silicone A (Zhong et al. 1992). The improvement was due to a reduction of foaming and cell adhesion to the vessel wall.

Estimation of cell growth rate through FW and DW measurements involves destructive sampling of the cell culture, which is time-consuming and can affect the performance of the remaining culture. To overcome this, Zhong et al. (1993b) have developed a laser turbidimeter that can measure cell concentrations of both uncoloured and coloured *P. frutescens* cell suspensions. They have shown that accurate results are obtained within the preferred growth parameters of these cell cultures in stirred bioreactors. The only parameter to significantly affect its accuracy was the expansion of the cells in the declining phase of the growth curve, which causes an apparent increase in cell concentration. Zhong et al. (1994c) have used this technique in conjunction with measurement of temperature, pH, dissolved oxygen, agitation speed, air flow and cultivation volume to develop a computer-aided on-line monitoring system for the analysis of plant cell cultures. Using this system they have shown that alteration of agitation speeds affects the physiological state of the cells. The growth phase of the cells could be identified, as well as the period of maximum anthocyanin production (stationary phase). The specific oxygen uptake rate increased along with the increase in anthocyanin production. This is curious, since the biosynthesis of anthocyanins does not directly use oxygen. In addition, the shear force produced by the impeller had a different effect over the time of the culture due to the growth state of the cells. In the stationary phase of culture, the cells expand significantly, thus increasing the viscosity of the medium and exacerbating the shear force (Zhong and Yoshida 1994). The monitoring system developed by these researchers will be useful to determine the dynamic growth status for use in feeding and control studies.

5.7.3 Conclusions

Perilla cell lines have been used extensively as a model to develop systems for the non-destructive monitoring of cell cultures. It is unlikely that these cell lines would be considered for commercial production because light is still a requirement for pigment synthesis and the cells seem particularly sensitive to shear stress.

5.8 *Vaccinium* Species

5.8.1 Types of Anthocyanins

Ohelo berry (*V. pahalae*) cell cultures have been extensively researched as a source of natural colourants and nutraceutical products. Callus and suspension cultures were established from leaf tissue and produced large amounts of anthocyanins under continuous irradiation (Smith et al. 1997). The cell lines contained similar anthocyanins to the parental tissue, and the major pigments were cyanidin-3-*O*-galactoside, cyanidin-3-*O*-arabinoside and traces of peonidin-3-*O*-galactoside. These cell lines have been successfully scaled up to production in 12 L bioreactors (Smith et al. 1997).

Cell cultures have been established from several other *Vaccinium* species. Cell lines established from a variety of cranberry (*V. macrocarpon)* explants and from bilberry (*V. myrtillus*) produced anthocyanins when exposed to light (Madhavi et al. 1995, 1998). The anthocyanin types in both cell lines were cyanidin-3-*O*-galactoside, cyanidin-3-*O*-glucoside and cyanidin-3-*O*-arabinoside and were the same for all explants. Cranberry and bilberry fruit also produce peonidin-based anthocyanins and the lack of these types in the cell cultures indicates a loss of methyltransferase activity. Analysis of PAL activity shows a concomitant increase with anthocyanin production upon exposure to light.

Analysis of cell cultures generated from rabbiteye blueberry (*V. ashei*) leaf tissue showed that, as with cell cultures from other *Vaccinium* species, the pigment profile was simpler than in the intact fruit (Nawa et al. 1993).

5.8.2 Phytohormones

Ohelo cultures were initially grown as colourless cultures in the dark on media containing NAA, 2,4-D and kinetin. Pigment production was induced by transferring the culture to continuous light in a media containing elevated sucrose (Smith et al. 1997) and was significantly improved by the use of BA rather than kinetin as the cytokinin (Fang et al. 1998). The use of BA improved both the cell growth rate and the portion of pigmented cells, but no change in the type and distribution of anthocyanins was observed, in contrast to strawberry cell cultures (Mori et al. 1994a). The presence of BA negated the osmotic stress effect of high sucrose levels, thus reverting to a loss of growth rate and further improving anthocyanin yield (Fang et al. 1998). In addition, the use of BA rather than kinetin significantly improved the consistency of anthocyanin production across many subcultures and restored colour to cell lines that had lost the capacity to make anthocyanins (Fang et al. 1998). The restorative effect of BA has enabled the use of a two-phase system in which the cell line is maintained in a high growth rate medium that generates little colour, then rapidly induced to make colour after transfer to BA medium (Fang et al. 1998). Anthocyanin biosynthesis in blueberry cell cultures was stimulated by small amounts of 2,4-D (0.1–0.2 mg/L) and 6% sucrose (Nawa et al. 1993).

5.8.3 Media Components and Elicitation

Modification of media components and addition of elicitors have been used to enhance anthocyanin production of ohelo cell cultures. Increasing the inoculum density (up to 30 mL inoculum/30 mL medium) and sucrose concentration (up to 7%) led to an increase in cell biomass and anthocyanin accumulation (Smith et al. 1997). This result is in line with that from other cell lines e.g. strawberry, grape and sweet potato. The alternative iron source FeEDDHA enhances anthocyanin production, perhaps through the greater availability of iron in a high light environment or through iron's ability to inhibit anthocyanin degradation (Fang et al. 1999). Elevated levels of $CuSO_4$ were also able to improve anthocyanin production. Purified β-glucan and chitosan stimulated anthocyanin production, as did MeJA (Fang et al. 1999). Surprisingly, when ibuprofen, an inhibitor of jasmonate biosynthesis, was added to cultures elicited with MeJA or β-glucan, anthocyanin production was dramatically enhanced rather than inhibited. Addition of ibuprofen to non-elicited cultures or cultures elicited with FeEDDHA causes a loss of anthocyanin production and growth rate. Addition of ibuprofen provides a means to rapidly increase anthocyanin production in ohelo cells even though the mechanism of action of ibuprofen is not yet known (Fang et al. 1999).

Shibli et al. (1997) showed that ethylene in the ohelo cell culture headspace slowed cell growth rate and anthocyanin production. Under this treatment a greater portion of cells were colourless, and anthocyanin production peaked earlier. The ethylene inhibitors $CoCl_2$ and $NiCl_2$ improved growth and anthocyanin accumulation. The use of vented closures largely alleviated the ethylene effect in these cultures, but even with a vented closure, ethylene inhibitors still enhanced productivity (Shibli et al. 1997).

5.8.4 Physical Parameters

As well as media, hormone and elicitor formulations, the physical environment needs to be optimised for maximum anthocyanin production. The ohelo cell line is a mixture of pigmented and unpigmented cells, with most of the anthocyanin production associated with small cell aggregates that may represent a later stage of differentiation than the larger more diffuse cell clumps (Fang et al. 1998). An effective image analysis tool has been developed for ohelo cell cultures, which enables the non-destructive on-line measurement of cell aggregate properties that are essential in the design and control of large-scale cell culture systems (Pepin et al. 1999). Results from this study confirm that most anthocyanin production is from small cell aggregates. Although the proportion of coloured aggregates in the larger sizes was greater than in the small class, the larger cell aggregates were only pigmented on the outer surface whereas the small aggregates were more intensely coloured throughout the multi-cell mass. The lack of colour inside large aggregates may be a result of the requirement for light induction. Maintenance of small cell aggregates in cell culture ensures a high level of anthocyanin production. This can be achieved through a low media pH (~4.5), low subculture frequency and low inoculum density (Meyer et al. 2002).

Ohelo cells have growth and anthocyanin accumulation characteristics common to many plant cell cultures. Growth rate assumes a sigmoidal curve and maximum anthocyanin biosynthesis is associated with a reduction in cell division at the end of the exponential growth phase. This result, coupled with the relationship between anthocyanin production and aggregate size, clearly shows a relationship between cell morphology and secondary metabolism. Increased irradiation can enhance anthocyanin biosynthesis without affecting cell growth rate. An increase in dissolved oxygen raises both cell biomass and anthocyanin production. When the ohelo cultures were transferred to a 14 L bioreactor, light became the major limiting factor in the production of anthocyanins. In addition, loss of cells from the medium through deposition on the bioreactor wall slowed growth rate (Meyer et al. 2002). To maximise anthocyanin production from ohelo cells in a bioreactor Meyer et al. used a high irradiance mercury light source and a magnetic scraper to remove cells from the bioreactor wall.

Ohelo cultures have been used to generate ^{14}C labelled anthocyanins and flavonoids for analysis of their uptake and metabolism in animal cell systems through the use of a fully enclosed shake flask system for the safe production of labelled metabolites (Grusak et al. 2004). In ohelo cultures a range of labelled flavonoid and anthocyanin compounds has been produced (Yousef et al. 2004). This system has also proven effective for the production of labelled compounds from grape cell cultures.

For commercial production of anthocyanins from cell cultures, a method for the storage of original cell lines is essential. Ohelo cells recovered from cryopreservation were able to grow and accumulate anthocyanins without loss of efficiency (Shibli et al. 1999).

5.8.5 Conclusions

Cell lines from the various *Vaccinium* species behave in a similar manner and all produce a similar range of anthocyanins. Most of the research to date has been done on the ohelo cell line and many useful data have been obtained on methods to maximise production of anthocyanins. The factors that influence anthocyanin production are very similar to other cell systems such as grape, strawberry and carrot, with the main difference being the dose rate of stimulants and the degree of stimulation. The principal drawback in using these cell lines for commercial production of anthocyanins is the need for light. It is possible that a mutant cell line could be generated, either through a random event or active mutagenesis, that could produce anthocyanins in the dark (e.g. for strawberry, Nakamura et al. 1999). Another approach is to modify the regulation of the anthocyanin biosynthetic pathway through genetic modification of the regulatory genes, thus overcoming the linkage between light and secondary metabolism. Deroles et al. (2002) have reported the genetic transformation of ohelo cell cultures using microprojectile bombardment as a first step towards this goal.

5.9 Other Plant Cell Lines

5.9.1 *Aralia Cordata*

Aralia cordata is an edible plant native to Japan, and is also used in traditional Chinese medicine. Sakamoto et al. (1994) developed two cell lines derived from stem and leaf tissue. One cell line can produce anthocyanins in the dark, and a second line requires exposure to light. The dark-grown cell line contains approximately 90% pigmented cells. This is much higher than other cell systems, which often report pigmented populations as low as 5%. The major anthocyanin identified in the cell lines is cyanidin 3-*O*-xylosylgalactoside and the minor pigment peonidin 3-*O*-xylosylgalactoside (Sakamoto et al. 1993; Asada et al. 1994). Conditions that enhance the production of anthocyanins in this cell line also alter the ratio of the two pigments, with a relative increase in peonidin (Asada et al. 1994). This indicates that conditions that enhance anthocyanin production also enhance methylation of the anthocyanin (Asada et al. 1994), as reported for grape cell lines (Do and Cormier 1991a). Analysis of phytohormone regimes showed that NAA/kinetin produced the best pigment yield using the dark-grown cell line and 2,4-D/kinetin was the best for the light-grown line, with anthocyanin yields of up to 13–15% DW (Sakamoto et al. 1993). Linsmaier and Skoog (LS) medium proved to be best for anthocyanin production and growth in the dark, with only 1 week needed for the cell line to reach the stationary phase. High sucrose concentrations reduced the growth rate and did not enhance anthocyanin production. Compared with sucrose, fructose significantly enhanced anthocyanin production, but growth rate was also reduced, so sucrose still gave the highest net yield of anthocyanins (Sakamoto et al. 1993). The effect of nitrogen supply was similar to many other cell lines. Anthocyanin production was enhanced by increasing the nitrate/ammonium ratio and by reducing the total amount of available nitrogen (Sakamoto et al. 1993, 1994). The dark-grown *A. cordata* cell line is one of the few cell lines that have been grown in large volumes (up to 500 L) to determine commercial viability (Kobayashi et al. 1993). With the addition of a continuous CO_2 supply to prevent browning of the culture, the cell line gave anthocyanin yields up to 17% DW in 500 L.

5.9.2 *Bupleurum falcatum*

One cell line has been isolated that produces anthocyanins in the light and one in the dark (Hiroaka et al. 1986). The dark-grown line was able to produce significantly more anthocyanins than the light-grown line. Anthocyanin production was associated with the onset of the stationary phase for both cell lines. The predominant pigments were malvidin *O*-glucosides.

5.9.3 *Callistephus chinensis* (China Aster)

Callus derived from leaf and stem tissue was able to produce anthocyanins under continuous light (Rau and Forkmann 1986). The most influential parameter for anthocyanin biosynthesis was the type and amount of auxin, with low concentrations of both IAA and NAA producing significantly more anthocyanin than 2,4-D. Increasing concentrations of any of the auxins led to a reduction in anthocyanin accumulation. Small amounts of cytokinin (e.g. kinetin) resulted in a loss of anthocyanin production. Alteration of nitrogen and phosphate levels showed a significant effect on growth rate but little effect on anthocyanin production. Callus derived from acyanic lines did not produce anthocyanins under any conditions tested.

5.9.4 *Campanula glomerata*

Hairy root cultures have been generated from *Campanula glomerata* after infection with *Agrobacterium rhizogenes* (Tanaka et al. 1999). The cell lines grew rapidly in the dark but not in continuous light. By contrast, illumination enabled the production of the anthocyanins cyanidin 3-*O*-glucoside and cyanidin 3-*O*-rutiniside. A two-stage culture system was developed involving a dark growth phase, followed by an illuminated phase to stimulate secondary metabolite production. Optimal accumulation of pigment occurred under low concentrations of BA. Increasing the BA content reduced pigment production, as did the addition of auxins.

5.9.5 *Camptotheca acuminata*

Cell cultures of *Camptotheca acuminata* accumulate cyanidin 3-*O*-galactoside and smaller amounts of cyanidin 3-*O*-glucoside (Pasqua et al. 2005). Optimum production was achieved using a mix of 2,4-D and kinetin rather than other hormones such as NAA and BAP. Increasing auxin levels reduced pigment accumulation. Anthocyanin production was dependent on light and was enhanced by high sucrose concentrations. Pigment accumulation was associated with the active growth phase of the cell line.

5.9.6 *Catharanthus roseus*

Callus and cell suspension cultures of *Catharanthus roseus* derived from hypocotyl tissue produce anthocyanins only when exposed to continuous light (Hall and Yeoman 1982; Knobloch et al. 1982). The anthocyanins present in both the callus tissue and cell suspension are the same as those found in the petal tissue: petunidin, malvidin and hirsutidin. However, as with other cell lines reported here, the level of derivitisation in the cell lines is less than in the parental tissue. Hirsutidin is the major pigment in the petal tissue, whereas the less methylated compounds petunidin and malvidin dominate in the cell lines (Knobloch et al. 1982; Hall and Yeoman 1986a).

This cell line responds to alterations in media content in a similar manner to many of the other cell lines described in this review. Reductions in inorganic nitrogen and phosphate, and increases in sucrose concentration all increase the production of anthocyanins and reduce growth rate at the same time (Knobloch et al. 1982; Hall and Yeoman 1986a, 1986b). The increase in anthocyanin yield under these conditions is due to the increase in cell size rather than an increase in intracellular anthocyanin concentration. Different growth conditions are able to alter the portion of coloured cells but only within a limited range (5–20%). Hall and Yeoman (1986b) suggest that there is a regulatory mechanism that limits the upper concentration of anthocyanins. Moreover, the constant portion of cells able to accumulate anthocyanins within the culture also indicates that this characteristic is actively regulated.

Exposure to continuous light was able to increase intracellular pigment concentration and the portion of coloured cells, but only to the upper limit described above (Hall and Yeoman 1986b). Accumulation of anthocyanins only occurred in the latter part of the cell cycle and when phosphate and nitrogen levels were reduced (Hall and Yeoman 1982). This is very similar to other cell lines discussed in this review (e.g carrot, strawberry and grape).

More recently a tumour cell line derived from *C. roseus* infected with *Agrobacterium* has been generated (Godoy-Hernandez and Loyola-Vargas 1997). This cell line is capable of rapid growth without the aid of growth regulators and produced anthocyanins under continuous light. Addition of acetylsalicylic acid (ASA) resulted in a dramatic increase (15-fold) in anthocyanin production. Addition of fungal elicitors and *trans*-cinnamic acid increased production by 3-fold. The mechanism of action of ASA on anthocyanin biosynthesis is unknown.

5.9.7 *Centaurea cyanus*

Stem tissue from the blue flowered *Centaurea cyanus* was used as the source tissue for the generation of a callus and suspension cell line that produces anthocyanins upon irradiation with UV light (Kakegawa et al. 1987). Maximum anthocyanin production was achieved by irradiating the cells at mid log phase, with anthocyanin production peaking at the start of the stationary phase. This shows that in this cell line anthocyanin production is associated with the active growth phase, not the stationary phase as in a number of other cell lines. The main anthocyanin produced was cyanidin 3-*O*- (6″-malonyl) glucoside, which is different from the anthocyanin in the blue flowers, which is cyanidin 3-*O*-(6″-succinyl glucoside)-5-*O*-glucoside, but the same as the anthocyanin produced in the stems after cold induction (Kakegawa et al. 1987). Anthocyanin biosynthesis was correlated with CHS activity, indicating that CHS is a primary site for the regulation of anthocyanin biosynthesis in this system (Kakegawa et al. 1991).

5.9.8 *Euphorbia millii*

An anthocyanin-producing cell line from *Euphorbia millii* was isolated from young leaves and the predominant anthocyanin was a cyanidin *O*-glycoside (Yamamoto et al. 1981). Successive subcultures and selection for pigmented cells yielded a stable anthocyanin producing cell line with a 3-fold improvement in anthocyanin yield (Yamamoto and Mizuguchi 1982). This cell line and its method of selection were used to develop a computer program for the tracing of cell line pedigrees (Yamamoto et al. 1983). The cell lines are maintained on a mix of 2,4-D and BA. Variations in the concentration of BA had little effect on anthocyanin production, whereas increasing levels of 2,4-D reduced anthocyanin output in a similar manner to many other cell lines (Yamamoto et al. 1989). Elevated sucrose levels improved anthocyanin production and cell growth with an optimal level of 5%. Analysis of growth conditions for this cell line showed that Gamborg's salts produced the highest anthocyanin yield compared with MS, LS, NN, and HE formulations. The ammonium level in Gamborgs media is significantly lower than in MS. Elevating ammonium significantly reduced anthocyanin production. In addition, the researchers showed that the most critical parameter for maximum anthocyanin production was the ammonium/nitrate ratio, with an optimum value of 1/16 (Yamamoto et al. 1989). The principal drawback in using these cells for large-scale production is their fragility. When transferred to a stirred bioreactor system the cell line showed significant reduction in growth rate and anthocyanin production (Yamamoto et al. 1989).

5.9.9 *Fagopyrum esculentum* (Buckwheat)

Cell cultures derived from hypocotyl segments produced anthocyanins only in the presence of light (Moumou et al. 1992). Increasing concentrations of 2,4-D increased cell growth rate but had no effect on anthocyanin production. Elevated sucrose concentrations reduced anthocyanin production. Under optimal conditions anthocyanin biosynthesis coincided with the period of maximum cell division, indicating that in this system secondary metabolism is not suppressed by cell division (Moumou et al. 1992).

5.9.10 *Glehnia littoralis*

Glehnia littoralis is a perennial herb grown in Japan, Korea and China. The young buds are edible and the roots and rhizomes are used in traditional medicine. A cell line was established from fast-growing and colourless petiole tissue and a dark-grown coloured version was isolated from small coloured loci (Kitamura 1998; Miura et al. 1998). Unlike many other cell lines derived from colourless parents, the growth rate of the coloured line was no less than the parental line both in solid and liquid culture. Analysis of the pigment content of this line showed the presence of five major anthocyanins, with cyanidin 3-*O*-(6-*O*-(6-*O*-E-feruloyl-β-D-glucopyranosyl)-2-*O*-β-D-xylopyranosyl-β-D-glucopyranside) comprising 62% of the total (Kitamura 1998; Miura et al. 1998). This compound has a high level of radical scavenging activity.

The initial cell line was maintained on 2,4-D and kinetin. Analysis of other phytohormone combinations showed that the optimum mix was NAA/kinetin for both anthocyanin production and cell growth. Different concentrations of NAA had little effect on growth rate but a significant effect on anthocyanin production, with 1 mg/L being the optimum level (Kitamura 1998). Analysis of carbon source showed that optimum growth and anthocyanin production was achieved using sucrose at 3%, followed by glucose, mannose and galactose (Kitamura 1998; Miura et al. 1998). Reduction of potassium levels did not significantly improve anthocyanin production and slowed growth rate.

The colourless cell line from the same source can produce furanocoumarin using fungal and yeast elicitors (Kitamura 1998). Biosynthesis of this compound was shown to be associated with an increase in PAL activity. Interestingly, the anthocyanin producing cell line (which also requires PAL activity) was unable to produce this compound. It is not known why the cell line is incapable of producing both products, which are derived from the same pathway. Treatment of the white and coloured cell lines with the radical generators H_2O_2 and AAPH showed quite different patterns of response. The coloured cell line showed a drop in PAL activity and concomitant drop in anthocyanin production, whereas the white cell line showed an increase in PAL activity, resulting in enhanced production of bergapten (Kitamura et al. 2002). This result further illustrates that PAL is under the control of very different regulatory mechanisms in the two cell lines. Treatment with yeast extracts also results in different responses from the two cell lines (Ishikawa et al. 2005). In the white line, PAL transcription and enzyme activity rise after treatment and are accompanied by the formation of umbelliferone, a precursor of furanocoumarin. No CHS activity was detected with this treatment. In the coloured line PAL and CHS transcription and activity are temporarily suppressed following elicitor treatment, but CHS suppression is more pronounced and is accompanied by an increase in umbelliferone levels (Ishikawa et al. 2005). The major difference between the two cell lines is the activity of CHS, which is absent in the white line and strongly expressed in the coloured line, indicating that this is a critical step in the ability to produce these pigments. The differences in the behaviour of PAL may be due to the differential response of genes encoding different isoforms. Initial analysis of PAL and CHS gene expression indicates the presence of up to five PAL genes and three CHS genes (Ishikawa et al.

2005). Since the coloured cell line can produce both anthocyanins and umbelliferone it provides a good model system for the study of the regulation of phenylpropanoid metabolism.

5.9.11 *Haplopappus gracilis*

Callus tissue proliferated rapidly in the dark without the aid of phytohormones. When this callus was exposed to light, growth was inhibited and anthocyanin biosynthesis was stimulated (Stickland and Sunderland 1972). Taking into account the growth rate and rate of pigment production, the researchers found that callus grown in the dark to stationary phase then exposed to light gave the greatest pigment yield. Two anthocyanin species were produced, cyanidin 3-*O*-glucoside and cyanidin 3-*O*-rutinoside. Addition of auxins inhibited both growth rate and anthocyanin production.

5.9.12 *Hibiscus sabdariffa* (Roselle)

Pigmented callus derived from seedling tissue contained two major pigments, cyanidin 3-*O*-xylosylglucoside and cyanidin 3-*O*-glucoside, whereas the host plant contains both cyanidin- and delphinidin-based pigments (Mizukami et al. 1988). 2,4-D was the best auxin for both enhancement of growth and pigment production. However, further elevated levels of 2,4-D caused a drop in both growth rate and anthocyanin yield. The addition of kinetin further enhanced both growth rate and pigment production, whereas GA_3 actively inhibited pigment formation without significantly affecting growth rate.

5.9.13 *Hyoscyamus muticus*

Anthocyanin producing callus was derived from cotyledonary and hypocotyl leaf pieces. Optimum pigmentation was achieved using a mix of 2,4-D and BAP (Basu and Chand 1996). Colour formation increased in the latter stages of the growth cycle and the predominant pigment was cyanidin-based. Light was required for colour formation. At the peak of colour production, 40–60% of the cells were pigmented.

5.9.14 *Leontopodium alpinum* (Edelweiss)

Callus, suspension and hairy root cultures (from infection with *A. rhizogenes*) were established from edelweiss (Comey et al. 1992; Hook 1994). All the cultures were able to produce anthocyanins but lost this ability with repeated subculturing. The addition of BAP increased anthocyanin production but reduced growth rate. Light was required for anthocyanin production in all three of the cell culture systems.

5.9.15 *Matthiola incana*

Anthocyanin-producing callus lines were established from 10 cultivars of *Matthiola incana* (Leweke and Forkmann 1982). The cyanidin-based pigments in the callus lines were similar to the flower pigments of the parental cultivar but without the acylation patterns. In addition, callus generated from the two white cultivars were incapable of producing anthocyanins under all conditions tested, suggesting genetic blocks in the anthocyanin biosynthetic pathway. Optimum pigment production was achieved using low levels of both 2,4-D and NAA together.

5.9.16 *Oxalis* sp.

Oxalis reclinata has green vegetative organs and only lightly coloured flowers (pink). In spite of the apparent lack of colour in the parent plant, a highly coloured callus line has been isolated from stem internode sections (Crouch et al. 1993). The major pigment in the callus line was cyanidin 3-*O*-glucosdie. The addition of auxins to the medium increased the growth rate, in particular with 2,4-D, but decreased the amount of pigment (Makunga et al. 1997). The addition of cytokinins decreased growth rate and stimulated pigment production. Anthocyanin-producing cell cultures of *O. linearis* showed a similar response to auxins and cytokinins (Meyer and van Staden 1995). Anthocyanin production is dependent on light (Makunga et al. 1997). Increasing sucrose concentrations also led to a decrease in growth rate and increase in pigment production (Meyer and van Staden 1995). Loss of anthocyanin production was associated with the onset of organogenesis (Crouch et al. 1993).

5.9.17 *Penstemon serrulatus*

An anthocyanin-producing cell suspension culture was isolated from a root explant after several cycles of selection for pigmented cells (Skrzypek et al. 1993). Anthocyanin production only occurred in the light and on NAA/BAP medium. The use of 2,4-D and kinetin halted pigment production. Two major pigments were produced, delphinidin 3-*O*-glucoside and peonidin 3-*O*-glucoside.

5.9.18 *Petunia hybrida*

Anthocyanin biosynthesis in petunia cell cultures was first reported by Colijn et al. (1981), who showed that the pigments (petunidin and malvidin) were identical to those found in the flowers of the host plant. A significant improvement in the level of anthocyanin production was achieved by subjecting the cells to dilution stress and reduced auxin levels (Hagendoorn et al. 1991a). However, a prolonged reduction in auxin level resulted in a loss of growth rate. PAL activity can be stimulated by a reduction in the activity of the plasma membrane H^+ATPase, causing a decrease in the proton gradient and a lowering of the cytoplasmic pH (Hagendoorn et al. 1991b). Auxins increase plasma membrane H^+ATPase activity, which matches the observation that reduced auxin stimulates anthocyanin accumulation (Hagendoorn et al. 1991b). However, further studies indicate that the reduction in cytoplasmic pH may not be part of the induction process leading to secondary metabolite production (Hagendoorn et al. 1994). Rather, it may be a function of secondary metabolite production, perhaps as a result of the transport of the metabolites into other cell compartments such as the vacuole.

5.9.19 *Plantanus acerifolia* (Plane Tree)

Callus and cell suspension cultures isolated from *Plantanus acerifolia* contained approximately 70% pigmented cells (Alami and Clerivet 2000). Anthocyanin production was associated with the exponential growth phase and only a single anthocyanin species, cyanidin 3-*O*-glucoside, was produced.

5.9.20 *Populus* (Poplar)

Anthocyanin production in poplar cell cultures has been studied for a considerable time. Callus and cell suspension cultures were first established from buds of *Populus* hybrids and grew rapidly in the dark, with anthocyanin production being induced by exposure to light (Matsumoto et al. 1973). NAA and IAA were more effective than 2,4-D in increasing anthocyanin production. Higher levels of 2,4-D and kinetin concentrations reduced anthocyanin accumulation. Anthocyanin formation and cell growth, in the presence of light, were enhanced by the addition of riboflavin and elevated sucrose (5%). None of these treatments was able to induce anthocyanin formation without the aid of light, indicating that light is an absolute requirement for pigment production (Matsumoto et al. 1973; Choi and Park 1997; Verma et al. 2000). Analysis of light frequencies showed that the blue light component was better at inducing anthocyanin than red, green and white light. In contrast, cell growth was most stimulated by red and green light.

Osmotic stress from the addition of mannitol causes a reduction in poplar cell growth rate and an increase in anthocyanin formation (Tholakalabavi et al. 1997). In conditions that induce anthocyanin formation, PAL activity is observed and precedes the accumulation of pigment (Tholakalabavi et al. 1997; Verma et al. 2000). Choi and Park (1997) found that pigmentation over successive subcultures was unstable and that constant selection for growth rate and anthocyanin production was needed to maintain the cell line. In addition, they observed that the growth rate was inversely related to anthocyanin production – a common feature of pigmented cell lines.

5.9.21 *Prunus cerasus* (Sour Cherry)

Cell lines were established from leaf petioles grown *in vitro* and roots from a cherry rootstock. The cell lines only produced pigment in the presence of light, sucrose and low concentrations of nitrates. Addition of urea (and hence production of ammonia) was able to mitigate the inhibitory effects of elevated nitrate levels (Durzan et al. 1991). The researchers used a novel technique to assay for anthocyanin production in cell suspensions in a non-destructive manner. Samples of cell suspension were aliquoted into 2 mL of media in 12-well plates. After each treatment, colour production was analysed by tri-stimulus reflectance colorimetry. Using this technique the researchers were able to analyse a large number of treatments within a single experiment.

Sour cherry cell cultures have also been established from leaf pieces grown *in vitro* (Blando et al. 2004). Cultures were maintained in the dark and assayed for anthocyanin production after transfer to light on media optimised for anthocyanin biosynthesis. Light is an absolute requirement for anthocyanin production in sour cherry. As a result the callus cultures were only capable of producing anthocyanins on the surface. Total anthocyanin production was significantly less than the parental fruit tissue. The anthocyanin profile in the parental fruit tissue included: cyanidin 3-*O*-glucosylrutinoside (the main pigment), cyanidin 3-*O*-rutinoside, cyanidin 3-*O*-sophoroside and cyanidin 3-*O*-glucoside. This differed markedly from the callus tissue in which only cyanidin 3-*O*-glucoside and cyanidin 3-*O*-rutinoside were present. This effect is common to several cell lines where the anthocyanin profile contains less evolved anthocyanin types, indicating a shut-down of anthocyanin modification activities such as methylation and acylation. The induction of anthocyanins did not alter the morphology or growth rate of the cells (Blando et al. 2005). Analysis of CHS activity showed an increase under continuous light. The addition of JA to the system resulted in an earlier onset of anthocyanin biosynthesis and a higher level of accumulation.

5.9.22 *Rosa* sp. (Paul's Scarlet Rose)

Suspension cultures of Paul's Scarlet Rose accumulate anthocyanins after exposure to light. Pigment accumulation occurs at the onset of the stationary phase and is enhanced by increasing concentrations of 2,4-D. These conditions also significantly inhibit growth rate (Davies 1972). It is possible that the enhanced anthocyanin biosynthesis is the result of the increased availability of the carbon source late in the cell cycle.

5.9.23 *Rudbeckia hirta*

Rudbeckia hirta is closely related to the *Echinacea* species and is used in traditional medicines. Callus isolated from cotyledonary tissue produced anthocyanins in the light (Luczkiewcz and Cisowski 2001). Analysis of different media formulations showed that maximum cell growth was associated with lower yields of pigments. Under most media formulations tested, only the surface layer of cells was pigmented. However, when cultured in Miller's medium, pigmentation occurred throughout the callus mass, but growth rate was low. As a result the authors established a two-phase system to maximise anthocyanin yield. Callus was first grown under conditions to maximise cell mass (SH medium) then transferred to Miller's medium for pigment production. Pigment yields of up to 5% DW were achieved, compared with 0.28% in the parental flowers. Analysis of phytohormones showed that a combination of NAA and zeatin provided the best pigment yield. Supplementation with amino acids showed that cysteine and phenylalanine both stimulated anthocyanin biosynthesis, with cysteine giving the greatest response. This is unusual as cysteine, unlike phenylalanine, is not a direct precursor of anthocyanin biosynthesis. Naringenin was the only anthocyanin precursor to enhance pigment production. The response to different carbon sources was similar to other cell lines. Sucrose in elevated concentrations (6%) enhanced pigmentation, probably due to osmotic stress. Glucose improved pigmentation, whereas fructose reduced it. Coconut water also enhanced pigmentation. Of all the components tested, the addition of cysteine enhanced pigmentation the most. The two main anthocyanins in the callus were cyanidin-3-O-(6-O-malonyl-β-D-glucopyranoside) and cyanidin-3-O-β-D-glucopyranoside.

5.9.24 *Solanum tuberosum* (Potato)

Highly coloured potato cell lines have been isolated from gamma-irradiated seedling tissue. The initial callus from this tissue was colourless with pigmented lines selected from randomly arising coloured loci (Zubko et al. 1992, 1993). Gamma irradiation significantly increased the frequency of colour breaks in the primary callus tissue. Analysis of PAL gene expression showed that in the coloured lines PAL is expressed constitutively, whereas it is silent in the colourless lines. The cell lines retained the ability to revert between the coloured or colourless states in both directions, indicating that the cell state required to produce pigments was not stable. The highly pigmented lines appeared to be insensitive to high concentrations of 2,4-D. In most other pigmented cell lines, high levels of 2,4-D enhance growth rate but inhibit anthocyanin accumulation. These pigmented potato lines show no reduction in anthocyanin biosynthesis in 2,4-D concentrations up to 3 mg/L (Zubko et al. 1992, 1993). In addition, the pigmented lines showed no reduction in growth rate compared with the original colourless lines. This is in contrast to other potato cell lines that show a significant reduction in growth rate upon selection for pigment production (S Deroles, unpublished data). The major pigments found in three of the coloured lines were peonidin and cyanidin, with the peonidin pigment identified as peonidin 3-O-[6-O-(4-O-E-p-coumaroyl-α-rhamnosyl)-β-glucoside]-5-O-β-glucoside. Pigmented potato lines established in our laboratories contain malvidin as the major pigment and reflect the pigment content of the parental tissue (S Deroles, unpublished data).

5.9.25 *Taraxacum officinale* (Dandelion)

Repeated selection for coloured loci in dandelion callus cultures has produced a stable highly pigmented cell line (Akashi et al. 1997). This cell line is only able to generate pigments in the presence of light and loss of pigmentation after dark culture is reversible. The main pigment is cyanidin 3-O-(6″-malonylglucoside). The accumulation of pigment after light irradiation is closely associated with an increase in CHS activity.

5.9.26 *Zea mays* (Maize)

Black Mexican Sweet (BMS) suspension cells do not accumulate anthocyanins due to the lack of expression of the appropriate anthocyanin regulatory genes. Transgenic lines carrying transgenes for the maize transcription factors R (bHLH) and C1 (MYB) under the 35SCaMV promoter can produce large amounts of anthocyanins in the dark (Grotewold et al. 1998). In addition, the transgenic line also responds to continuous light with increased anthocyanin biosynthesis. The anthocyanins produced were cyanidin O-glycosides containing arabinose residues. This is in contrast to the parental tissue in which the cyanidin pigments are glycosylated with glucose molecules. The induction of anthocyanin biosynthesis through the use of anthocyanin regulatory genes highlights a useful technique for the development of cell cultures as biofactories for the production of secondary metabolites. In all the cell lines discussed in this chapter, no commercially viable production has been achieved primarily due to the failure to increase intracellular anthocyanin concentrations to a high enough level without compromising growth rate. The use of anthocyanin regulatory genes offers a promising way to overcome this problem.

5.10 Conclusions

It is clear from the information available that anthocyanin production is controlled by several common factors and that to a large extent the response to these factors is similar across different plant species.

Anthocyanin production in plant cell cultures is the result of the activation of regulatory systems that control anthocyanin biosynthesis under defined conditions in the intact plant. Based on the results common to most of the cell systems described above, anthocyanin biosynthesis in these cultures is activated by regulatory systems associated with light stimulus and/or the stress response. Most of the cell lines (22 out of 34) described here are unable to make anthocyanins without the aid of light. In these cultures the response to other stimuli can improve the yield of anthocyanins but none overcame the primary need for light. However, in some carrot cell cultures the light response is dependent on the hormone regime (low auxin/high cytokinin). The other class of cell lines does not require light, which indicates that the primary control of the biosynthetic pathway is via another mechanism or group of regulatory genes. Dark-grown cultures are often derived from light-grown lines. Dark-grown cultures also respond to other stimuli in a similar manner to the light-grown cell lines. Owing to the simple and relatively uniform nature of cell cultures compared with whole plants, they provide an ideal platform to explore the different regulatory systems and signal transduction pathways that control secondary metabolism. One common feature that is emerging is that the control of anthocyanin biosynthesis by light is via expression of the early biosynthetic genes, PAL and CHS. The specific regulatory genes acting on these targets in cell culture systems are not yet determined.

Phytohormone composition in the media is another factor that affects anthocyanin production. The optimum composition involves the ratio of auxin to cytokinin as well as the absolute values and types of hormone (e.g. 2,4-D or NAA). Owing to the variability in response across different plant species, it is difficult to settle on an optimum phytohormone composition for the production of anthocyanins in plant cell cultures. However, there are some general trends. Anthocyanin production is inversely related to cell growth in most systems, so phytohormone levels that promote growth often result in reduced anthocyanin production. In general, low concentrations of auxins and cytokinins, with more auxin than cytokinin, are best for anthocyanin production. A popular combination is 2,4-D and kinetin. However, a number of cell lines respond better to NAA or IAA as the auxin, due to their reduced ability to accelerate cell growth. Some cell lines (e.g. strawberry) are able to grow on hormone-free media. This makes extraction of the pigment easier due to the lack of contamination by potentially toxic phytohormones (e.g. 2,4-D). Of course, the effect of phytohormone combinations is often interconnected with other growth conditions such as the presence of light and nutrient supply. High auxin levels increase the transmembrane proton gradient through up-regulation of H^+ATPase activity. Anthocyanin production is associated with low proton gradients through regulation of PAL expression in petunia cell cultures, thus indicating a mechanism of action between auxin levels and anthocyanin production. Optimisation for maximum anthocyanin yield using phytohormone regimes is problematic owing to the correlation between cell growth rate and anthocyanin production. In most cell lines the optimum level is a compromise between these two factors. This highlights the need to uncouple these processes through altered regulation of the anthocyanin biosynthetic pathway in order to generate commercially viable cell lines for colour production.

Most of the cell lines described increase anthocyanin biosynthesis in response to a reduction in nitrogen, alteration in the ratio of ammonia/nitrate and a reduction in phosphate. All these treatments reduce the growth rate of the cell line. The exact mechanism controlling this effect is unknown, but it does involve activation of PAL and/or CHS. A suggestion from several authors is that a slowing of the cell growth rate alters the balance between primary and secondary metabolism. At a fast growth rate, substrates such as phenylalanine are used by primary metabolic processes, leaving little available for secondary metabolism. Once growth rate is slowed (e.g. via low nitrogen) more substrate becomes available and an increase in anthocyanin biosynthesis is observed. Feeding experiments with phenylalanine support this theory in some cases. However, it is still possible that stimulation of anthocyanin biosynthesis under these conditions is controlled by an active regulatory process, as well as by a passive response to substrate supply.

The response to increased carbon supply seems to be mostly a response to increased osmotic pressure, a subsequent reduction in cell growth rate and (like conditions of reduced nitrogen etc.) a stimulation of anthocyanin biosynthesis. This response could again be a result of the shift in substrate supply and/or activation of a regulatory mechanism that up-regulates PAL and CHS.

In general the pigments in cell cultures are similar to those found in the parent plants but often are less substituted though loss of acylation, methylation and in some cases F3′H/F3′5′H activity. The final pigment type can be altered through the feeding of alternative organic acids, thus altering the acylation pattern through substrate supply.

Most treatments that stimulate anthocyanin biosynthesis also reduce cell growth rate. This is a fundamental problem in developing a commercially viable culture system for the production of anthocyanins; since any gain in anthocyanin production (per g FW), is offset by a reduction in growth rate. A second significant drawback for many of these cell lines is the need for light stimulation. This is very difficult to achieve in large-scale systems. Even in cell lines capable of anthocyanin production in the dark, an association remains between the ability to make anthocyanins and cell differentiation state. The constant conflict between cell growth rate and the ability to produce pigments is the main reason why no cell lines have successfully been brought into commercial production, even though there is an increasing demand to establish such systems for the reliable supply of custom made anthocyanin-based pigments for the processed food industry.

The results for these cell lines show that a likely way to achieve commercially viable production of anthocyanin pigments is to control the regulation of anthocyanin biosynthesis, in particular the regulation of PAL and CHS. We have taken the first step towards this goal through the enhancement of anthocyanin biosynthesis via overexpression of an anthocyanin regulatory gene (S Deroles, unpublished data). The transgenic cell line is capable of high levels of anthocyanin production and maintains the high growth rates of its uncoloured parent line. Only by controlling the regulatory process, either by genetic modification or random/induced mutation, will it be possible to produce pigment by cell culture at a commercial scale.

References

Afifi, M., El-Kereamy, A., Legrand, V., Chervin, C., Monje, M.C., Nepveu, F. and Roustan, J.P. (2003) Control of anthocyanin biosynthesis pathway gene expression by eutypine, a toxin from *Eutypa lata*, in grape cell tissue cultures. J. Plant Physiol. 160, 971–975.

Akashi, T., Saito, N., Hirota, H. and Ayabe, S.I. (1997) Anthocyanin-producing dandelion callus as a chalcone synthase source in recombinant polyketide reductase assay. Phytochemistry 46, 283–287.

Alami, I. and Clerivet, A. (2000) Cyanidin 3-glucoside accumulation in plane tree (*Platanus acerifolia*) cell-suspension cultures. Biotech. Lett. 22, 87–89.

Andersen, O.M. and Jordhein, M. (2006) The anthocyanins. In: O.M. Andersen and K.R. Markham (Eds.). *Flavonoids: Chemistry, Biochemistry and Applications*. London, CRC Press: 471–552.

Asada, Y., Sakamoto, K. and Furuya, T. (1994) A minor anthocyanin from cultured cells of *Aralia cordata*. Phytochemistry 35, 1471–1473.

Asano, S., Ohtsubo, S., Nakajima, M., Kusunoki, M., Kaneko, K., Katayama, H. and Nawa, Y. (2002) Production of anthocyanins by habituated cultured cells of Nyoho strawberry (*Fragaria ananassa* Duch.). Food Sci. Tech. Res. 8, 64–69.

Aumont, V., Larronde, F., Richard, T., Budzinski, H., Decendit, A., Deffieux, G., Krisa, S. and Mérillon, J.M. (2004) Production of highly ^{13}C-labeled polyphenols in *Vitis vinifera* cell bioreactor cultures. J. Biotech. 109, 287–294.

Bailly, C., Cormier, F. and Do, C.B. (1997) Characterization and activities of S-adenosyl-l-methionine:cyanidin 3-giucoside 3'-O-methyltransferase in relation to anthocyanin accumulation in *Vitis vinifera* cell suspension cultures. Plant Sci. 122, 81–89.

Baker, D.C., Dougall, D.K., Gläßgen, W.E., Johnson, S.C., Metzger, J.W., Rose, A. and Seitz, H.U. (1994) Effects of supplied cinnamic acids and biosynthetic intermediates on the anthocyanins accumulated by wild carrot suspension cultures. Plant Cell Tissue Organ Cult. 39, 79–91.

Basu, P. and Chand, S. (1996) Anthocyanin accumulation in *Hyoscyamus muticus* L. tissue cultures. J. Biotech. 52, 151–159.

Blando, F., Gerardi, C. and Nicoletti, I. (2004) Sour cherry (*Prunus cerasus* L) anthocyanins as ingredients for functional foods. J. Biomed. Biotech. 2004, 253–258.

Blando, F., Scardino, A.P., De Bellis, L., Nicoletti, I. and Giovinazzo, G. (2005) Characterization of *in vitro* anthocyanin-producing sour cherry (*Prunus cerasus* L.) callus cultures. Food Res. Int. 38, 937–942.

Bridle, P. and Timberlake, C.F. (1996) Anthocyanins as natural food colourants – selected aspects. Food Chem. 58, 103–109.

Calderon, A.A., Pedreno, M.A., Munoz, R. and Ros Barcelo, A. (1993) Evidence for non-vacuolar localization of anthocyanoplasts (anthocyanin-containing vesicles) in suspension cultured grapevine cells. Phyton 54, 91–98.

Callebaut, A., Decleire, M. and Vandermeiren, K. (1993). I *Ajuga repens* (Bugle): *in vitro* production of anthocyanins. In: Y.P.S. Bajaj (Ed.) *Biotechnology in Agriculture and Forestry*. Berlin, Heidelberg, Springer-Verlag, 24: 1–22.

Callebaut, A., Hendrickx, G., Voets, A.M. and Motte, J.C. (1990a) Anthocyanins in cell cultures of *Ajuga reptans*. Phytochemistry 29, 2153–2158.

Callebaut, A., Hendrickx, G., Voets, A.M. and Perwez, C. (1988) Production of anthocyanins by cell cultures of *Ajuga reptans*. Mededelingen van de Faculteit Landbouwwetenschappen, Rijksuniversiteit Gent 53, 1713–1715.

Callebaut, A., Terahara, N., De Haan, M. and Decleire, M. (1997) Stability of anthocyanin composition in *Ajuga reptans* callus and cell suspension cultures. Plant Cell Tissue Organ Cult. 50, 195–201.

Callebaut, A., Terahara, N. and Decleire, M. (1996) Anthocyanin acyltransferases in cell cultures of *Ajuga reptans*. Plant Sci. 118, 109–118.

Callebaut, A., Voets, A.M. and Motte, J.C. (1990b) Anthocyanin production by plant cell cultures on media based on milk whey. Biotech. Lett. 12, 215–218.

Ceoldo, S., Levi, M., Marconi, A.M., Baldan, G., Giarola, M. and Guzzo, F. (2005) Image analysis and in vivo imaging as tools for investigation of productivity dynamics in antho-cyanin-producing cell cultures of *Daucus carota*. New Phytologist 166, 339–352.

Choi, M. and Park, Y. (1997) Selection of a high anthocyanin-producing cell line from callus cultures of hybrid poplar (*Populus alba* L. × *P. glandulosa* Uyeki). Forest Gen. 4, 253–257.

Colijn, C.M., Jonsson, L.M.V., Schram, A.W. and Kool, A.J. (1981) Synthesis of malvidin and petunidin in pigmented tissue cultures of *Petunia hybrida*. Protoplasma 107, 63–68.

Comey, N., Hook, I. and Sheridan, H. (1992) Enhancement of anthocyanin production in cell cultures and hairy roots of *Leontopodium alpinum*. Planta Medica 58 A605–A606.

Conn, S., Zhang, W. and Franco, C. (2003) Anthocyanic vacuolar inclusions (AVIs) selectively bind acylated anthocyanins in *Vitis vinifera* L. (grapevine) suspension culture. Biotech. Lett. 25, 835–839.

Cormier, F., Couture, R., Do, C.B., Pham, T.Q. and Tong, V.H. (1997) Properties of anthocyanins from grape cell culture. J. Food Sci. 62, 246–248.

Cormier, F., Crevier, H.A. and Do, C.B. (1990) Effects of sucrose concentration on the accumulation of anthocyanins in grape (*Vitis vinifera*) cell suspension. Canadian J. Bot. 68, 1822–1825.

Cormier, F., Do, C.B. and Moresoli, C. (1992) Anthocyanin release from grape (*Vitis vinefera* L.) cell suspension. Biotech. Lett. 14, 1029–1034.

Cormier, F., Do, C.B. and Nicolas, Y. (1994) Anthocyanin production in selected cell lines of grape (*Vitis vinifera* L.). In Vitro Cell. Dev. Biol. Plant 30, 171–173.

Crouch, N.R., Staden, L.F.v., Staden, J.v., Drewes, F.E., Drewes, S.E. and Meyer, H.J. (1993) Accumulation of cyanidin-3-glucoside in callus and cell cultures of *Oxalis reclinata*. J. Plant Physiol. 142, 109–111.

Curtin, C., Zhang, W. and Franco, C. (2003) Manipulating anthocyanin composition in *Vitis vinifera* suspension cultures by elicitation with jasmonic acid and light irradiation. Biotech. Lett. 25, 1131–1135.

Davies, K.M. (Ed.) (2004) Plant pigments and their manipulation. *Annual Plant Reviews*. Blackwell, Oxford, UK.

Davies, K.M. and Schwinn, K.E. (2006) Molecullar biology and biotechnology of flavonoid biosynthesis. In: O.M. Andersen and K.R. Markham (Eds.) *Flavonoids: Chemistry, Biochemistry and Applications*. London, CRC Press: 143–218.

Davies, M.E. (1972) Polyphenol synthesis in cell suspension cultures of Paul's Scarlet rose. Planta 104, 50–65.

Decendit, A. and Mérillon, J.M. (1996) Condensed tannin and anthocyanin production in *Vitis vinifera* cell suspension cultures. Plant Cell Rep. 15, 762–765.

Decendit, A., Ramawat, K.G., Waffo, P., Deffieux, G., Badoc, A. and Mérillon, J.M. (1996) Anthocyanins, catechins, condensed tannins and piceid production in *Vitis vinifera* cell bioreactor cultures. Biotech. Lett. 18, 659–662.

Dedaldechamp, F. and Uhel, C. (1999) Induction of anthocyanin synthesis in nonpigmented grape cell suspensions by acting on DFR substrate availability or precursors level. Enz. Microb. Tech. 25, 316–321.

Dedaldechamp, F., Uhel, C. and Macheix, J.J. (1995) Enhancement of anthocyanin synthesis and dihydroflavonol reductase (DFR) activity in response to phosphate deprivation in grape cell suspensions. Phytochemistry 40, 1357–1360.

Delgado-Vargas, F., Jimenez, A.R. and Paredes-Lopez, O. (2000) Natural pigments: Carotenoids, anthocyanins, and betalains - characteristics, biosynthesis, processing, and stability. Crit. Rev. Food Sci. Nut. 40, 173–289.

Deroles, S., Smith, M.A.L. and Lee, C. (2002) Factors affecting transformation of cell cultures from three dicotyledonous pigment-producing species using microprojectile bombardment. Plant Cell Tissue Organ Cult. 70, 69–76.

Do, C.B. and Cormier, F. (1990) Accumulation of anthocyanins enhanced by a high osmotic potential in grape (*Vitis vinifera* L.) cell suspensions. Plant Cell Rep. 9, 143–146.

Do, C.B. and Cormier, F. (1991a) Accumulation of peonidin 3-glucoside enhanced by osmotic stress in grape (*Vitis vinifera* L.) cell suspension. Plant Cell Tissue Organ Cult. 24, 49–54.

Do, C.B. and Cormier, F. (1991b) Effects of low nitrate and high sugar concentrations on anthocyanin content and composition of grape (*Vitis vinifera* L.) cell suspension. Plant Cell Rep. 9, 500–504.

Do, C.B. and Cormier, F. (1991c) Effects of high ammonium concentrations on growth and anthocyanin formation in grape (*Vitis vinifera* L.) cell suspension cultured in a production medium. Plant Cell Tissue Organ Cult. 27, 169–174.

Do, C.B., Cormier, F. and Nicolas, Y. (1995) Isolation and characterization of a UDP-glucose:cyanidin 3-*O*-glucosyltransferase from grape cell suspension cultures (*Vitis vinifera* L.). Plant Sci. 112, 43–51.

Dougall, D.K. (1989) Sinapic acid stimulator of anthocyanin accumulation in carrot cell cultures. Plant Sci. 60, 259–262.

Dougall, D.K., Baker, D.C., Gakh, E.G., Redus, M.A. and Whittemore, N.A. (1998) Studies on the stability and conformation of monoacylated anthocyanins part 2 – Anthocyanins from wild carrot suspension cultures acylated with supplied carboxylic acids. Carbohydrate Res. 310, 177–189.

Dougall, D.K. and Frazier, G.C. (1989) Nutrient utilization during biomass and anthocyanin accumulation in suspension cultures of wild carrot cells. Plant Cell Tissue Organ Cult. 18, 95–104.

Downham, A. and Collins, P. (2000) Colouring our foods in the last and next millennium. Int. J. Food Sci. Tech. 35, 5–22.

Durzan, D.J., Hansen, K. and Peng, C. (1991) Anthocyanin production in cell suspension cultures of *Prunus cerasus* cv. Vladimir. Adv. Hort. Sci. 5, 3–10.

Edahiro, J. and Seki, M. (2006) Phenylpropanoid metabolite supports cell aggregate formation in strawberry cell suspension culture. J. Biosci. Bioeng. 102, 8–13.

Edahiro, J., Yamada, M., Seike, S., Kakigi, Y., Miyanaga, K., Nakamura, M., Kanamori, T. and Seki, M. (2005b) Separation of cultured strawberry cells producing anthocyanins in aqueous two-phase system. J. Biosci. Bioeng. 100, 449–454.

Edahiro, J.I., Nakamura, M., Seki, M. and Furusaki, S. (2005a) Enhanced accumulation of anthocyanin in cultured strawberry cells by repetitive feeding of L-phenylalanine into the medium. J. Biosci. Bioeng. 99, 43–47.

Endress, R. (1994) Plant cells as producers of secondary compounds. In: R. Endress (Ed.) *Plant Cell Biotechnology.* Berlin, Springer-Verlag: 121–255.

Fang, Y., Smith, M.A.L. and Pépin, M.F. (1998) Benzyl adenine restores anthocyanin pigmentation in suspension cultures of wild *Vaccinium pahalae.* Plant Cell Tissue Organ Cult. 54, 113–122.

Fang, Y., Smith, M.A.L. and Pépin, M.F. (1999) Effects of exogenous methyl jasmonate in elicited anthocyanin-producing cell cultures of ohelo (*Vaccinium pahalae*). In Vitro Cell. Dev. Biol. Plant 35, 106–113.

Francis, F.J. (1989) Food colourants: Anthocyanins. Crit. Rev. Food Sci. Nut. 28, 273–314.

Frankel, E.N., Kanner, J., German, J.B., Parks, E. and Kinsella, J.E. (1993) Inhibition of oxidation of human low-density lipoprotein by phenolic substances in red wine. Lancet 341, 454–457.

Frankel, E.N., Waterhouse, A.L. and Teissedre, P.L. (1995) Principal phenolic phytochemicals in selected California wines and their antioxidant activity in inhibiting oxidation of human low-density lipoproteins. Ag. Food Chem. 43, 890–894.

Gläßgen, W.E., Rose, A., Madlung, J., Koch, W., Gleitz, J. and Seitz, H.U. (1998) Regulation of enzymes involved in anthocyanin biosynthesis in carrot cell cultures in response to treatment with ultraviolet light and fungal elicitors. Planta 204, 490–498.

Gläßgen, W.E., Seitz, H.U. and Metzger, J.W. (1992b) High-performance liquid chromatography/electrospray mass spectrometry and tandem mass spectrometry of anthocyanins from plant tissues and cell cultures of *Daucus carota* L. Biol. Mass Spec. 21, 271–277.

Gläßgen, W.E., Wray, V., Strack, D., Metzger, J.W. and Seitz, H.U. (1992a) Anthocyanins from cell suspension cultures of *Daucus carota.* Phytochemistry 31, 1593–1601.

Gleitz, J., Schnitzler, J.P., Steimle, D. and Seitz, H.U. (1991) Metabolic changes in carrot cells in response to simultaneous treatment with ultraviolet light and a fungal elicitor. Planta 184, 362–367.

Gleitz, J. and Seitz, H.U. (1989) Induction of chalcone synthase in cell suspension cultures of carrot (*Daucus carota* L. spp. *sativus*) by ultraviolet light: evidence for two different forms of chalcone synthase. Planta 179, 323–330.

Godoy-Hernandez, G. and Loyola-Vargas, V.M. (1997) Effect of acetylsalicylic acid on secondary metabolism of *Catharanthus roseus* tumor suspension cultures. Plant Cell Rep. 16, 287–290.

Gould, K.S. and Lister, C. (2006). Flavonoid functions in plants. In: O.M. Andersen and K.R. Markham (Eds.) *Flavonoids: Chemistry, Biochemistry and Applications.* London, CRC Press: 397–442.

Gronbaek, M., Deis, A., Sorensen, T., Becker, U., Schnor, P. and Jensen, G. (1995) Mortality associated with moderate intakes of wine, beer, or spirits. British Med. J. 310, 1165–1169.

Grotewold, E., Chamberlin, M., Snook, M., Siame, B., Butler, L., Swenson, J., Maddock, S., St. Clair, G. and Bowen, B. (1998) Engineering secondary metabolism in maize cells by ectopic expression of transcription factors. Plant Cell 10, 721–740.

Grusak, M.A., Rogers, R.B., Yousef, G.G., Erdman Jr, J.W. and Lila, M.A. (2004) An enclosed-chamber labeling system for the safe ^{14}C-enrichment of phytochemicals in plant cell suspension cultures. In Vitro Cell. Dev. Biol. Plant 40, 80–85.

Guardiola, J., Iborra, J.L. and Canovas, M. (1995) A model that links growth and secondary metabolite production in plant cell suspension cultures. Biotech. Bioeng. 46, 291–297.

Hagendoorn, M.J.M., Poortinga, A.M., Wong Fong Sang, H.W., van der Plas, L.H.W. and van Walraven, H.S. (1991b) Effect of elicitors on the plasmamembrane of *Petunia hybrida* cell suspensions. Plant Physiol. 96, 1261–1267.

Hagendoorn, M.J.M., Wagner, A.M., Segers, G., van der Plas, L.H.W., Oostdam, A. and van Walraven, H.S. (1994) Cytoplasmic acidification and secondary metabolite production in different plant cell suspensions. Plant Physiol. 106, 723–730.

Hagendoorn, M.J.M., Zethof, J.L.M., Hunnik, E.v. and Plas, L.H.W.v.d. (1991a) Regulation of anthocyanin and lignin synthesis in *Petunia hybrida* cell suspensions. Plant Cell Tissue Organ Cult. 27, 141–147.

Hall, R.D. and Yeoman, M.M. (1982) Anthocyanin production in cell cultures of *Catharanthus roseus*. In: A. Fujiwara (Ed.) *Plant Tissue Culture 1982*, Tokyo, Japanese Association for Plant Tissue Culture: 281–282.

Hall, R.D. and Yeoman, M.M. (1986a) Factors determining anthocyanin yield in cell cultures of *Catharanthus roseus* (L.)G. Don. New Phytologist 103, 33–43.

Hall, R.D. and Yeoman, M.M. (1986b) Temporal and spatial heterogeneity in the accumulation of anthocyanins in cell cultures of *Catharanthus roseus* (L.) G. Don. J. Exp. Bot. 37, 48–60.

Harborne, J.B., Mayer, A.M. and Bar-Nun, N. (1983) Identification of the major anthocyanin of carrot cells in tissue culture as cyanidin 3-(sinapoylxylosylglucosylgalactoside). Z. Naturforsch. 38c, 1055–1056.

Heinzmann, U. and Seitz, U. (1977) Synthesis of phenylalanine ammonia-lyase in anthocyanin-containing and anthocyanin-free callus cells of *Daucus carota* L. Planta 135, 63–67.

Hemingson, J.C. and Collins, R.P. (1982) Anthocyanins present in cell cultures of *Daucus carota*. J. Nat. Prod. 45, 385–389.

Hinderer, W., Petersen, M. and Seitz, H.U. (1984) Inhibition of flavonoid biosynthesis by gibberellic acid in cell suspension cultures of *Daucus carota* L. Planta 160, 544–549.

Hirasuna, T.J., Shuler, M.L., Lackney, V.K. and Spanswick, R.M. (1991) Enhanced anthocyanin production in grape cell cultures. Plant Sci. 78, 107–120.

Hirner, A.A., Veit, S. and Seitz, H.U. (2001) Regulation of anthocyanin biosynthesis in UV-A-irradiated cell cultures of carrot and in organs of intact carrot plants. Plant Sci. 161, 315–322.

Hiroaka, N., Kodama, T. and Tomita, K. (1986) Selection of *Bulpeurum falcatum* callus line producing anthocyanins in darkness. J. Nat. Prod. 49, 470–474.

Hirose, M., Yamakawa, T., Kodama, T. and Komamine, A. (1990) Accumulation of betacyanin in *Phytolacca americana* cells and of anthocyanin in *Vitis* sp. cells in relation to cell division in suspension cultures. Plant Cell Physiol. 31, 267–271.

Honda, H., Hiraoka, K., Nagamori, E., Omote, M., Kato, Y., Hiraoka, S., Kobayashi, T. (2002) Enhanced anthocyanin production from grape callus in an air-lift type bioreactor using a viscous additive-supplemented medium. J. Biosci. Bioeng. 94, 135–139.

Hong, Y.C., Read, P.E., Harlander, S.K. and Labuza, T.P. (1989) Development of a tissue culture system from immature strawberry fruits. J. Food Sci. 54, 388–392.

Hook, I. (1994) Secondary metabolites in hairy root cultures of *Leontopodium alpinum* Cass. (edelweiss). Plant Cell Tissue Organ Cult. 38, 321–326.

Hopp, W. and Seitz, H.U. (1987) The uptake of acylated anthocyanin into isolated vacuoles from a cell suspension culture of *Daucus carota*. Planta 170, 74–85.

How, F. and Smith, M.A. (2003) Effect of light/dark cycling on growth and anthocyanin production of *Ajuga reptans* in callus culture. J. Ag. Res. China 52, 291–296.

Iborra, J.L., Guardiola, J., Montaner, S., Canovas, M. and Manjon, A. (1994) Enhanced accumulation of anthocyanins in *Vitis vinifera* cells immobilized in polyurethane foam. Enz. Microb. Tech. 16, 416–419.

Ilan, A. and Dougall, D.K. (1992) The effect of growth retardants on anthocyanin production in carrot cell suspension cultures. Plant Cell Rep. 11, 304–309.

Ilan, A. and Dougall, D.K. (1994) Effects of gibberellic acid and uniconazole on the activities of some enzymes of anthocyanin biosynthesis in carrot cell cultures. J. Plant Growth Reg. 13, 213–220.

Ilan, A., Zanewich, K.P., Rood, S.B. and Dougall, D.K. (1994) Gibberellic acid decreases anthocyanin accumulation in wild carrot cell suspension cultures but does not alter 3'-nucleotidase activity. Physiol. Plant. 92, 47–52.

Ishikawa, A., Kitamura, Y., Ozeki, Y., Itoh, Y., Yamada, A. and Watanabe, M. (2005) Post-stress metabolism involves umbelliferone production in anthocyanin-producing and non-producing cells of *Glehnia littoralis* suspension cultures. J. Plant Physiol. 162, 703–710.

Jang, J.C., Leon, P., Zhou, L. and Sheen, J. (1997) Hexokinase as a sugar sensor in higher plants. Plant Cell 9, 5–19.

Kakegawa, K., Hattori, E., Koike, K. and Takeda, K. (1991) Induction of anthocyanin synthesis and related enzyme activities in cell cultures of *Centaurea cyanus* by UV-light irradiation. Phytochemistry 30, 2271–2273.

Kakegawa, K., Kaneko, Y., Hattori, E., Koike, K. and Takeda, K. (1987) Cell cultures of *Centaurea cyanus* produce malonated anthocyanin in UV light. Phytochemistry 26, 2261–2263.

Kakegawa, K., Suda, J., Sugiyama, M. and Komamine, A. (1995) Regulation of anthocyanin biosynthesis in cell suspension cultures of *Vitis* in relation to cell division. Physiol. Plant. 94, 661–666.

Kanabus, J., Bressen, R.A. and Carpita, N.C. (1986) Carbon assimilation in carrot cells in liquid culture. Plant Physiol. 82, 363–368.

Kinnersley, A.M. and Dougall, D.K. (1980) Increase in anthocyanin yield from wild-carrot cell cultures by a selection system based on cell-aggregate size. Planta 149, 200–204.

Kitamura, Y. (1998) The production of anthocyanin and furanocoumarin defense compounds by cultured cells of *Glehnia littoralis*. Recent Res. Dev. Phytochem. 2, 397–412.

Kitamura, Y., Ohta, M., Ikenaga, T. and Watanabe, M. (2002) Responses of anthocyanin-producing and non-producing cells of *Glehnia littoralis* to radical generators. Phytochemistry 59, 63–68.

Knobloch, K.H., Bast, G. and Berlin, J. (1982) Medium- and light-induced formation of serpentine and anthocyanins in cell suspension cultures of *Catharanthus roseus*. Phytochemistry 21, 591–594.

Kobayashi, Y., Akita, M., Sakamoto, K., Liu, H., Shigeoka, T., Koyano, T., Kawamura, M. and Furuya, T. (1993) Large-scale production of anthocyanin by *Aralia cordata* cell suspension cultures. App. Microb. Biotech. 40, 215–218.

Kokubo, T., Ambe-Ono, Y., Nakamura, M., Ishida, H., Yamakawa, T. and Kodama, T. (2001) Promotive effect of auxins on UDP-glucose: flavonol glucosyltransferase activity in *Vitis* sp cell cultures. J. Biosci. Bioeng. 91, 564–569.

Konczak, I., Okuno, S., Yoshimoto, M. and Yamakawa, O. (2004) Caffeoylquinic acids generated *in vitro* in a high-anthocyanin-accumulating sweet potato cell line. J. Biomed. Biotech. 2004, 287–292.

Konczak, I., Terahara, N., Yoshimoto, M., Nakatani, M., Yoshinaga, M. and Yamakawa, O. (2005) Regulating the composition of anthocyanins and phenolic acids in a sweetpotato cell culture towards production of polyphenolic complex with enhanced physiological activity. Trends Food Sci. Tech. 16, 377–388.

Konczak-Islam, I., Nakatani, M., Yoshinaga, M. and Yamakawa, O. (2001) Effect of ammonium ion and temperature on anthocyanin composition in sweet potato cell suspension culture. Plant Biotech. 18, 109–117.

Konczak-Islam, I., Okuno, S., Yoshimoto, M. and Yamakawa, O. (2003a) Composition of phenolics and anthocyanins in a sweet potato cell suspension culture. Biochem. Eng. J. 14, 155–161.

Konczak-Islam, I., Yoshimoto, M., Hou, D.X., Terahara, N. and Yamakawa, O. (2003b) Potential chemopreventive properties of anthocyanin-rich aqueous extracts from *in vitro* produced tissue of sweetpotato (*Ipomoea batatas* L.). J. Ag. Food Chem. 51(20), 5916–5922.

Konczak-Islam, I., Yoshinaga, M., Nakatani, M., Terahara, N. and Yamakawa, O. (2000) Establishment and characteristics of an anthocyanin-producing cell line from sweet potato storage root. Plant Cell Rep. 19, 472–477.

Krisa, S., Vitrac, X., Decendit, A., Larronde, F., Deffieux, G. and Mérillon, J.M. (1999a) Obtaining *Vitis vinifera* cell cultures producing higher amounts of malvidin-3-*O*-ß-glucoside. Biotech. Lett. 21, 497–500.

Krisa, S., Waffo Téguo, P., Decendit, A., Deffieux, G., Vercauteren, J. and Mérillon, J.M. (1999b) Production of ^{13}C-labelled anthocyanins by *Vitis vinifera* cell suspension cultures. Phytochemistry 51, 651–656.

Kurata, H., Mochizuki, A., Okuda, N., Seki, M. and Furusaki, S. (2000) Intermittent light irradiation with second- or hour-scale periods controls anthocyanin production by strawberry cells. Enz. Microb. Tech. 26, 621–629.

Larronde, F., Krisa, S., Decendit, A., Chèze, C., Deffieux, G. and Mérillon, J.M. (1998) Regulation of polyphenol production in *Vitis vinifera* cell suspension cultures by sugars. Plant Cell Rep. 17, 946–950.

Leweke, B. and Forkmann, G. (1982) Genetically controlled anthocyanin synthesis in cell cultures of *Matthiola incana*. Plant Cell Rep. 1, 98–100.

Lila, M.A. (2004) Anthocyanins and human health: an *in vitro* investigative approach. J. Biomed. Biotech. 2004, 306–313.

Luczkiewcz, M. and Cisowski, W. (2001) Optimisation of the second phase of a two phase growth system for anthocyanin accumulation in callus cultures of *Rudbeckia hirta*. Plant Cell Tissue Organ Cult. 65, 57–68.

Madhavi, D.L., Bomser, J., Smith, M.A.L. and Singletary, K. (1998) Isolation of bioactive constituents from *Vaccinium myrtillus* (bilberry) fruits and cell cultures. Plant Sci. 131, 95–103.

Madhavi, D.L., Juthangkoon, S., Lewen, K., Berber-Jimenez, M.D. and Smith, M.A.L. (1996) Characterization of anthocyanins from *Ajuga pyramidalis* metallica crispa cell cultures. J. Ag. Food Chem. 44, 1170–1176.

Madhavi, D.L., Smith, M.A. and Berber-Jimenez, M.D. (1995) Expression of anthocyanins in callus cultures of cranberry (*Vaccinium macrocarpon* Ait). J. Food Sci. 60, 351–355.

Madhusudhan, R. and Ravishankar, G.A. (1996) Gradient of anthocyanin in cell aggregates of *Daucus carota* in suspension cultures. Biotech. Lett. 18, 1253–1256.

Makunga, N.P., van Staden, J. and Cress, W.A. (1997) The effect of light ad 2,4-D on anthocyanin production in *Oxalis reclinata* callus. Plant Growth Reg. 23, 153–158.

Markham, K.R., Gould, K.S., Winefield, C.S., Mitchell, K.A., Bloor, S.J. and Boase, M.R. (2000) Anthocyanic vacuolar inclusions - their nature and sifnificance in flower colouration. Phytochemistry 55, 327–336.

Marshall, G.B., Smith, M.A.L., Lee, C.K.C., Deroles, S.C. and Davies, K.M. (2002) Differential gene expression between pigmented and non-pigmented cell culture lines of *Daucus carota*. Plant Cell Tissue Organ Cult. 70, 91–97.

Matsumoto, T., Nishida, K., Noguchi, M. and Tamaki, E. (1973) Some factors affecting the anthocyanin formation by *Populus* cells in suspension culture. Ag. Biol. Chem. 37, 561–567.

Meyer, H.J. and van Staden, J. (1995) The *in vitro* production of anthocyanin from callus cultures of *Oxalis linearis*. Plant Cell Tissue Organ Cult. 40, 55–58.

Meyer, J.E., Pépin, M.F. and Smith, M.A.L. (2002) Anthocyanin production from *Vaccinium pahalae*: limitations of the physical microenvironment. J. Biotech. 93, 45–57.

Miura, H., Kitamura, Y., Ikenaga, T., Mizobe, K., Shimizu, T., Nakamura, M., Kato, Y., Yamada, T., Maitani, T. and Goda, Y. (1998) Anthocyanin production of *Glehnia littoralis* callus cultures. Phytochemistry 48, 279–283.

Miyanaga, K., Seki, M. and Furusaki, S. (2000a) Quantitative determination of cultured strawberry-cell heterogeneity by image analysis: effects of medium modification on anthocyanin accumulation. Biochem. Eng. J. 5, 201–207.

Miyanaga, K., Seki, M. and Furusaki, S. (2000b) Analysis of pigmentation in individual cultured plant cells using an image processing system. Biotech. Letters 22, 977–981.

Mizukami, H., Tomita, K., Ohashi, H. and Hiraoka, N. (1988) Anthocyanin production in callus cultures of roselle (*Hibiscus sabdariffa* L.). Plant Cell Rep. 7, 553–556.

Mori, T. and Sakura, M. (1999) Preparation of conditioned medium to stimulate anthocyanin production using suspension cultures of *Fragaria ananassa* cells. World J. Microb. Biotech. 15, 635–637.

Mori, T. and Sakurai, M. (1994) Production of anthocyanin from strawberry cell suspension cultures: effects of sugar and nitrogen. J. Food Sci. 59, 588–593.

Mori, T. and Sakurai, M. (1995) Effects of riboflavin and increased sucrose on anthocyanin production in suspended strawberry cell cultures. Plant Sci. 110, 147–153.

Mori, T. and Sakurai, M. (1996) Riboflavin affects anthocyanin synthesis in nitrogen culture using strawberry suspended cells. J. Food Sci. 61, 698–702.

Mori, T. and Sakurai, M. (1998) Conditioned medium from heterogeneous plants (rose and grape) on cell growth and anthocyanin synthesis of *Fragaria ananassa*. Biotech. Lett. 20, 73–75.

Mori, T., Sakurai, M. and Sakuta, M. (2000) Changes in PAL, CHS, DAHP synthase (DS-Co and DS-Mn) activity during anthocyanin synthesis in suspension culture of *Fragaria ananassa*. Plant Cell Tissue and Organ Cult. 62, 135–139.

Mori, T., Sakurai, M. and Sakuta, M. (2001) Effects of conditioned medium on activities of PAL, CHS, DAHP synthase (DS-Co and DS-Mn) and anthocyanin production in suspension cultures of *Fragaria ananassa*. Plant Sci. 160, 355–360.

Mori, T., Sakurai, M., Seki, M. and Furusaki, S. (1994a) Use of auxin and cytokinin to regulate anthocyanin production and composition in suspension cultures of strawberry cell. J. Sci. Food Agric. 65, 271–276.

Mori, T., Sakurai, M., Seki, M. and Furusaki, S. (1994b) Effects of conditioning factor on anthocyanin production in strawberry suspension cultures. J. Sci. Food Agric. 66, 381–388.

Mori, T., Sakurai, M., Shigeta, J.I., Yoshida, K. and Kondo, T. (1993) Formation of anthocyanins from cells cultured from different parts of strawberry plants. J. Food Sci. 58, 788–792.

Moumou, Y., Trotin, F., Dubois, J., Vasseur, J. and El-Boustani, E. (1992) Influence of culture conditions on polyphenol production by *Fagopyrum esculentum* tissue cultures. J. Nat. Prod. 55, 33–38.

Nagamori, E., Hiraoka, K., Honda, H. and Kobayashi, T. (2001) Enhancement of anthocyanin production from grape (*Vitis vinifera*) callus in a viscous additive-supplemented medium. Biochem. Eng. J. 9, 59–65.

Nagarajan, R.P., Keshavarz, E. and Gerson, D.F. (1989) Optimization of anthocyanin yield in a mutated carrot cell line (*Daucus carota*) and its implications in large scale production. J. Ferm. Bioeng. 68, 102–106.

Nakamura, M., Seki, M. and Furusaki, S. (1998) Enhanced anthocyanin methylation by growth limitation in strawberry suspension culture. Enz. Micro. Tech. 22, 404–408.

Nakamura, M., Takeuchi, Y., Miyanaga, K., Seki, M. and Furusaki, S. (1999) High anthocyanin accumulation in the dark by strawberry (*Fragaria ananassa*) callus. Biotech. Lett. 21, 695–699.

Narayan, M.S., Thimmaraju, R. and Bhagyalakshmi, N. (2005) Interplay of growth regulators during solid-state and liquid-state batch cultivation of anthocyanin producing cell line of *Daucus carota*. Proc. Biochem. 40, 351–358.

Narayan, M.S. and Venkataraman, L.V. (2000) Characterisation of anthocyanins derived from carrot (*Daucus carota*) cell culture. Food Chem. 70, 361–363.

Narayan, M.S. and Venkataraman, L.V. (2002) Effect of sugar and nitrogen on the production of anthocyanin in cultured carrot (*Daucus carota*) cells. J. Food Sci. 67, 84–86.

Nawa, Y., Asano, S., Motoori, S. and Ohtani, T. (1993) Production of anthocyanins, carotenoids, and proanthocyanidins by cultured cells of rabbiteye blueberry (*Vaccinium ashei* Reade). Biosci. Biotech. Biochem. 57, 770–774.

Noe, W. and Seitz, H.U. (1982) Induction of de novo synthesis of phenylalanine ammonialyase by l-alpha-aminooxy-beta-phenylpropionic acid in suspension cultures of *Daucus carota* L. Planta 154, 454–458.

Nozue, M., Baba, S., Kitamura, Y., Xu, W., Kubo, H., Nogawa, M., Shioiri, H. and Kojima, M. (2003) VP24 found in anthocyanic vacuolar inclusions (AVIs) of sweet potato cells is a member of a metalloprotease family. Biochem. Eng. J. 14, 199–205.

Nozue, M., Kawai, J. and Yoshitama, K. (1987) Selection of a high anthocyanin-producing cell line of sweet potato cell cultures and identification of pigments. J. Plant Physiol. 129, 81–88.

Nozue, M., Kubo, H., Nishimura, M., Katou, A., Hattori, C., Usuda, N., Nagata, T. and Yasuda, H. (1993) Characterization of intravacuolar pigmented structures in anthocyanin-containing cells of sweet potato suspension cultures. Plant Cell Physiol. 34, 803–808.

Nozue, M., Kubo, H., Nishimura, M. and Yasuda, H. (1995) Detection and characterization of a vacuolar protein (VP24) in anthocyanin-producing cells of sweet potato in suspension culture. Plant Cell Physiol. 36, 883–889.

Nozue, M., Yamada, K., Nakamura, T., Kubo, H., Kondo, M. and Nishimura, M. (1997) Expression of a vacuolar protein (VP24) in anthocyanin-producing cells of sweet potato in suspension culture. Plant Physiol. 115, 1065–1072.

Nozue, M. and Yasuda, H. (1985) Occurrence of anthocyanoplasts in cell suspension cultures of sweet potato. Plant Cell Rep. 4, 252–255.

Ozeka, Y. and Komamine, A. (1982). Induction of anthocyanin synthesis in a carrot suspension culture. Correlation of metabolic differentiation with morphological differentiation. Plant Tissue Cult. 1982, 355–356.

Ozeki, Y. (1996) Regulation of anthocyanin synthesis in carrot suspension cultured cells. J. Plant Res. 109(1095), 343–351.

Ozeki, Y., Davies, E. and Takeda, J. (1993) Structure and expression of chalcone synthase gene in carrot suspension cultured cells regulated by 2,4-D. Plant Cell Physiol. 34, 1029–1037.

Ozeki, Y., Ito, Y., Sasaki, N., Oyanagi, M., Akimoto, H., Chikagawa, Y. and Takeda, J. (2000) Phenylalanine ammonia-lyase genes involved in anthocyanin synthesis and the regulation of its expression in suspension cultured carrot cells. J. Plant Res. 113, 319–326.

Ozeki, Y. and Komamine, A. (1981) Induction of anthocyanin synthesis in relation to embryogenesis in a carrot suspension culture: correlation of metabolic differentiation with morphological differentiation. Physiol. Plant. 53, 570–577.

Ozeki, Y. and Komamine, A. (1985) Changes in activities of enzymes involved in general phenylpropanoid metabolism during the induction and reduction of anthocyanin synthesis in a carrot suspension culture as regulated by 2,4-D. Plant Cell Physiol. 26, 903–911.

Ozeki, Y. and Komamine, A. (1986) Effects of growth regulators on the induction of anthocyanin synthesis in carrot suspension cultures. Plant Cell Physiol. 27, 1361–1368.

Ozeki, Y., Komamine, A., Noguchi, T. and Sankawa, U. (1987) Changes in activities of enzymes involved in flavonoid metabolism during the initiation and suppression of anthocyanin synthesis in carrot suspension cultures regulated by 2,4-dichlorophenoxyacetic acid. Physiol. Plant. 69, 123–128.

Ozeki, Y., Komamine, A. and Tanaka, Y. (1990) Induction and repression of phenylalanine ammonia-lyase and chalcone synthase enzyme proteins and mRNAs in carrot cell suspension cultures regulated by 2,4-D. Physiol. Plant. 78, 400–408.

Ozeki, Y. and Takeda, J. (1994) Regulation of phenylalanine ammonia-lyase genes in carrot suspension cultured cells. Plant Cell Tissue Organ Cult. 38, 221–225.

Park, H.-H., Kang, S.K., Lee, J.H., Choi, J.Y., Lee, Y.S., Kwon, I.B. and Yu, J.H. (1989) Production of anthocyanins by Vitis hybrid cell culture. Korean J. Microbiol. Bioeng. 17, 257–262.

Pasqua, G., Monacelli, B., Mulinacci, N., Rinaldi, S., Giaccherini, C., Innocenti, M. and Vinceri, F.F. (2005) The effect of growth regulators and sucrose on anthocyanin production in Camptotheca acuminata cell cultures. Plant Physiol. Biochem. 43, 293–298.

Pepin, M.F., Archambault, J., Chavarie, C. and Cormier, F. (1995) Growth kinetics of Vitis vinifera cell suspension cultures: I. Shake flask cultures. Biotech. Bioeng. 47, 131–138.

Pepin, M.F., Smith, M.A. and Reid, J.F. (1999) Application of imaging tools to plant cell culture: relationship between plant cell aggregation and flavonoid production. In Vitro Cell. Dev. Biol. Plant 35, 290–295.

Petersen, M. and Seitz, H.U. (1986) UDP-glucose:cyanidin 3-O-glucosyltransferase in anthocyanin-containing cell cultures from Daucus carota L. J. Plant Physiol. 125, 383–390.

Plata, N., Konczak-Islam, I., Jayram, S., McClelland, K., Woolford, T. and Franks, P. (2003) Effect of methyl jasmonate and p-coumaric acid on anthocyanin composition in a sweet potato cell suspension culture. Biochem. Eng. J. 14, 171–177.

Qu, J., Zhang, W., Yu, X. and Jin, M. (2005) Instability of anthocyanin accumulation in Vitis vinifera L. var. Gamay Fréaux suspension cultures. Biotech. Bioproc. Eng. 10, 155–161.

Rajendran, L., Ravishankar, G.A., Venkataraman, L.V. and Prathiba, K.R. (1992) Anthocyanin production in callus cultures of Daucus carota as influenced by nutrient stress and osmoticum. Biotech. Lett. 14, 707–712.

Rajendran, L., Suvarnalatha, G., Ravishankar, G.A. and Venkataraman, L.V. (1994) Enhancement of anthocyanin production in callus cultures of Daucus carota L. under the influence of fungal elicitors. Appl. Microbiol. Biotechnol. 42, 227–231.

Ramachandra Rao, S. and Ravishankar, G.A. (2002) Plant cell cultures: chemical factories of secondary metabolites. Biotech. Adv. 20, 101–153.

Ramachandra Rao, S., Sarada, R. and Ravishankar, G.A. (1996) Phycocyanin, a new elicitor for capsaicin and anthocyanin accumulation in plant cell cultures. Appl. Microbiol. Biotechnol. 46(5–6), 619–621.

Rau, D. and Forkmann, G. (1986) Anthocyanin synthesis in tissue cultures of Callistephus chinensis (China aster). Plant Cell Rep. 5, 435–438.

Renaud, S. and De Lorgeril, M. (1992) Wine, alchohol, platelets, and the French paradox for coronary heart disease. Lancet 339, 1523–1526.

Rose, A., Gläßgen, W.E., Hopp, W. and Seitz, H.U. (1996) Purification and characterization of glycosyltransferases involved in anthocyanin biosynthesis in cell-suspension cultures of Daucus carota L. Planta 198, 397–403.

Saigne-Soulard, C., Richard, T., Mérillon, J.M. and Monti, J.P. (2006) ^{13}C NMR analysis of polyphenol biosynthesis in grape cells: impact of various inducing factors. Analytica Chimica Acta 563(1–2 SPEC. ISS.), 137–144.

Sakamoto, K., Iida, K., Sawamura, K., Hajiro, K., Asada, Y., Yoshikawa, T. and Furuya, T. (1993) Effects of nutrients on anthocyanin production in cultured cells of Aralia cordata. Phytochemistry 33, 357–360.

Sakamoto, K., Iida, K., Sawamura, K., Hajiro, K., Asada, Y., Yoshikawa, T. and Furuya, T. (1994) Anthocyanin production in cultured cells of Aralia cordata Thunb. Plant Cell Tissue Organ Cult. 36, 21–26.

Sakurai, M. and Mori, T. (1996) Stimulation of anthocyanin synthesis by conditioned medium produced by strawberry suspension cultures. J. Plant Physiol. 149, 599–604.

Sakurai, M., Mori, T., Seki, M. and Furusaki, S. (1996) Changes of anthocyanin composition by conditioned medium and cell inoculum size using strawberry suspension culture. Biotech. Lett. 18, 1149–1154.

Sakurai, M., Mori, T., Seki, M. and Furusaki, S. (1997b) Influence of conditioned medium on cyanidin and peonidin synthesis. J. Chem. Eng. Japan 30, 951–953.

Sakurai, M., Ozeki, Y. and Mori, T. (1997a) Induction of anthocyanin accumulation in rose suspension-cultured cells by conditioned medium of strawberry suspension cultures. Plant Cell Tissue Organ Cult. 50, 211–214.

Sakuta, M., Hirano, H., Kakegawa, K., Suda, J., Hirose, M., Joy, R.W.I., Sugiyama, M. and Komamine, A. (1994) Regulatory mechanisms of biosynthesis of betacyanin and anthocyanin in relation to cell division activity in suspension cultures. Plant Cell Tissue Organ Cult. 38, 167–169.

Sato, K., Nakayama, M. and Shigeta, J.I. (1996) Culturing conditions affecting the production of anthocyanin in suspended cell cultures of strawberry. Plant Sci. 113, 91–98.

Schwinn, K.E. and Davies, K.M. (2004). Flavonoids. In: K.M. Davies (Ed.) Plant Pigments and Their Maniupulation. Oxford, Blackwell Publishing, 14: 92–149.

Seitz, H.U., Bach, M., Richter, S., Schnitzler, J.P. and Steimle, D.E. (1994) Elicitor-induced changes in the phenol metabolism of suspension-cultured carrot cells. Acta Hort. 381, 113–120.

Shibli, R.A., Smith, M.A.L. and Kushad, M. (1997) Headspace ethylene accumulation effects on secondary metabolite production in Vaccinium pahalae cell culture. Plant Growth Reg. 23, 201–205.

Shibli, R.A., Smith, M.A.L. and Shatnawi, M.A. (1999) Pigment recovery from encapsulated-dehydrated Vaccinium pahalae (ohelo) cryopreserved cells. Plant Cell Tissue Organ Cult. 55, 119–123.

Skrzypek, Z., Swiatek, L. and Wysokinska, H. (1993) Investigations of the anthocyanins of Penstemon serrulatus cell suspension cultures. Planta Medica 59A565.

Smith, M.A.L., Madhavi, D.L., Fang, Y. and Tomczak, M.M. (1997) Continuous cell culture and product recovery from wild Vaccinium pahalae germplasm. J. Plant Physiol. 150, 462–466.

Smith, M.A.L., Reid, H.F., Hansen, A.C., Li, Z. and Madhavi, D.L. (1995) Non-destructive machine vision analysis of pigment-producing cell cultures. J. Biotech. 40, 1–11.

Stickland, R.G. and Sunderland, N. (1972) Production of anthocyanins, flavonols and chlorogenic acids by cultured callus tissues of Haplopappus gracilis. Ann. Bot. 36, 443–457.

Suda, I., Furuta, S., Nishiba, Y., Yamakawa, O., Matsugano, K. and Sugita, K. (1997) Hepatoprotective activity of purple coloured sweet potato juice. Sweetpotato Res. Front 4, 3.

Sudha, G. and Ravishankar, G.A. (2003a) Elicitation of anthocyanin production in callus cultures of Daucus carota and involvement of calcium channel modulators. Cur. Sci. 84, 775–779.

Sudha, G. and Ravishankar, G.A. (2003b) Influence of putrescine on anthocyanin production in callus cultures of Daucus carota mediated through calcium ATPase. Acta Physiol. Plant. 25, 69–75.

Sudha, G. and Ravishankar, G.A. (2003c) Elicitation of anthocyanin production in callus cultures of Daucus carota and the involvement of methyl jasmonate and salicylic acid. Acta Physiol. Plant. 25, 249–256.

Suvarnalatha, G., Rajendran, L., Ravishankar, G.A. and Venkataraman, L.V. (1994) Elicitation of anthocyanin production in cell cultures of carrot (Daucus carota L) by using elicitors and abiotic stress. Biotech. Lett. 16, 1275–1280.

Suzuki, M. (1995) Enhancement of anthocyanin accumulation by high osmotic stress and low pH in grape cells (*Vitis* hybrids). J. Plant Physiol. 147, 152–155.

Takeda, J. (1988) Light-induced synthesis of anthocyanin in carrot cells in suspension. I. The factors affecting anthocyanin production. J. Exp. Bot. 39, 1065–1077.

Takeda, J. (1990) Light-induced synthesis of anthocyanin in carrot cells in suspension. II. Effects of light and 2,4-D on induction and reduction of enzyme activities related to anthocyanin synthesis. J. Exp. Bot. 41(227), 749–755.

Takeda, J. and Abe, S. (1992) Light-induced synthesis of anthocyanin in carrot cells in suspension-IV. The action spectrum. Photochem. Photobiol. 56, 69–74.

Takeda, J., Abe, S., Hirose, Y. and Ozeki, Y. (1993) Effect of light and 2,4-dichlorophenoxyacetic acid on the level of mRNAs for phenylalanine ammonia-lyase and chalcone synthase in carrot cells cultured in suspension. Physiol. Plant. 89, 4–10.

Takeda, J., Obi, I. and Yoshida, K. (1994) Action spectra of phenylalanine ammonia-lyase and chalcone synthase expression in carrot cells in suspension. Physiol. Plant. 91, 517–521.

Takeda, T., Inomata, M., Matsuoka, H., Hikuma, M. and Furusaki, S. (2003) Release of anthocyanin from strawberry cultured cells with heating treatment. Biochem. Eng. J. 15, 205–210.

Tamura, H., Kumaoka, Y. and Sugisawa, H. (1989) Identification and quantitative variation of anthocyanins produced by cultured callus tissue of *Vitis* sp. Ag. Biol. Chem. 53, 1969–1970.

Tanaka, N., Matsuura, E., Terahara, N. and Ishimaru, K. (1999) Secondary metabolites in transformed root cultures of *Campanula glomerata*. J. Plant Physiol. 155, 251–254.

Terahara, N., Callebaut, A., Ohba, R., Nagata, T., Ohnishi-Kameyama, M. and Suzuki, M. (1996) Triacylated anthocyanins from *Ajuga reptans* flowers and cell cultures. Phytochemistry 42, 199–203.

Terahara, N., Callebaut, A., Ohba, R., Nagata, T., Ohnishi-Kameyama, M. and Suzuki, M. (2001) Acylated anthocyanidin 3-sophoroside-5-glucosides from *Ajuga reptans* flowers and the corresponding cell cultures. Phytochemistry 58, 493–500.

Terahara, N., Konczak, I., Ono, H., Yoshimoto, M. and Yamakawa, O. (2004) Characterization of acylated anthocyanins in callus induced from storage root of purple-fleshed sweet potato, *Ipomoea batatas* L. J. Biomed. Biotech. 2004, 279–286.

Tholakalabavi, A., Zwiazek, J.J. and Thorpe, T.A. (1997) Osmotically-stressed poplar cell cultures: anthocyanin accumulation, deaminase activity, and solute composition. J. Plant Physiol. 151, 489–496.

Tsukaya, H., Ohshima, T., Naito, S., Chino, M. and Komeda, Y. (1991) Sugar-dependent expression of the CHS-A gene for chalcone synthase from *Petunia* in transgenic Arabidopsis. Plant Physiol. 97, 1414–1421.

Verma, A.K., Chaudhary, U., Rakesh, K., Pant, A.K., Gaur, A.K. and Lakhchaura, B.D. (2000) anthocyanin, bisabolol and phenylammonialyase activity in cell cultures of *Populus deltoides*. Indian J. Exp. Biol. 38, 1050–1053.

Verpoorte, R., van der Heijden, R., ten Hoopen, H.J.G. and Memelink, J. (1999) Metabolic engineering of plant secondary metabolite pathways for the production of fine chemicals. Biotech. Lett. 21, 467–479.

Vitrac, X., Krisa, S., Decendit, A., Vercauteren, J., Nührich, A., Monti, J.P., Deffieux, G. and Mérillon, J.M. (2002) Carbon-14 biolabelling of wine polyphenols in *Vitis vinifera* cell suspension cultures. J. Biotech. 95, 49–56.

Vitrac, X., Larronde, F., Krisa, S., Decendit, A., Deffieux, G. and Mérillon, J.M. (2000) Sugar sensing and Ca^{2+}-calmodulin requirement in *Vitis vinifera* cells producing anthocyanins. Phytochemistry 53, 659–665.

Vogelien, D.L., Hrazdina, G., Reeves, S. and Dougall, D.K. (1990) Phenotypic differences in anthocyanin accumulation among clonally related cultured cells of carrot. Plant Cell Tissue Organ Cult. 22, 213–222.

Wang, J.W., Xia, Z.H., Chu, J.H. and Tan, R.X. (2004) Simultaneous production of anthocyanin and triterpenoids in suspension cultures of *Perilla frutescens*. Enz. Micro. Tech. 34, 651–656.

Yamakawa, T., Ishida, H., Kato, S., Kodama, T. and Minoda, Y. (1983b) Formation and identification of anthocyanins in cultured cells of *Vitis* sp. Ag. Biol. Chem. 47, 997–1001.

Yamakawa, T., Kato, S., Ishida, H., Kodama, T. and Minoda, Y. (1983a) Production of anthocyanins by *Vitis* cells in suspension culture. Ag. Biol. Chem. 47, 2185–2191.

Yamakawa, T., Ohtsuka, H., Onomichi, K., Kodama, T. and Minoda, Y. (1982) Production of anthocyanin pigment by grape cell culture. In: A. Fujiwara (Ed.) *Plant Tissue Culture 1982*, Tokyo, Japanese Association for Plant Tissue Culture: 273–274.

Yamakawa, T., Onomichi, K., Kodama, T. and Minoda, Y. (1985) Application of feeder layer method for improved colony formation of grape cells and protoplasts a t low cell-density. Ag. Biol. Chem. 49, 3583–3585.

Yamamoto, Y., Kadota, N., Mizuguchi, R. and Yamada, Y. (1983) Computer tracing of the pedigree of cultured *Euphorbia millii* cells that produce high levels of anthocyanin. Ag. Biol. Chem. 47, 1021–1026.

Yamamoto, Y., Kinoshita, Y., Watanabe, S. and Yamada, Y. (1989) Anthocyanin production in suspension cultures of high-producing cells of *Euphorbia millii*. Ag. Biol. Chem. 53, 417–423.

Yamamoto, Y. and Mizuguchi, R. (1982) Selection of a high and stable pigment-producing strain in cultured *Euphorbia millii* cells. Theo. App. Gen. 61, 113–116.

Yamamoto, Y., Mizuguchi, R. and Yamada, Y. (1981) Chemical constituents of cultured cells of *Euphorbia tirucalli* and *E. millii*. Plant Cell Rep. 1, 29–30.

Yeoman, M.M., Lindsey, K., Miedzybrodzka, M.B. and McLauchlan, W.R. (1982) Accumulation of secondary products as a facet of differentiation in plant cell and tissue cultures. In: M.M. Yeoman and D.E.S. Truman (Eds.) *Differentiation In Vitro. British Society for Cell Biology, Symposium 4*, Cambridge, Cambridge University Press, pp. 65–82.

Yousef, G.G., Seigler, D.S., Grusak, M.A., Rogers, R.B., Knight, C.T.G., Kraft, T.F.B., Erdman Jr, J.W. and Lila, M.A. (2004) Biosynthesis and characterization of ^{14}C-enriched flavonoid fractions from plant cell suspension cultures. J. Ag. Food Chem. 52, 1138–1145.

Zhang, H., Wang, L., Deroles, S.C., Bennett, R. and Davies, K.M. (2006) New insight into the structures and formation of anthocyanic vacuolar inclusions in flower petals. BMC Plant Biol. 6, 29.

Zhang, W., Curtin, C., Kikuchi, M. and Franco, C. (2002a) Integration of jasmonic acid and light irradiation for enhancement of anthocyanin biosynthesis in *Vitis vinifera* suspension cultures. Plant Sci. 162, 459–468.

Zhang, W. and Furusaki, S. (1997) Regulation of anthocyanin synthesis in suspension cultures of strawberry cell by pH. Biotech. Lett. 19, 1057–1061.

Zhang, W., Jin, M.F., Yu, X.J. and Yuan, Q. (2001) Enhanced anthocyanin production by repeated-batch culture of strawberry cells with medium shift. App. Microbiol. Biotech. 55, 164–169.

Zhang, W., Seki, M. and Furusaki, S. (1997) Effect of temperature and its shift on growth and anthocyanin production in suspension cultures of strawberry cells. Plant Sci. 127, 207–214.

Zhang, W., Seki, M., Furusaki, S. and Middelberg, A.P.J. (1998) Anthocyanin synthesis, growth and nutrient uptake in suspension cultures of strawberry cells. J. Ferm. Bioeng. 86, 72–78.

Zhong, J.J., Fujiyama, K., Seki, T. and Yoshida, T. (1993b) On-line monitoring of cell concentration of *Perilla frutescens* in a bioreactor. Biotech. Bioeng. 42, 542–546.

Zhong, J.J., Fujiyama, K., Seki, T. and Yoshida, T. (1994b) A quantitative analysis of shear effects on cell suspension and cell culture of *Perilla frutescens* in bioreactors. Biotech. Bioeng. 44, 649–654.

Zhong, J.J., Konstantinov, K.B. and Yoshida, T. (1994c) Computer-aided on-line monitoring of physiological variables in suspended cell cultures of *Perilla frutescens* in a bioreactor. J. Ferm. Bioeng. 77, 445–447.

Zhong, J.J., Seki, M., Kinoshita, S.-I. and Yoshida, T. (1992) Effects of surfactants on cell growth and pigment production in suspension cultures of *Perilla frutescens*. World J. Microbiol. Biochem. 8, 106–109.

Zhong, J.J., Seki, T., Kinoshita, S.I. and Yoshida, T. (1991) Effect of light irradiation on anthocyanin production by suspended culture of *Perilla frutescens*. Biotech. Bioeng. 38, 653–658.

Zhong, J.J., Xu, W.X. and Yoshida, T. (1994a) Effects of initial sucrose concentration on excretion of anthocyanin pigments in suspended cultures of *Perilla frutescens* cells. World J. Microbiol. Biochem. 10, 590–592.

Zhong, J.J., Yoshida, M., Fujiyama, K., Seki, T. and Yoshida, T. (1993a) Enhancement of anthocyanin production by *Perilla frutescens* cells in a stirred bioreactor with internal light irradiation. J. Ferm. Bioeng. 75, 299–303.

Zhong, J.J. and Yoshida, T. (1993) Effects of temperature on cell growth and anthocyanin production in suspension cultures of *Perilla frutescens*. J. Ferm. Bioeng. 76, 530–531.

Zhong, J.J. and Yoshida, T. (1994). Rheological characteristics of suspended cultures of *Perilla frutescens* and their implications in bioreactor operation for anthocyanin production. Adv. Plant Biotech., 255–279.

Zhong, J.J. and Yoshida, T. (1995) High-density cultivation of *Perilla frutescens* cell suspensions for anthocyanin production: effects of sucrose concentration and inoculum size. Enzyme Microb. Tech. 17, 1073–1079.

Zubko, M.K., Muradov, A.Z., Patskovskii, Y.V. and Voronin, V.V. (1992) Selection of potato cell lines with constitutive anthocyan biosynthesis. Sov. Biotech. (Biotekhnologiya), 5, 106–110.

Zubko, M.K., Schmeer, K., Gläßgen, W.E., Bayer, E. and Seitz, H.U. (1993) Selection of anthocyanin-accumulating potato (*Solanum tuberosum* L.) cell lines from calli derived from seedlings produced by gamma-irradiated seeds. Plant Cell Rep. 12, 555–558.

Zwayyed, S.K., Frazier, G.C. and Dougall, D.K. (1991) Growth and anthocyanin accumulation in carrot cell suspension cultures growing on fructose, glucose, or their mixtures. Biotech. Prog. 7, 288–290.

6

Modification and Stabilization of Anthocyanins

Keiko Yonekura-Sakakibara[1], Toru Nakayama[2], Mami Yamazaki[3], and Kazuki Saito[4]

[1] RIKEN Plant Science Center, keikoys@psc.riken.jp
[2] Tohoku University, Department of Biomolecular Engineering, nakayama@seika.che.tohoku.ac.jp
[3] Chiba University, Graduate School of Pharmaceutical Science, mamiy@p.chiba-u.ac.jp
[4] RIKEN Plant Science Center/Chiba University, Graduate School of Pharmaceutical Science, ksaito@faculty.chiba-u.jp

Abstract. Modification of anthocyanins is responsible for the stabilization of vacuolar anthocyanins. Cytosolic modification reactions include glycosylation, methylation, and acylation immediately following the synthesis of anthocyanidin aglycones. Knowledge of the biochemistry and molecular biology of these modification reactions, and the enzymes involved, has increased dramatically due to the molecular cloning of the genes encoding these enzymes from a variety of plant species in the last decade. Recent advances in the modification of anthocyanins are summarized and discussed in terms of fundamental biochemistry and application for the modification and stabilization of anthocyanins.

6.1 Introduction

Anthocyanins are flavonoid pigments that are stabilized by modification of the aglycone forms (anthocyanidins) by glycosylation, methylation, and acylation. Mutants lacking the gene responsible for the initial glycosylation of cyanidin accumulate no anthocyanins despite the presence of an intact anthocyanidin biosynthetic pathway, indicating that modification of anthocyanidins is necessary for the stable storage of colored anthocyanin pigments (Boss et al. 1996; Tohge et al. 2005). Stacking of polyaromatic moieties of acylated anthocyanins has been shown to play a role in stability and blue shift of anthocyanin color (Tanaka and Brugliera 2006). Several lines of evidence suggest that these anthocyanin modifications take place in the cytosol just after formation of relatively unstable anthocyanidin (Nakajima et al. 2001; Davies and Schwinn 2006). However, some serine carboxylpeptidase-like anthocyanin acyltransferases may localize in the vacuole (Hause et al. 2002). In this chapter, we describe the modification of anthocyanins and its effects on stabilization.

K. Gould et al. (eds.), *Anthocyanins*, DOI: 10.1007/978-0-387-77335-3_6,
© Springer Science+Business Media, LLC 2009

6.2 Glycosyltransferases

6.2.1 Glycosylation of Anthocyanidins/Anthocyanins

In general, naturally occurring anthocyanins are found as a glycosylated form with one or more sugar moieties. A few unusual anthocyanidins are exceptions, such as the 3-desoxyanthocyanidins and pyranoanthocyanidins, which are detected in both unglycosylated and glycosylated forms (Anderson and Jordheim 2006). Glycosylation usually replaces hydroxyl groups at the C-3, C-5, C-7, C-3', C-4', and C-5' positions, with the C-3 position being the most frequently glycosylated, followed by C-5 (Anderson and Jordheim 2006). Glycosyl groups (i.e. glucose, galactose, xylose, glucuronic acid, and arabinose) linked to anthocyanidins are frequently further glycosylated and/or acylated. Anthocyanidin C-glycosylation is also observed at the C-8 position in *Tricyrtis formosana* as a rare case (Saito et al. 2003).

6.2.2 Stabilization of Anthocyanidins by Glycosylation

Glycosylation is an important modification for increasing the hydrophilicity and stability of hydrophobic flavonoids. The accumulation of anthocyanins in plants which are deficient in anthocyanidin 3-O-glucosylation activity is significantly reduced (Tohge et al. 2005) because anthocyanidins are inherently unstable under physiological conditions. Glycosylation is also essential for color stability because the aromatic acylation that plays a key role in color stability is generally linked to glycosyl groups of anthocyanins. The color of glycosylated anthocyanidins is unstable in aqueous solutions at mildly acidic pH values unless aromatic acyl groups are added. Glycosylation is also thought to be a signal for vacuolar transport (Marrs et al. 1995; Ono et al. 2006).

6.2.3 Anthocyanidin/Anthocyanin Glycosyltransferases

Anthocyanidin/anthocyanin O-glycosylation is catalyzed by family 1 glycosyltransferases (UGTs), which use flavonoids as the sugar acceptor and UDP-sugars as the sugar donor. UGTs recognize the hydroxyl groups of a wide variety of flavonoid aglycones including anthocyanidins. The structures of anthocyanins indicate that UGTs recognize UDP-glucose, UDP-galactose, UDP-rhamnose, UDP-xylose, UDP-glucuronic acid, and UDP-arabinose as sugar donors.

An anthocyanidin/anthocyanin UGT gene was first identified in maize by transposon tagging (Dooner and Nelson 1977; Fedoroff et al. 1984). Since then, a number of cDNAs encoding for anthocyanidin/anthocyanin glycosyltransferases have been isolated by several methods. These clones include anthocyanidin 3-O-glucosyltransferase (A3GlcT) (Furtek et al. 1998; Wise et al. 1990; Tanaka et al. 1996; Gong et al. 1997; Ford et al. 1998; Tohge et al. 2005), anthocyanidin 3-O-galactosyltransferase (A3GalT) (Kubo et al. 2004), anthocyanin 5-O-glucosyltransferase (A5GlcT) (Yamazaki et al. 1999; Yamazaki et al. 2002; Tohge et al. 2005), anthocyanidin 5,3 -O-glucosyltransferase (A5,3GlcT) (Ogata et al. 2005), anthocyanin 3'-O-glucosyltransferase (A3'GlcT) (Fukuchi-Mizutani, et al. 2003), anthocyanin 3',5' -O-glucosyltransferase (A3',5'GlcT) (Noda et al. 2004), anthocyanidin 3-O-glucoside

Table 6.1 Cloned anthocyanidin or anthocyanin glycosyltransferases

function	plant species	substrate flavonoid	UDP-sugar	GenBank accession# (AGI)	references
A3GlcT	Gentiana trifbra	A/F	UDP-Glc	D85186	Tanaka et al. 1996
	Perilla frutescens	A/F	UDP-Glc	AB002818	Gong et al. 1997
	Vitis vinifera	A/F	UDP-Glc	AF000371	Ford et al. 1998
	Petunia hybrida	A/F	UDP-Glc	AB027454	Yamazaki et al. 2002
	Arabidopsis thaliana	A/F	UDP-Glc	NM_121711 (At5g17050)	Tohge et al. 2005
	Zea mays	ND	ND	X13501	Furtek et al. 1988
	Hordeum vulgare	ND	ND	X15694	Wise et al. 1990
	Solanum melongena	ND	ND	X77369	–
A3GalT	Aralia cordata	A/F	UDP-Gal	AB103471	Kubo et al. 2004
A5GlcT	Perilla frutescens	A	UDP-Glc	AB013596	Yamazaki et al. 1999
	Petunia hybrida	A	UDP-Glc	AB027455	Yamazaki et al. 2002
	Arabidopsis thaliana	A	UDP-Glc	NM_117485 (At4g14090)	Tohge et al. 2005
	Torenia hybrida	ND	ND	AB076698	–
	Verbena hybrida	ND	ND	AB013598	–
A5,3GlcT	Rosa hybrida	A	UDP-Glc	AB201048	Ogata et al. 2005
A3'GlcT	Gentiana trifbra	A	UDP-Glc	AB076697	Fukuchi-Mizutani et al. 2003
A3',5'GlcT	Clitoria ternatea	A	UDP-Glc	DD148309	Noda et al. 2004
A3G-2''GlcAT	Bellis perennis	A	UDP-GlcA	AB190262	Sawada et al. 2005
A3G-2''GlcT	Ipomoea nil	A	UDP-Glc	AB192314	Morita et al. 2005
	Ipomoea purpurea	ND	UDP-Glc	AB192315	Morita et al. 2005
A3G-RhaT	Petunia hybrida	A	UDP-Rha	X71059	Brugliera et al. 1994 Kroon et al. 1994

A3GlcT; anthocyanidin 3-O-glucosyltransferase, A3GalT; anthocyanidin 3-O-galactosyltransferase, A5GlcT; anthocyanin 5-O-glucosyltransferase, A5,3GlcT; anthocyanidin 5,3-O-glucosyltransferase, A3'GlcT; anthocyanin 3'-O-glucosyltransferase, A3',5'GlcT; anthocyanin 3',5'-O-glucosyltransferase, A3G-2''GlcAT; anthocyanidin 3-O-glucoside 2''-O-glucuronosyltransferase, A3G-2''GlcT; anthocyanidin 3-O-glucoside 2''-O-glucosyltransferase, A3G-RhaT; anthocyanidin 3-O-glucoside rhamnosyltransferase, A; anthocyanin or anthocyanidin, F; flavonol, ND; not determined.

2''-O-glucuronosyltransferase (A3G-2''GlcAT) (Sawada et al. 2005), anthocyanidin 3-O-glucoside 2''-O-glucosyltransferase (A3G-2''GlcT) (Morita et al. 2005), and anthocyanidin 3-O-glucoside rhamnosyltransferase (A3G-RhaT) (Brugliera et al. 1994; Kroon et al. 1994) (Table 6.1).

Plant UGTs contain a conserved sequence called "plant secondary product glycosyltransferase" (PSPG) motif (Gachon et al. 2005). The PSPG motif conserved among anthocyanin UGTs is Trp-(Ala/Cys/Val)-(Pro/Ser/Gln)-Gln-Xaa$_3$-Leu-Xaa-His-Xaa-(Ala/Ser)-Xaa-Gly-Xaa-Phe-Val-(Thr/Ser)-His-Cys-Gly-Trp-Asn-Ser-Xaa$_2$-Glu-Xaa$_4$-Gly-Val-Pro-Xaa-(Ile/Val)-Xaa$_2$-Pro-Xaa$_3$-Asp-Gln (Amino acids conserved among all UGTs listed in Table 6.1 are underlined.). The motif has been thought to be the binding site of a nucleotide-diphosphate-sugar. With the exception of this region, overall sequence identity is low among UGTs.

A phylogenetic tree of deduced amino acid sequences of flavonoid UGTs reveals that UGTs form distinct clusters by the regioselectivity for glycosylation (Fig. 6.1). Clustering by bond formation preference suggests that differentiation of UGT functions occurred prior to divergence of plant species. An exception is *Clitoria ternatea* anthocyanin 3',5'GlcT which falls into the same cluster as A3GlcT and A3GalT (Fig. 6.1, Cluster I) (Noda et al. 2004; Tanaka and Brugliera 2006).

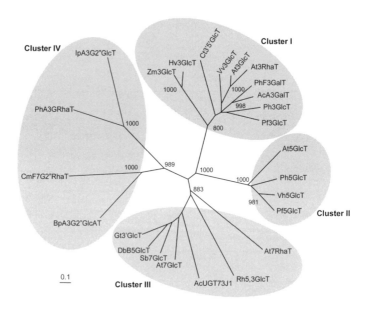

Fig. 6.1 Phylogenetic tree of flavonoid UGT deduced amino acid sequences.   The tree was constructed using the neighbor joining method of TREEVIEW program (Page 1996). Numbers indicate bootstrap values greater than 800. Bar = 0.1 amino acid substitutions per site. GenBank accession numbers are: At3RhaT (NM_102790); At3GlcT (NM_121711); Vv3GlcT (AF000371); Ph3GalT (AF316552); Ph3GlcT (AB027454); Pf3GlcT (AB002818); Hv3GlcT (X15694); Zm3GlcT (X13501); Ct3'5'GlcT (DD148309); At5GlcT (NM_117485); Pf5GlcT (AB013596); Vh5GlcT (BAA36423); Ph5GlcT (AB027455); At7GlcT (NM_129234); AcUGT73J1 (AY62063); DbB5GlcT (Y18871); Sb7GlcT (BAA83484); Gt3'GlcT (AB076697); CmF7G2″RhaT (AY048882); BpA3G2″GlcAT (AB190262); IpA3G2″GlcT (AB192315); PhA3G2″RhaT (Z25802). 3GlcT; UDP-glucose: flavonoid 3-*O*-glucosyltransferase, 3'GlcT; UDP-glucose: flavonoid 3'-*O*-glucosyltransferase, 3'5'GlcT; UDP-glucose: flavonoid 3',5'-*O*-glucosyltransferase, 3GalT; UDP-galactose: flavonoid 3-*O*-galactosyltransferase, 3RhaT; UDP-rhamnose: flavonoid 3-*O*-rhamnosyltransferase, 5GlcT; UDP-glucose: flavonoid 5-*O*-glucosyltransferase, 7GlcT; UDP-glucose: flavonoid 7-*O*-glucosyltransferase, 7RhaT; UDP-rhamnose: flavonoid 7-*O*-rhamnosyltransferase, F7G2″RhaT; UDP-rhamnose: flavanone 7-*O*-glucoside 2″-*O*-rhamnosyltransferase, A3G2″GlcAT; UDP-glucuronic acid: anthocyanin 3-*O*-glucoside 2″-*O*-glucuronosyltransferase, A3G2″GlcT; UDP-glucose: anthocyanin 3-*O*-glucoside 2″-*O*-glucosyltransferase, A3G2″RhaT; UDP-rhamnose: anthocyanin 3-*O*-glucoside 2″-

O-rhamnosyltransferase, B5GlcT; UDP-glucose: betanidin 5-*O*-glucosyltransferase. Abbreviations for species: Ac; *Allium cepa*, At; *Arabidopsis thaliana*, Bp; *Bellis perennis*, Cm; *Citrus maxima*, Ct; *Clitoria ternatea*, Db; *Dorotheanthus bellidiformis*, Gt; *Gentiana triflora*, Hv; *Hordeum vulgare*, Ip; *Ipomoea purpurea*, Pf; *Perilla frutescens*, Ph; *Petunia hybrida*, Rh; *Rosa hybrida*, Sb; *Scutellaria baicalensis*, Vh; *Verbena hybrida*, Vv; *Vitis vinifera*, Zm; *Zea mays*.

The A3GlcT enzymes that have been tested for substrate specificity have the ability to utilize anthocyanidins and flavonols as substrates. A3GalT from *Aralia cordata* has a similar ability. Although petunia flavonol 3GalT (F3GalT) belongs in the same cluster, F3GalT recognizes flavonols but not anthocyanidins, and catalyzes the reverse reaction with comparable efficiency to the forward reaction (Miller et al. 1999). Petunia *F3GalT* is expressed exclusively in pollen, suggesting that it has evolved for a specific function from an ancestral gene.

In the majority of plants, glycosylation of a hydroxyl group at the C-3 position of anthocyanidin is the primary modification step. Interestingly, rose A5,3GlcT catalyzes C-5 glucosylation prior to C-3 glucosylation, and an intermediate, anthocyanidin 5-*O*-glucoside, rapidly converts to anthocyanidin 3,5-*O*-diglucoside without detection (Ogata et al. 2005). Despite of its function, rose A5,3GlcT belongs to neither 3UGT- nor 5UGT- subfamily, and seems to be classified into flavonoid 7UGT subfamily (Fig. 6.1, Cluster III).

Gentian A3'GlcT is also a member of the flavonoid 7UGT subfamily. Thus far, no genes encoding an anthocyanin 7-*O*-glycosyltransferase have been reported. Betanidin 5GlcT from *Dorotheanthus bellidiformis* has both flavonol 7-*O*- and 4'-*O*-glucosyltransferase activity (Vogt et al. 1999).

Both A3G-2''GlyT and A5GlcT recognize anthocyanidin 3-*O*-glucoside as a substrate. However, they fall into different clusters (Fig. 6.1, Clusters IV and II, respectively). Citrus flavanone 7-*O*-glucoside2''-RhaT which catalyze the conversion of flavanone 7-*O*-glucoside to flavanone 7-*O*-neohesperidoside (Frydman et al. 2004), is in the same cluster as A3G-2''GlcT, suggesting that this group consists of UGTs that recognize the glucose moieties of flavonoids as a substrate.

6.2.4 Enzymatic Characteristics of Anthocyanidin/Anthocyanin UGTs

General aspects – Anthocyanidin/anthocyanin UGTs are approximately 50–55 kDa monomeric proteins localized in the cytosol with an optimum pH in the range of 8.0–8.5.

Substrate specificity – Anthocyanidin/anthocyanin UGTs have a broad substrate range for sugar acceptors (i.e. flavonoids). By contrast, UGTs show a distinct specificity for sugar donors (Ford et al. 1998; Fukuchi-Mizutani et al. 2003). A3GlcTs recognize anthocyanidin and flavonol as acceptor substrates (Table 6.1). There is no data for the ability of flavonoids to act as substrates, except for anthocyanin, for other anthocyanin UGTs. Grape A3GlcT catalyzes the glycosylation of quercetin, cyanidin, delphinidin, and malvidin. However, as a sugar donor, it recognizes only UDP-glucose, specifically excluding UDP-xylose, UDP-galactose, UDP-glucuronic acid, UDP-mannose, ADP-glucose, CDP-glucose, GDP-glucose, and TDP-glucose.

Kinetic parameters – Biochemical analyses with recombinant proteins show that the Km values for quercetin, cyanidin, delphinidin, malvidin, and UDP-glucose of grape A3GlcT are 15, 30, 16, 35.7, and 1.88 mM, respectively (Ford et al. 1998). Km values for cyanidin 3-*O*-glucoside and UDP-glucose of perilla A5GlcT are 31 and 940 µM, respectively (Yamazaki et al. 1999), and for delphinidin 3,5-*O*-diglucoside of gentian A3′GlcT is 120 µM (Fukuchi-Mizutani et al. 2003). In the case of daisy A3G-2″GlcAT, the Km of cyanidin 3-*O*-6″-*O*-malonylglucoside is 19 µM (native protein) and 32 µM (recombinant protein), and for UDP-glucuronic acid is 476 µM (native protein) and 497 µM (recombinant protein) (Sawada et al. 2005). In contrast, citrus flavanone UGT has a higher affinity for the substrates, naringenin 7-*O*-glucoside (Km = 2.4 µM) and UDP-rhamnose (Km = 1.3 µM) (Bar-Peled et al. 1991).

Inhibition studies – The effects of divalent metal ions and metal chelators have been examined in many reports (reviewed in Heller and Forkmann 1993). For example, A3GlcT enzyme activity is completely inhibited by 1 mM Cu^{2+}, Mn^{2+}, and Zn^{2+} (Ford et al. 1998), and A3G-2″GlcAT is inhibited by 0.1 mM Cu^{2+} and Hg^{2+}(Sawada et al. 2005). It should be noted that the inhibitory effect may due to the destruction of substrate anthocyanins by these metal ions (Ford et al. 1998). EDTA has a negligible effect on the activities of A5GlcT, A3′GlcT, and A3G-2″GlcAT (Yamazaki et al. 1999; Fukuchi-Mizutani et al. 2003; Sawada et al. 2005).

6.2.5 Structure of Anthocyanidin/Anthocyanin Glycosyltransferases

The crystal structures of non-plant GT proteins are divided into two GT groups, GT-A and GT-B, and UGTs from the Actinomycete *Amycolatopsis orientalis* (GtfA, GtfB, and GtfD) have a GT-B structure (Mulichak et al. 2001). The predicted structures of UGTs from *Amycolatopsis* and plants are similar, despite their low sequence identity (Lim and Bowles 2004). Site-directed mutagenesis and 3D-structure homology modeling suggest that His22, Glu378, and Glu394 in betanidin 5GlcT are essential for its catalytic activity (Hans et al. 2004). His22 and Glu378 are highly conserved amino acid residues in flavonoid UGTs. The crystal structure of a triterpene/flavonoid UGT from *Medicago truncatula* (Shao et al. 2005) suggests His22 as the catalytic base and Asp121 as a residue for deprotonation for the acceptor. His22 and Asp121 are well conserved among flavonoid UGTs and A3GlyTs, respectively. The results of His22 to Ala, and Asp121 to Ala or Asn substitutions support this hypothesis.

Recently, the crystal structure of authentic flavonoid UGT from grape (*Vitis vinifera*) has been solved (Offen et al. 2006). The structure indicates that Asp374, Gln375, and Thr141 are crucial residues for direct interaction with hydroxyl groups on the glucose moiety of UDP-glucose. Asp374 in grape A3GlcT corresponds to Glu394 in betanidin 5GlcT. Gln375 is conserved among anthocyanidin/anthocyanin GlcT enzymes. The corresponding His residues in GalTs may be involved in UDP-galactose binding. However, Gln375His mutagenesis of grape A3GlcT did not result in GalT activity (Kubo et al. 2004). Further structural studies of flavonoid UGTs based on crystal structure data, site-directed mutagenesis, and 3D-modelling will be useful in identifying substrate and regiospecificity of these enzymes, and could provide the critical information necessary for production of valuable compounds by metabolic engineering.

6.3 Methyltransferases

Methylation is also a frequently observed modification of flavonoids. Most of the reported anthocyanins (90%) with properly identified structures are based on six common anthocyanidins (pelargonidin, cyanidin, delphinidin, peonidin, petunidin, and malvidin). The three anthocyanidins (peonidin, petunidin, and malvidin) that are methylated comprise 20% of the reported anthocyanins. Rare methylated anthocyanidins include 5-methylcyanidin, 5-methyldelphinidin, 5-methylpetunidin, 5-methylmalvidin, 7-methylpeonidin, 7-methylmalvidin, and methylated 3-desoxyanthocyanidins (Anderson and Jordheim 2006).

S-adenosyl-L-methionine (SAM) dependent O-methyltransferases (OMTs) catalyze the methylation of many natural plant compounds. Plant OMTs are classified into two groups (Joshi and Chiang 1998). Class I OMTs have a molecular weight of approximately 23–27 kDa and require divalent ions such as Mg^{2+} for the activity. Class II includes Mg^{2+}-independent OMTs with a molecular weight of approximately 38–43 kDa. Both OMTs have three conserved motifs (motifs A through C), though the distances between motifs are slightly different from one class to another (Joshi and Chiang 1998). The motifs are thought to be SAM binding domains (motif A; (Val/Ile/Leu)-(Val/Leu)-(Asp/Lys)-(Val/Ile)-Gly-Gly-Xaa$_2$-(Gly/Ala), motif B; (Val/Ile/Phe)-(Ala/Pro/Glu)-Xaa-(Ala/Pro/Gly)-Asp-Ala-Xaa3-Lys-(Trp/Tyr/Phe), motif C; (Ala/Pro/Gly/Ser)-(Leu/Ile/Val)-(Ala/Pro/Gly/Ser)-Xaa$_2$-(Ala/Pro/Gly/Ser)-(Lys/Arg)-(Val/Ile)-(Glu/Ile)-(Leu/Ile/Val)). Class I OMTs include caffeoyl coenzyme A 3-OMTs. Class II OMTs consist of flavonoid OMTs (flavonol 3'-OMT, flavonoid 7-OMT, isoflavone OMT, and isoliquiritigenin 2'-OMT), caffeic acid 3-OMTs, catechol OMTs, *myo*-inositol OMTs, and others (Joshi and Chiang 1998).

6.3.1 Anthocyanin Methyltransferases

The anthocyanin biosynthetic pathway has been extensively studied using petunia, snapdragon, and maize as model plants, with most of the information about anthocyanin OMT (AOMT) coming from petunia. AOMT is predominantly localized in the cytosol (Jonsson et al. 1983), and anthocyanin 3'-OMT and the anthocyanin 3',5'-OMT are encoded by genes, *Mt1/Mt2* and *Mf1/Mf2*, respectively. The enzyme activity of both types of AOMTs is correlated with the *An1* and *An2* genotypes (Gerats et al. 1984; Quattrocchio et al. 1993). cDNAs encoding AOMTs were first isolated from *Petunia hybrida* by differential screening using cDNAs from two genetic lines, V26 (*An1*) and W162 (*an1*), as probes (Brugliera et al. 2003). AOMT cDNAs from *Torenia hybrida* and *Fuchsia* were also cloned. Interestingly, these three anthocyanin OMTs are in the class I OMT family, rather than in the class II family, whereas flavonoid OMTs are in the class II family (Fig. 6.2). AOMTs from petunia and torenia catalyze the methylation of delphinidin derivatives (delphinidin 3-O-glucoside,

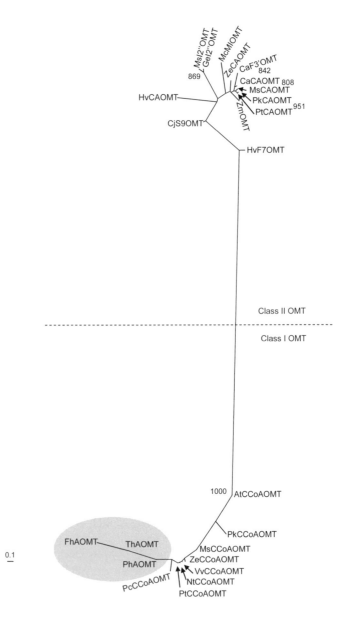

Fig. 6.2 Phylogenetic tree of deduced amino acid sequences of cloned OMTs. The tree was constructed using the neighbor joining method of TREEVIEW program (Page 1996). Numbers indicate bootstrap values greater than 800. Bar = 0.1 amino acid substitutions per site. Anthocyanin OMTs are shaded. GenBank accession numbers are in parentheses: AtCCoAOMT (L40031); CaCAOMT (U16793); CaF3′OMT

(U16794); CjS9OMT (D29809); GeI2″OMT (D88742); HvCAOMT (U54767); HvF7OMT (X77467); McMIOMT (M87340); MsCAOMT (M63853); MsCCoAOMT (U20736); MsI2″OMT (L10211); NtCCoAOMT (Z56282); PeC-CoAOMT (A22706); PkCAOMT (D49710); PkCCoAOMT (AB000408); PtCAOMT (X62096); PtCCoAOMT (U27116); VvCCoAOMT (Z54233); Ze-CAOMT (U19911); ZeCCoAOMT (U13151); ZmOMT (M73235). AOMT; anthocyanin OMT, CAOMT; Caffeic acid OMT, CCoAOMT; cafferoyl-CoA OMT, F7OMT; flavonoid 7-OMT, I2″OMT; isoliquiritigenin 2″-OMT, MIOMT; myoinositol OMT; S9OMT; scoulerine 9-OMT. Abbreviations for species: Ca; *Chrysosplenium americanum*, Cj; *Coptis japonica*, Fh; *Fuchsia spp*, Ge; *Glycyrrhiza echinata*, Hv; *Hordeum vulgare*, Mc; *Mesembryanthemum crystallinum*, Ms; *Medicago sativa*, Nt; *Nicotiana tabacum*, Pc; *Petroselium crispum*, Ph; *Petunia hybrida*, Pk; *Populus kitakamiensis*, Pt; *Populus tremuloides*, Th; *Torenia hybrida*, Vv; *Vitis vinifera*, Ze; *Zinnia elegans*, Zm; *Zea mays*

delphinidin 3-*O*-rutinoside, and delphinidin 3,5-*O*-diglucoside) to petunidin and malvidin glycoside using SAM as a methyl donor. However, the predominant OMT activity of petunia AOMT is anthocyanin 3′-OMT and that for torenia AOMT is anthocyanin 3′,5′-OMT. Sequence identity of the petunia and torenia AOMTs at the amino acid level is 56%, but additional details about these AOMTs are limited. Identification of the AOMT-encoding genes and biochemical characterization of their products from various plants will be required to provide a better understanding of anthocyanin methylation.

6.4 Acyltransferases

6.4.1 Acylation of Anthocyanins

Acylation is one of the most common modifications of plant secondary metabolites, including anthocyanins. More than 65% of the reported anthocyanins whose structures are adequately characterized are acylated (Anderson and Jordheim 2006). The structural diversity of anthocyanins is greatly increased by acylation with aromatic and/or aliphatic substituents, which are generally linked to the glycosyl moieties of anthocyanins (Fig. 6.3). Aromatic acyl substituents of acylated anthocyanins are generally hydroxycinnamoyl groups, such as *p*-coumaryl, caffeyl, feruryl, and sinapyl groups, with some exceptions (e.g. gallyl groups). Aliphatic acyl substituents include malonyl, acetyl, succinyl, malyl, oxalyl, and tartaryl groups, among which the malonyl group is the most widely found. For anthocyanins with acylated glucosyl group(s), both the aromatic and aliphatic acyl groups are, in many cases, linked to the 6-positions of glucosy groups. In cultured cells of carrot (*Daucas carota*) and parsley (*Petroselium hortense*), the acylation of anthocyanins and other flavonoids was shown to be important for selective transport of these flavonoids into vacuoles (Hopp and Seitz 1987; Matern et al. 1986).

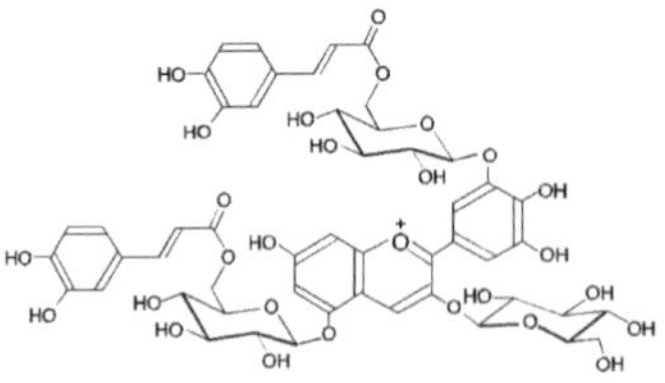

Fig. 6.3 Examples of polyacylated anthocyanins. **1**, gentiodelphin (*Gentiana triflora*); **2**, cinerarin (*Senecio cruentus*); **3**, ternatin A₁ (*Clitoria ternatea*)

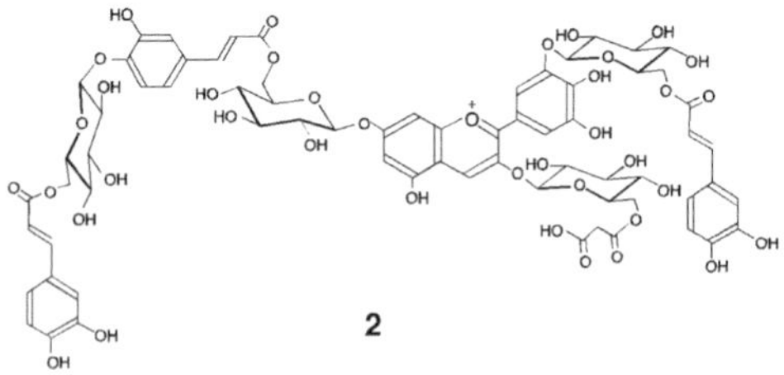

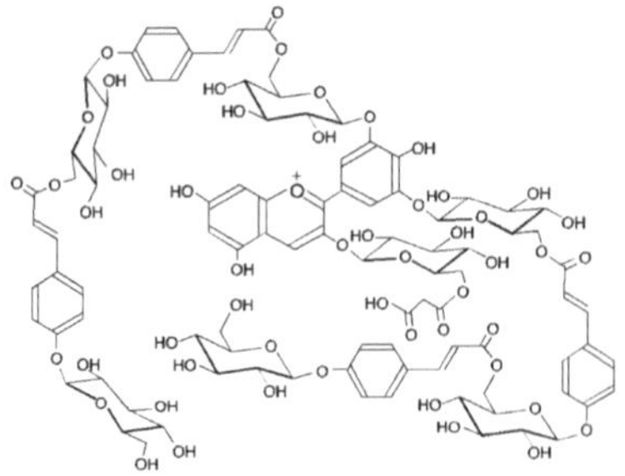

6.4.2 Stabilization of Anthocyanins by Acylation

Anthocyanins, without acyl groups are quickly decolorized in neutral or weakly acidic aqueous solutions. This decoloration arises from hydration at the C-2 position of the anthocyanidin flavylium nucleus (Brouillard and Dangles 1994). Generally, the coloration of acylated anthocyanins is more stable than non-acylated forms under the same cellular conditions. In particular, modification with multiple aromatic acyl groups (i.e., polyacylation; see Fig. 6.3) makes anthocyanin coloration highly stable and bluer (Brouillard and Dangles 1994), thus playing a very important role in the stable coloration of many blue flowers (Honda and Saito 2002) such as gentian (*Gentiana triflora*), cineraria (*Senecio cruentus*), and butterfly pea (*Clitoria ternatea*). These effects of polyacylation arise from the intramolecular face-to-face stacking of aromatic acyl groups and the anthocyanidin nucleus (Brouillard and Dangls 1994). Both the bluing and stabilization effects of polyacylation depend on the number as well as positions of aromatic acyl groups. Anthocyanins with aromatic acylated glycosyl groups in both the 7- and 3′-positions appear to provide the most stable blue flower colors (Honda and Saito 2002; Anderson and Jordheim 2006). Aliphatic acylation does not appear to intrinsically alter the absorption spectra of anthocyans *in vitro* or *in vivo* (Suzuki et al. 2002). However, aliphatic acylation does enhance the stability of anthocyanin coloration. For example, the stability of cyanidin 3-*O*-6″-*O*-malonylglucoside coloration in aqueous solution at pH 5 and 7 was higher than those of cyanidin 3-*O*-glucoside and cyanidin aglycone (Suzuki et al. 2002). It must be mentioned that the modification of anthocyanins with dicarboxylic acid(s) (e.g., malonylation and malylation) is also important for aqueous solubility. Moreover, the terminal carboxy group of dicarboxylic acyl substituents should serve as a buffer component that maintains the acidity of vacuolar contents, thus making a contribution to flower color stabilization.

Finally, acylation should also confer biochemical or catabolic stability on stored anthocyanins in plant cells, as acylation prevents the indiscriminate degradation of anthocyanin storage forms by microbial glycosidases, most of which are unable to act on acylated glycosides (Suzuki et al. 2002).

6.4.3 Anthocyanin Acyltransferases

The formation of acylated anthocyanins is mediated by anthocyanin acyltransferases (AATs), which are anthocyanin-specific and catalyze the transfer of acyl groups from activated acyl donors to the acceptor anthocyanins. Two distinct types of AATs have been identified based on the acyl donors: the BAHD superfamily of enzymes, which use acyl coenzyme A thioesters (acyl-CoA) as acyl donors (Nakayama et al. 2003), and the serine carboxypeptidase-like (SCPL) group, which use acyl-activated sugars (i.e., β-acetal esters or 1-O-β-glucose esters) as acyl donors (Milkowski and Strack 2004). A large number of acyl-CoA-dependent AATs have been identified in many plant species and extensively characterized at the molecular level. By contrast, acyl-1-O-β-glucose-dependent AAT activities have been found only in a few plant species, until recently.

6.4.4 Acyl-CoA-Dependent AATs

General aspects – Studies on acyl-CoA-dependent AAT activities with different acyl-CoA specificity, acyl-acceptor specificity, and acyl transfer regiospecificity have been pursued in cell culture as well as in flowers and other plant organs where acylated anthocyanins accumulate (Nakayama et al. 2003). To date, many acyl-CoA-dependent AATs have been cloned, sequenced, and biochemically characterized in their native and/or recombinant forms (Fig. 6.4) (Nakayama et al. 2003). These AATs are monomeric proteins with an approximate molecular mass of 50 kDa and are predicted to be located in the cytoplasm (or cytoplasmic surface of the ER), judging from the absence of any transit sequences in their primary structures, and their optimum pH for catalytic activity being at neutral to slightly alkaline (7.0–8.5). Results of immunocytochemical analyses of the subcellular localization of Gt5AT confirm this prediction (Fujiwara et al. 1998). Analyses of spatial and temporal expression of some of these AATs show that they are specifically expressed in pigmented organs in a temporally-regulated manner (Yonekura-Sakakibara et al. 2000; Suzuki et al. 2002), as are other anthocyanin biosynthetic enzymes, though some acyl-CoA-dependent AATs are expressed in acyanic organs (Suzuki et al. 2004b). Recent integrated metabolome/transcriptome analyses of *Arabidopsis* plants that overexpress a MYB-related transcription factor (PAP1*)* led to the identification of three AAT genes that are likely to be involved in the biosynthesis of an anthocyanin carrying one each of sinapyl, *p*-coumaryl, and malonyl groups in *A. thaliana* (Tohge et al. 2005). Expression of these AAT genes is up-regulated by PAP1 along with those of another set of flavonoid biosynthetic genes.

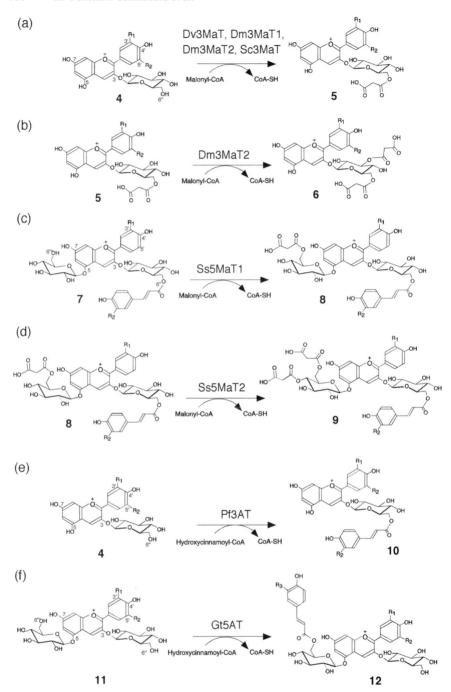

Fig. 6.4 Regiospecific acyl transfer reactions catalyzed by acyl-CoA-dependent AATs. Dv3MaT, malonyl-CoA: anthocyanidin 3-*O*-glucoside 6″-*O*-malonyltransferase of *Dahlia*

variabilis; Dm3MaT1, malonyl-CoA: anthocyanidin 3-*O*-glucoside 6''-*O*-malonyltransferase of *Dendranthema* x *morifolium*; Dm3MaT2, malonyl-CoA: anthocyanidin 3-*O*-glucoside 3'',6''-*O*-dimalonyltransferase of *D. morifolium*; Sc3MaT, malonyl-CoA: anthocyanidin 3-*O*-glucoside 6''-*O*-malonyltransferase of *Senecio cruentus*; Ss5MaT1, malonyl-CoA: anthocyanin 5-*O*-glucoside 6'''-*O*-malonyltransferase of *Salvia splendens*; Ss5MaT2, malonyl-CoA: anthocyanin 5-*O*-glucoside 4'''-*O*-malonyltransferase of *S. splendens*; Pf3AT, hydroxycinnamoyl-CoA: anthocyanidin 3-*O*-glucoside 6''-*O*-hydroxycinnamoyltransferase of *Perilla frutescens*; Gt5AT, hydroxycinnamoyl-CoA: anthocyanin 3,5-*O*-diglucoside 6'''-*O*-hydroxycinnamoyltransferase of *Gentiana triflora*.

Acyl donor specificity – Acyl-CoA-dependent AATs can be classified into two distinct categories on the basis of their acyl-donor specificity; i.e., aliphatic acyl-transferases or aromatic acyltransferases. Aliphatic AATs do not act on aromatic acyl-CoAs, and aromatic AATs do not utilize aliphatic acyl-CoAs (Nakayama et al. 2003). Aromatic AATs that have been characterized so far can use *p*-coumaroyl- as well as caffeoyl-CoAs. All of the aliphatic-acyl-CoA-dependent AATs that have been cloned and characterized are malonyltransferases involved in the biosynthesis of malonylated anthocyanins. These enzymes can use other aliphatic acyl-CoAs, such as acetyl-CoA, methylmalonyl-CoA, and succinyl-CoA, as weak acyl donors.

Regiospecificity of acyl transfer – Acyl-CoA-dependent AATs generally display strict specificity for the position of anthocyanin acylation. In most cases, AATs catalyze a single regiospecific acyl-transfer reaction, and the resulting product no longer serves as an acyl acceptor of that enzyme (Nakayama et al. 2003). Thus, during the biosynthesis of anthocyanins with multiple acyl groups, a series of acylations is completed by the actions of a series of AATs with different acyl donor specificity and regiospecificity, each furnishing the necessary precursor for subsequent modifications. During biosynthesis of the dominant anthocyanin of scarlet sage flowers, (salvianin, Fig. 6.4, **9**) for example, 6''-*O*-caffeylation takes place first, followed by 6'''-*O*-malonylation and 4'''-*O*-malonylation. The 6''-*O*-caffeylation is catalyzed by a 3-aromatic AAT (Fig. 6.4e), and the 6'''-*O*-malonylation (Fig. 6.4c) and subsequent 4'''-*O*-malonylation steps (Fig. 4d) are catalyzed by two distinct malonyltransferases called Ss5MaT1 (Suzuki et al. 2001) and Ss5MaT2 (Suzuki et al. 2004b), respectively. It must be mentioned, however, that an AAT capable of catalyzing multiple, consecutive acylations of anthocyanins has recently been identified in the flowers of chrysanthemum (*Dendranthema* x *morifolium*) (Suzuki et al. 2004a). This enzyme, Dm3MaT2, malonylates cyanidin 3-*O*-glucoside to produce cyanidin 3-*O*-6''-*O*-malonylglucoside (Fig. 6.4a), which subsequently serves as a substrate for the same enzyme to produce cyanidin 3-*O*-3'',6''-*O*-dimalonylglucoside (Fig. 6.4b), a dominant anthocyanin in red chrysanthemum flowers .

Acyl acceptor specificity – It has been generally observed that the number of B-ring hydroxy functions of substrate anthocyanin does not affect reactivity to AATs, which, therefore, can act equally on any of the anthocyanidin types – pelargonidin, cyanidin, and delphinidin types – of anthocyanins (Nakayama et al. 2003). Such a specificity for anthocyanidin types contributes to the formation of a "metabolic grid" in the anthocyanin biosynthetic pathways (Yamazaki et al. 1999). Moreover, some AATs have even been shown to act on flavonoid glycosides, such as quercetin 3-*O*-glucoside, though weakly (Suzuki et al. 2002). Depending on the enzyme, the extent and positions of glycosylation and acylation of substrate anthocyanin may affect their reactivity to AATs. For example, Ss5MaT1 acts on anthocyanidin 3-*O*-(6″-*O*-hydroxycinnamylglucoside)-5-*O*-glucosides (Fig. 6.4, **7**) but not on either anthocyanidin 3,5-*O*-diglucosides (**11**) (Suzuki et al. 2001). *p*-Coumaric acid, which mimics the 6″-*O*-hydroxycinnamyl moiety of **7**, is a dead-end competitive inhibitor of Ss5MaT1 with respect to substrate anthocyanins (Suzuki et al. 2003), suggesting the absolute requirement for a 6″-*O*-hydroxycinnamyl group at the 3-*O*-glucosyl moiety of the substrate for its efficient binding to Ss5MaT1 (Suzuki et al. 2001). In contrast, Gt5AT acts on **11**, but not on **7** (Fujiwara et al. 1997, 1998).

Phylogenetics of acyl-CoA-dependent AATs – Amino acid sequences of known AATs from different plant species share only 30-60% sequence identity, despite their biochemical similarity. However, they share the consensus sequences His-Xaa$_3$-Asp (motif 1) and Asp-Phe-Gly-Trp-Gly (motif 3), which have been specifically identified in members of the BAHD superfamily (or versatile acyltransferase (VAT) family) of enzymes (Suzuki et al. 2001; Nakayama et al. 2003). This indicates a close phylogenetic relationship of these AATs to this family. The BAHD family is a large group of acyltransferases that are involved in plant secondary metabolism, such as the biosynthesis of volatile esters, waxes, alkaloids, and terpenoids (D'Auria 2006). Molecular phylogenetic analyses of BAHD family members yield a cladogram with several major clades (Fig. 6.5). Most AATs may be categorized, along with other BAHD members involved in the modification of other flavonoids and phenolic glucosides into a single clade (Taguchi et al. 2005) (Fig. 6.5; shown with gray background). BAHD members of this clade share the Tyr-Phe-Gly-Asn-Cys sequence (motif 2), which may serve as a cladic signature sequence (Suzuki et al. 2001). cDNA clones for Dv3MaT and Sc3MaT were isolated by a homology-based strategy taking full advantage of the specific conservation of motif 2 (Suzuki et al. 2002). This phylogenetic tree also suggests that AATs may be further sub-clustered on the basis of plant species, rather than acyl-donor specificity (i.e., aliphatic vs. aromatic). It must be noted, however, that not all AATs are in this clade. Ss5MaT2 (see above and Fig. 6.4d) lacks the signature sequence of the other AAT members and is in a clade that contains enzymes that accept a wide range of alcohol substrates to produce volatile esters, alkaloids, and other compounds (Fig. 6.5) (Suzuki et al. 2004b).

Structure and proposed catalytic mechanism of AAT – AAT catalysis was kinetically predicted to proceed with the formation of a ternary complex consisting of acyl-CoA, an acyl acceptor, and an enzyme, prior to chemical catalysis (Suzuki et al. 2003). It was also proposed that, in the ternary complex, a general base deprotonates a hydroxyl of the acyl acceptor to facilitate its nucleophilic attack on the carbonyl carbon of acyl-CoA. Several lines of biochemical evidence suggest the involvement of His (in motif 1) or Asp (in motif 3) residues during AAT catalysis (Suzuki et al. 2003). The crystal structure of a member of the BAHD family, vinorine synthase of *Rauvolfia serpentina*, was solved in 2005 (Ma et al. 2005). In 2007, the crystal structures of Dm3MaT3, an AAT homolog of Dm3MaT1 and Dm3MaT2, were determined at 1.8~2.2-Å resolutions (Unno et al. 2007), providing the first crystal structure of a BAHD member complexed with acyl-CoA. The crystal structures, along with the results of mutagenesis studies of Dm3MaT1 and Dm3MaT2, allowed to unambiguously identify the acyl-CoA and anthocyanin binding sites in AATs,

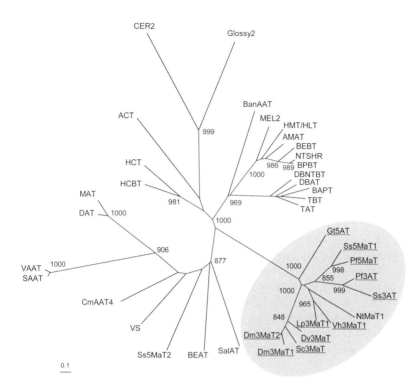

Fig. 6.5 Non-rooted molecular phylogenetic tree of the BAHD family constructed by the neighbor-joining method. The lengths of the lines indicate the relative distances between nodes. Numbers indicate bootstrap values greater than 800. Bar = 0.1 amino acid substitutions per site. The AAT-related clade is shown with gray background. Enzymes used for alignment are as follows: ACT, agmatine coumaroyltransferase of *Hordeum vulgare*; AMAT, an-thraniloyl-CoA:methanol acyltransferase of *Vitis labrusca*; BanAAT, alcohol acyltransferase of banana; BAPT, baccatin III *O*-phenylpropanoyltransferase of *Taxus cuspidata*; BEAT, benzylalcohol *O*-acetyltransferase of *Clarkia breweri*; BEBT, benzoyl-CoA:benzylalcohol *O*-benzoyltransferase of *C. breweri*; BPBT, benzoyl-CoA:bezylalcohol/phenylethanol benzoyl-transferase of *Petunia hybrida*; CER2, an acyltransferase involved in wax biosynthesis in *Arabidopsis thaliana*; CmAAT4, alcohol acyltransferase of *Cucumis melo*; DAT, deacetylvin-doline 4-*O*-acetyltransferase of *Catharanthus roseus*; DBAT, 10-deacetylbaccatin III-10-*O*-acetyltransferase of *T. cuspidata*; DBNTBT, 3′-*N*-debenzoyl-2′-deoxytaxol *N*-benzoyltransferase of *T. cuspidata*; Dm3MaT1, malonyl-CoA: anthocyanin 3-*O*-glucoside 6″-*O*-malonyltransferase of *Dendranthema x morifolium*; Dm3MaT2, malonyl-CoA: anthocyanin 3-*O*-glucoside 3″,6″-*O*-dimalonyltransferase of *D. morifolium*; Dv3MaT, malonyl-CoA: anthocyanin 3-*O*-glucoside 6″-*O*-malonyltransferase of *Dahlia variabilis*; Glossy2, an acyl-transferase involved in wax biosynthesis in *Zea mays*; Gt5AT, hydroxycinnamoyl-CoA: an-thocyanin 3,5-*O*-diglucoside 6″′-*O*-hydroxycinnamoyltransferase of *Gentiana triflora*; HCBT, anthranilate *N*-hydroxycinnamoyl/benzoyltransferase of *Dianthus caryophyllus*; HCT, hy-droxycinnamoyltransferase from *Nicotiana tabacum*; HMT/HLT, tigloyl-CoA:13α-

hydroxymultifluorine/13α-hydroxylupanine *O*-tigloyltransferase of *Lupinus albus*; Lp3MaT1, flavonol 3-*O*-glucoside-6″-*O*-malonyltransferase of *Lamium purpureum*; MAT, minovincinine-19-hydroxy-*O*-acetyltransferase of *C. roseus*; MEL2, an acyltransferase homolog of *C. melo*; NTSHR, Hsr201 protein of *N. tabacum*; NtMaT1, malonyl-CoA:flavonoid/naphthol glucoside acyltransferase of *N. tabacum*; Pf3AT, hydroxycinnamoyl-CoA: anthocyanin 3-*O*-glucoside 6″-*O*- hydroxycinnamoyltransferase of *Perilla frutescens*; Pf5MaT, malonyl-CoA: anthocyanin 5-*O*-glucoside 6‴-*O*-malonyltransferase of *P. frutescens*; SAAT, alcohol acyltransferase of *Fragaria x ananassa*; SalAT, salutaridinol 7-*O*-acetyltransferase of *Papaver somniferum*; Sc3MaT, malonyl-CoA: anthocyanin 3-*O*-glucoside 6″-*O*-malonyltransferase of *Senecio cruentus*; Ss5MaT1, malonyl-CoA: anthocyanin 5-*O*-glucoside 6‴-*O*-malonyltransferase of *Salvia splendens*; Ss5MaT2, malonyl-CoA: anthocyanin 5-*O*-glucoside 4‴-*O*-malonyltransferase of *S. splendens*; Ss3AT, hydroxycinnamoyl-CoA: anthocyanin 3-*O*-glucoside 6″-*O*- hydroxycinnamoyltransferase of *S. splendens*; TAT, taxa-4(20),11(12)-dien-5α-ol-*O*-acetyltransferase of *T. cuspidata*; TBT, taxane benzoyltransferase of *T. cuspidata*; VAAT, alcohol acyltransferase of *Fragaria vesca*; Vh3MaT1, flavonol 3-*O*-glucoside-6″-malonyltransferase of *Verbena hybrida*; VS, vinorine synthase of *Rauvolfia serpentina*

providing important information about the catalytic mechanism and the roles of the shared motifs of BAHD family enzymes (Unno et al. 2007). The active site of Dm3MaT3 is located in a solvent channel that runs between two domains, and the His residue of motif 1 that is available from both sides of the channel is predicted to serve as a catalytic residue. Motif 3 is located away from the active site of the enzyme and does not seem to participate in the postulated catalytic mechanism. These observations suggest the importance of the His residue in motif 1 in the general acid/base mechanism of AAT catalysis.

6.4.5 Serine Carboxypeptidase-like (SCPL)-AATs

In plant secondary metabolism, there is an alternative ester formation pathway that is catalyzed by acyltransferases utilizing 1-*O*-β-acetal esters (1-*O*-β-glucose esters) as acyl donors rather than acyl-CoA thioesters (Li and Steffens 2000). Sequence and phylogenetic comparisons of these acyltransferases revealed their strong homology with serine carboxypeptidases belonging to the α/β hydrolase superfamily, thus defining them as serine carboxypeptidase-like (SCPL) acyltransferases (Milkowski and Strack 2004). 1-*O*-acylglycoside-dependent AAT activity was identified in cell cultures of *Daucus carota* in 1992 (Gläßgen and Seitz 1992). This enzyme catalyzes acyl transfer from hydroxycinnamoyl-1-*O*-glucosides to position 6 of the glucose moiety of cyanidin 3-(6″-*O*-glucosido-2″-*O*-xylosido) galactoside. 1-*O*-malylglucoside-dependent AAT activity was also identified in red flowers of carnations (*Dianthus caryophyllus*), whose color mainly arises from malylated anthocyanins. The malylation of anthocyanins in red carnation is catalyzed by 1-*O*-malylglucose:pelargonidin 3-*O*-glucose-6″-*O*-malyltransferase, which utilizes 1-*O*-β-d-malylglucose as the malyl donor (Abe et al. 2008). Although primary structures of these AATs remain to be determined, it is likely that these enzymes are SCPL proteins. In 2006, an AAT catalyzing the aromatic acyl transfer from 1-*O*-acylglucose to an anthocyanin that is involved in the biosynthesis of ternatin A_1 (Fig. 6.3, **3**) was purified from petals of butterfly pea (*C. ternatea*) (Noda et al. 2006). Molecular cloning of a cDNA encoding this enzyme, delphinidin 3-(6″-O-malonyl)glucoside-3,5-diglucoside-6‴-p-coumaroyltransferase (3′-AT), revealed that it encoded a presumptive SCPL protein with an N-terminal signal sequence, three potential N-glycosylation sites, and a Ser-His-Asp putative catalytic triad. Cellular localization of SCPL-AATs remains to be established. However, some other SCPL acyltransferases appear to be localized in vacuoles (Hause et al. 2002), unlike the acyl-CoA-dependent AATs (see above). Thus, SCPL-AATs could also be vacuolar enzymes.

6.4.6 Potential Application of AATs

Flower color modification – The accumulation of polyacylated anthocyanins in the flowers of transgenic plants may serve as one strategy for the creation of blue flowers (Tanaka et al. 2005). Therefore, genes encoding for aromatic AATs with different specificities will be important tools for flower color modification. Although aliphatic acylation does not intrinsically alter the absorption spectra of anthocyanins, it enhances the stability of anthocyanin coloration (see above), thus would also be important for controlling flower color. Overexpression of AAT-encoding genes in heterologous plants may provide epistatic anthocyanin modifications and result in color-modified plants.

Modification of the functional properties of anthocyanins – Because anthocyanin acylation may also improve their biological activities (Yoshimoto et al. 2001), biological half-lives, membrane permeability, and intestinal absorption, there has been a growing interest among the food and biomedical industries in using specific acylations to modify the functional properties of anthocyanins. The regioselective acylation of cyanidin 3-O-glucoside with aromatic acids has been attained in cultured cells of *Ipomoea batatas* that have aromatic AAT activity (Nakajima et al. 2000).

6.5 Conclusions and Perspectives

In the last decade, our understanding of the molecular genetics and biochemistry of anthocyanin modification has been greatly advanced by the cloning of cDNAs encoding glycosylation, methylation and acylation enzymes. With the aim of further stabilizing anthocyanins, a more rationalized system of engineering must be based upon a fundamental understanding of enzymatic mechanisms. Recent advances in three-dimensional structural analyses of key enzymes such as UDP-sugar-dependent glycosyltransferase (Offen et al. 2006) and BAHD-family acyltransferases (Ma et al. 2005; Unno et al. 2007) should facilitate the rational design of engineered proteins for anthocyanin modification. Functional identification of genes involved in the modification of anthocyanins in exotic plant species with diverse anthocyanin molecules can be achieved by integrating extensive gene expression profiling (transcriptomics) and detailed metabolic profiling (metabolomics). A successful example of this approach is the comprehensive annotation of the genes responsible for anthocyanin modification in *Arabidopsis* (Tohge et al. 2005) and tomato (Mathews et al. 2003). The combination of structural biology and functional genomics will allow us to realize the rational design of anthocyanins as demonstrated by the recent triumph of a "blue rose" (Tanaka and Brugliera 2006).

References

Abe, Y., Tera, M., Sasaki, N., Okamura, M., Umemoto, N., Momose, M., Kawahara, N., Kamakura, H., Goda, Y., Nagasawa, K., Ozeki, Y. (2008) Detection of 1-*O*-malylglucose: pelargonidin 3-*O*-glucose-6"-*O*-malyltransferase activity in carnation (Dianthus caryophyllus). Biochem. Biophys. Res. Commun. DOI: 10.1016/j.bbrc.2008.04.153

Anderson, Ø.M. and Jordheim, M. (2006) The Anthocyanins. In: Ø.M. Anderson and K.R. Markham (Eds.), *Flavonoids: Chemistry, Biochemistry and Applications.* CRC/Taylor & Francis, Boca Raton, pp. 471–551.

Bar-Peled, M., Lewinsohn, E., Fluhr, R. and Gressel, J. (1991) UDP-rhamnose: flavanone-7-*O*-glucoside-2"-*O*-rhamnosyltransferase. Purification and characterization of an enzyme catalyzing the production of bitter compounds in citrus. J. Biol. Chem. 266, 20953–20959.

Boss, P.K., Davies, C. and Robinson, S.P. (1996) Expression of anthocyanin biosynthesis pathway genes in red and white grapes. Plant Mol. Biol. 32, 565–569

Brouillard, R. and Dangles, O. (1994) Flavonoids and flower colour. In: J.B. Harbone (Ed.), *The Flavonoids.* Chapman & Hall/CRC, Washington DC, pp. 565–599.

Brugliera, F., Holton, T.A., Stevenson, T.W., Farcy, E., Lu, C.Y. and Cornish, E.C. (1994) Isolation and characterization of a cDNA clone corresponding to the *Rt* locus of *Petunia hybrida.* Plant J. 5, 81–92.

Brugliera, F., Demelis, L., Koes, R. and Tanaka, Y. (2003) Genetic sequences having methyltransferase activity and uses therefore. International publication number WO 03/062428 A1.

D'Auria, J.C. (2006) Acyltransferases in plants: a good time to be BAHD. Curr. Opin. Plant Biol. 9, 331–340.

Davies, K.M. and Schwinn, K.E. (2006) Molecular Biology and Biotechnology of Flavonoid Biosynthesis. In: Ø.M. Anderson and K.R. Markham (Eds.), *Flavonoids: Chemistry, Biochemistry and Applications.* CRC/Taylor & Francis, Boca Raton, pp. 143–218.

Dooner, H.K. and Nelson, O.E. (1977) Controlling element-induced alterations in UDPglucose:flavonoid glucosyltransferase, the enzyme specified by the *bronze* locus in maize. Proc. Natl. Acad. Sci. USA 74, 5623–5627.

Fedoroff, N.V., Furtek, D.B. and Nelson, O.E. (1984) Cloning of the *bronze* locus in maize by a simple and generalizable procedure using the transposable controlling element *Activator* (*Ac*). Proc. Natl. Acad. Sci. USA 81, 3825–3829.

Ford, C.M., Boss, P.K. and Hoj, P.B. (1998) Cloning and characterization of *Vitis vinifera* UDP-glucose: flavonoid 3-*O*-glucosyltransferase, a homologue of the enzyme encoded by the maize *Bronze-1* locus that may primarily serve to glucosylate anthocyanidins *in vivo.* J. Biol. Chem. 273, 9224–9233.

Frydman, A., Weisshaus, O., Bar-Peled, M., Huhman, D.V., Sumner, L.W., Marin, F.R., Lewinsohn, E., Fluhr, R., Gressel, J. and Eyal, Y. (2004) Citrus fruit bitter flavors: isolation and functional characterization of the gene Cm1,2RhaT encoding a 1,2 rhamnosyltransferase, a key enzyme in the biosynthesis of the bitter flavonoids of citrus. Plant J. 40, 88–100.

Fujiwara, H., Tanaka, Y., Fukui, Y., Nakao, M., Ashikari, T. and Kusumi, T. (1997) Anthocyanin 5-aromatic acyltransferase from *Gentiana triflora.* Purification, characterization, and its role in anthocyanin biosysnthesis. Eur. J. Biochem. 249, 45–51.

Fujiwara, H., Tanaka, Y., Yonekura-Sakakibara, K., Fukuchi-Mizutani, M., Nakao, M., Fukui, Y., Yamaguchi, M., Ashikari, T. and Kusumi, T. (1998) cDNA cloning, gene expression and subcellular localization of anthocyanin 5-aromatic acyltransferase from *Gentiana triflora.* Plant J. 16, 421–431.

Fukuchi-Mizutani, M., Okuhara, H., Fukui, Y., Nakao, M., Katsumoto, Y., Yonekura-Sakakibara, K., Kusumi, T., Hase, T. and Tanaka, Y. (2003) Biochemical and molecular characterization of a novel UDP-glucose: anthocyanin 3'-O-glucosyltransferase, a key enzyme for blue anthocyanin biosynthesis, from gentian. Plant Physiol. 132, 1652–1663.

Furtek, D., Schiefelbein, J.W., Johnston, F. and Nelson Jr., O.E. (1988) Sequence comparisons of three wild-type *Bronze-1* alleles from *Zea mays*. Plant Mol. Biol. 11, 473–481.

Gachon, C.M.M., Langlois-Meurinne M. and Saindrenan P. (2005) Plant secondary metabolism glycosyltransferases: the emerging functional analysis. Trends Plant Sci. 10, 542–549.

Gerats, A.G.M., Farcy, E., Wallroth, M., Groot, S.P.C. and Schram, A. (1984) Control of anthocyanin biosynthesis in *Petunia hybrida* by multiple allelic series of the genes *An1* and *An2*. Genetics 106, 501–508.

Gläßgen, W.E. and Seitz, H.U. (1992) Acylation of anthocyanins with hydroxycinnamic acids via 1-O-acylglucosides by protein preparations from cells cultures of *Daucus carota* L. Planta 186, 582–585

Gong, Z., Yamazaki, M., Sugiyama, M., Tanaka, Y. and Saito, K. (1997) Cloning and molecular analysis of structural genes involved in anthocyanin biosynthesis and expressed in a forma-specific manner in *Perilla frutescens*. Plant Mol. Biol. 35, 915–927.

Hans, J., Brandt, W. and Vogt, T. (2004) Site-directed mutagenesis and protein 3D-homology modelling suggest a catalytic mechanism for UDP-glucose-dependent betanidin 5-O-glucosyltransferase from *Dorotheanthus bellidiformis*. Plant J. 39, 319–333.

Hause, B., Meyer, K., Viitanen, P.V., Chapple, C. and Strack, D. (2002) Immunolocalization of 1- O-sinapoylglucose: malate sinapoyltransferase in *Arabidopsis thaliana*. Planta 215, 26–32.

Heller, W. and Forkmann, G. (1993) Biosynthesis of Flavonoids. In: J.B. Harbone (Ed.), *The Flavonoids: Advances in Research Since 1986*. Chapman & Hall/CRC, London, pp. 499–535.

Honda, T. and Saito, N. (2002) Recent progress in the chemistry of polyacylated anthocyanins as flower color pigments. Heterocycles 56, 633–692

Hopp, W. and Seitz, H.U. (1987) The uptake of acylated anthocyanin into isolated vacuoles from a cell suspension culture of *Daucus carota*. Planta 170, 74–85.

Jonsson, L.M., Donker-Koopman, W.E., Uitslager, P. and Schram, A.W. (1983) Subcellular localization of anthocyanin methyltransferase in flowers of *Petunia hybrida*. Plant Physiol. 72, 287–290.

Joshi, C.P. and Chiang, V.L. (1998) Conserved sequence motifs in plant S-adenosyl-L-methionine-dependent methyltransferases. Plant Mol. Biol. 37, 663–674.

Kroon, J., Souer, E., de Graaff, A., Xue, Y., Mol, J. and Koes, R. (1994) Cloning and structural analysis of the anthocyanin pigmentation locus *Rt* of *Petunia hybrida*: characterization of insertion sequences in two mutant alleles. Plant J. 5, 69–80.

Kubo, A., Arai, Y., Nagashima, S. and Yoshikawa, T. (2004) Alteration of sugar donor specificities of plant glycosyltransferases by a single point mutation. Arch. Biochem. Biophys. 429, 198–203.

Li, A. X. and Steffens, J. C. (2000) An acyltransferase catalyzing the formation of diacylglucose is a serine carboxypeptidase-like protein. Proc. Natl. Acad. Sci. USA 97, 6902–6907.

Lim, E.K. and Bowles, D.J. (2004) A class of plant glycosyltransferases involved in cellular homeostasis. EMBO J. 23, 2915–2922.

Ma, X., Koepke, J., Panjikar, S., Fritzsch, G. and Stöckigt, J. (2005) Crystal structure of vindroine synthase, the first representative of the BAHD superfamily. J. Biol. Chem. 280, 13576–13583.

Marrs, K.A., Alfenito, M.R., Lloyd, A.M. and Walbot, V. (1995) A glutathione S-transferase involved in vacuolar transfer encoded by the maize gene *Bronze-2*. Nature 375, 397–400.

Matern, U., Reichenbach, C. and Heller, W. (1986) Efficient uptake of flavonoids into parsley (*Petroselinum hortense*) vacuoles requires acylated glycosides. Planta 167, 183–189.

Mathews, H., Clendennen, S.K., Caldwell, C.G., Liu, X.L., Connors, K., Matheis, N., Schuster, D.K., Menasco, D.J., Wagoner, W., Lightner, J. and Wagner, D.R. (2003) Activation tagging in tomato identifies a transcriptional regulator of anthocyanin biosynthesis, modification, and transport. Plant Cell 15, 1689–1703.

Milkowski, C. and Strack, D. (2004) Serine carboxypeptidase-like acyltransferases. Phytochemistry 65, 517–524.

Miller, K.D., Guyon, V., Evans, J.N., Shuttleworth, W.A. and Taylor, L.P. (1999) Purification, cloning, and heterologous expression of a catalytically efficient flavonol 3-O-galactosyltransferase expressed in the male gametophyte of Petunia hybrida. J. Biol. Chem. 274, 34011–34019.

Morita, Y., Hoshino, A., Kikuchi, Y., Okuhara, H., Ono, E., Tanaka, Y., Fukui, Y., Saito, N., Nitasaka, E., Noguchi, H. and Iida, S. (2005) Japanese morning glory dusky mutants displaying reddish-brown or purplish-gray flowers are deficient in a novel glycosylation enzyme for anthocyanin biosynthesis, UDP-glucose: anthocyanidin 3-O-glucoside-2"-O-glucosyltransferase, due to 4-bp insertions in the gene. Plant J. 42, 353–363.

Mulichak, A.M., Losey, H.C., Walsh, C.T. and Garavito, R.M. (2001) Structure of the UDP-glucosyltransferase GtfB that modifies the heptapeptide aglycone in the biosynthesis of vancomycin group antibiotics. Structure 9, 547–557.

Nakajima, J., Tanaka, Y., Yamazaki, M. and Saito, K. (2001) Reaction mechanism from leucoanthocyanidin to anthocyanidin 3-glucoside, a key reaction for coloring in anthocyanin biosynthesis. J. Biol. Chem. 276, 25797–25803.

Nakajima, N., Ishihara, K., Hamada, H., Kawabe, S. and Furuya, T. (2000) Regioselective acylation of flavonoid glucoside with aromatic acid by an enzymatic reaction system from cultured cells of Ipomoea batatas. J. Biosci. Bioengin. 90, 347–349.

Nakayama, T., Suzuki, H. and Nishino, T. (2003) Anthocyanin acyltransferases: specificities, mechanism, phylogenetics, and applications. J. Mol. Cat. B: Enzymatic. 23, 117–132.

Noda, N., Kato, N., Kogawa, K., Kazuma, K. and Suzuki M (2004) Cloning and characterization of the gene encoding anthocyanin 3', 5'-O-glucosyltransferase involved in ternatin biosynthesis from blue petals of butterfly pea (Clitoria ternatea). Plant Cell Physiol. 45, s132.

Noda, N., Kazuma, K., Sasaki, T., Kogawa, K. and Suzuki, M. (2006) Molecular cloning of 1-O-acylglucose dependent anthocyanin aromatic acyltransferase in ternatin biosynthesis of butterfly pea (Clitoria ternatea). Plant Cell Physiol. 47, S109.

Offen, W., Martinez-Fleites, C., Yang, M., Kiat-Lim, E., Davis, B.G., Tarling, C.A., Ford, C.M., Bowles, D.J. and Davies, G.J. (2006) Structure of a flavonoid glucosyltransferase reveals the basis for plant natural product modification. EMBO J. 25, 1396–1405.

Ogata, J., Kanno, Y., Itoh, Y., Tsugawa, H. and Suzuki, M. (2005). Plant biochemistry: anthocyanin biosynthesis in roses. Nature 435, 757–758.

Ono, E., Fukuchi-Mizutani, M., Nakamura, N., Fukui, Y., Yonekura-Sakakibara, K., Yamaguchi, M., Nakayama, T., Tanaka, T., Kusumi, T. and Tanaka, Y. (2006). Yellow flowers generated by expression of the aurone biosynthetic pathway Proc. Natl. Acad. Sci. USA 103, 11075–11080.

Page, R. (1996). TreeView: an application to display phylogenetic trees on personal computers. Comput. Appl. Biosci. 12, 357–358.

Quattrocchio, F., Wing, J.F., Leppen, H., Mol, J. and Koes, R.E. (1993) Regulatory genes controlling anthocyanin pigmentation are functionally conserved among plant species and have distinct sets of target genes. Plant Cell 5, 1497–1512.

Saito, N., Tatsuzawa, F., Miyoshi, K., Shigihara, A. and Honda, T. (2003) The first isolation of C-glycosylanthocyanin from the flowers of Tricyrtis formosana. Tetrahedron Lett. 44, 6821–6823.

Sawada, S., Suzuki, H., Ichimaida, F., Yamaguchi, M.A., Iwashita, T., Fukui, Y., Hemmi, H., Nishino, T. and Nakayama, T. (2005) UDP-glucuronic acid: anthocyanin glucuronosyltransferase from red daisy (*Bellis perennis*) flowers. Enzymology and phylogenetics of a novel glucuronosyltransferase involved in flower pigment biosynthesis. J. Biol. Chem. 280, 899–906.

Shao, H., He, X., Achnine, L., Blount, J.W., Dixon, R.A. and Wang, X. (2005) Crystal structures of a multifunctional triterpene/flavonoid glycosyltransferase from *Medicago truncatula*. Plant Cell 17, 3141–3154.

Suzuki, H., Nakayama, T., Yonekura-Sakakibara, K., Fukui, Y., Nakamura, N., Nakao, M., Tanaka, Y., Yamaguchi, M.-A., Kusumi, T. and Nishino, T. (2001) Malonyl-CoA:anthocyanin 5-*O*-glucoside-6'''-*O*-malonyltransferase from scarlet sage (*Salvia splendens*) flowers. J. Biol. Chem. 276, 49013–49019.

Suzuki, H., Nakayama, T., Yonekura-Sakakibara, K., Fukui, Y., Nakamura, N., Yamaguchi, M.-A., Tanaka, Y., Kusumi, T. and Nishino, T. (2002) cDNA cloning, heterologous expressions, and functional characterization of malonyl-coenzyme A: anthocyanidin 3-*O*-glucoside-6'''-*O*-malonyltransferase from dahlia flowers. Plant Physiol. 130, 2142–2151.

Suzuki, H., Nakayama, T. and Nishino, T. (2003) Proposed mechanism and functional amino acid residues of malonyl-CoA: anthocyanin 5-*O*-glucoside-6'''-*O*-malonyltransferase from flowers of *Salvia splendens*, a member of the versatile plant acyltransferase family. Biochemistry 42, 1764–1771.

Suzuki, H., Nakayama, T., Yamaguchi, M. and Nishino, T. (2004a) cDNA cloning and characterization of two *Dendranthema* x *morifolium* anthocyanin malonyltransferases with different functional activities. Plant Sci. 166, 89–96.

Suzuki, H., Sawada, S., Watanabe, K., Nagae, S., Yamaguchi, M.A., Nakayama, T. and Nishino, T. (2004b) Identification and characterization of a novel anthocyanin malonyltransferase from scarlet sage (*Salvia splendens*) flowers: an enzyme that is phylogenetically separated from other anthocyanin acyltransferases. Plant J. 38, 994–1003.

Taguchi, G., Shitchi, Y., Shirasawa, S., Yamamoto, H. and Hayashida, N. (2005) Molecular cloning, characterization, and downregulation of an acyltransferase that catalyzes the malonylation of flavonoid and naphthol glucosides in tobacco cells. Plant J. 42, 481–491.

Tanaka, Y., Yonekura, K., Fukuchi-Mizutani, M., Fukui, Y., Fujiwara, H., Ashikari, T. and Kusumi, T. (1996) Molecular and biochemical characterization of three anthocyanin synthetic enzymes from *Gentiana triflora*. Plant Cell Physiol. 37, 711–716.

Tanaka, Y., Katsumoto, Y., Brugliera, F. and Mason, J. (2005) Genetic engineering in floriculture. Plant Cell Tissue Org. Cult. 80, 1–24.

Tanaka, Y. and Brugliera, F. (2006) Flower Colour. In: C. Ainsworth (Ed.), *Flowering and Its Manipulation*. Blackwell Publishing, Oxford, pp. 201–239.

Tohge, T., Nishiyama, Y., Hirai, M.Y., Yano, M., Nakajima, J., Awazuhara, M., Inoue, E., Takahashi, H., Goodenowe, D.B., Kitayama, M., Noji, M., Yamazaki, M. and Saito, K. (2005) Functional genomics by integrated analysis of metabolome and transcriptome of Arabidopsis plants over-expressing an MYB transcription factor. Plant J. 42, 218–235.

Unno, H., Ichimaida, F., Suzuki, H., Takahashi, S., Tanaka, Y., Saito, A., Nishino, T., Kusunoki, M., and Nakayama, T. (2007) Structural and mutational studies of anthocyanin malonyltransferases establish the features of BAHD enzyme catalysis. J. Biol. Chem. 282, 15812-15822.

Vogt, T., Grimm, R. and Strack, D. (1999) Cloning and expression of a cDNA encoding betanidin 5-*O*-glucosyltransferase, a betanidin- and flavonoid-specific enzyme with high homology to inducible glucosyltransferases from the Solanaceae. Plant J. 19, 509–519.

Wise, R.P., Rohde, W. and Salamini, F. (1990) Nucleotide sequence of the *Bronze-1* homologous gene from *Hordeum vulgare*. Plant Mol. Biol. 14, 277–279.

Yamazaki, M., Gong, Z., Fukuchi-Mizutani, M., Fukui, Y., Tanaka, Y., Kusumi, T. and Saito, K. (1999) Molecular cloning and biochemical characterization of a novel anthocyanin 5-O-glucosyltransferase by mRNA differential display for plant forms regarding anthocyanin. J. Biol. Chem. 274, 7405–7411.

Yamazaki, M., Yamagishi, E., Gong, Z., Fukuchi-Mizutani, M., Fukui, Y., Tanaka, Y., Kusumi, T., Yamaguchi, M. and Saito, K. (2002) Two flavonoid glucosyltransferases from *Petunia hybrida*: molecular cloning, biochemical properties and developmentally regulated expression. Plant Mol. Biol. 48, 401–411.

Yonekura-Sakakibara, K., Tanaka, Y., Fukuchi-Mizutani, M., Fujiwara, H., Fukui, Y., Ashikari, T., Murakami, Y., Yamaguchi, M. and Kusumi T. (2000) Molecular and biochemical characterization of a novel hydroxycinnamoyl-CoA: anthocyanin 3-O-glucoside -6"-O-acyltransferase from *Perilla frutescens*. Plant Cell Physiol. 41, 495–502.

Yoshimoto, M., Okuno, S., Yamaguchi, M.-A. and Yamakawa, O. (2001) Antimutagenicity of deacylated anthocyanins in purple-fleshed sweet potato. Biosci. Biotech. Biochem. 65, 1652–1655.

7

Flavonoid Biotransformations in Microorganisms

Joseph A. Chemler[1], Effendi Leonard[2] and Mattheos A.G. Koffas[3]

State University of New York at Buffalo, Chemical and Biological Engineering
[1] jchemler@buffalo.edu
[2] eleonard@buffalo.edu
[3] mkoffas@buffalo.edu

Abstract. Flavonoids are a diverse group of secondary metabolites found ubiquitously in the plant kingdom. Their associated health benefits have gained these fascinating compounds an increasing amount of attention towards their use as medicinal agents, supplements and natural colorants. With the rapid progress in unraveling the flavonoid biosynthetic pathways, the first part of this chapter presents the recent advances and challenges in utilizing recombinant bacteria and yeast to produce a number of different classes of flavonoid compounds including stilbenes, flavanones, isoflavones, flavones and anthocyanins. The second part presents a review on the biomodifications of flavonoids by non-recombinant microorganisms that result in an array of natural products, with special emphasis on the metabolism of flavonoids by intestinal microflora.

7.1 Microbes as Flavonoid Production Platforms

A growing number of scientific evidence demonstrates the potent health promoting activities of many flavonoids (Block et al. 1992; Harborne and Williams 2000). However, flavonoid availability for human consumption is a concerning problem. Even though many flavonoids are contained in edible plant products, diets in many parts of the world are low in fruits and vegetables, therefore preventing sufficient intake of flavonoids (Birt et al. 2001). Moreover, some of the very promising flavonoids only exist in minute quantities in minor food groups such as herbs, which exacerbate the low bioavailability of these plant chemicals.

Today, some flavonoid compounds are available as nutraceutical supplements such as isoflavone containing food commodities that are popular for treating hormone related disorders. Flavonoids are traditionally derived from plant extraction to meet general public needs. However, due to the low flavonoid concentration *in planta,* abundant natural resources are required for large-scale production for nutraceutical supplements. To resolve this problem, certain plants have been genetically engineered by increasing the activity of flavonoid biosynthetic enzymes. However, the existence of competing pathways in plants complicates the substantial in-crease of content of specific flavonoid compounds (Liu et al. 2002). For that reason, blocking of competing pathways had to be implemented in order to further increase flavonoid content (Yu et al. 2003). Plant cultivation also depends heavily on environmental, seasonal and geological conditions. Therefore, consistent quality and quantity of plant resources could present a rate-limiting step to

K. Gould et al. (eds.), *Anthocyanins,* DOI: 10.1007/978-0-387-77335-3_7,

large-scale production. In the down-stream processing line, flavonoid extraction and purification is also inefficient due to contamination of numerous plant small molecules and the loss of products due to processing conditions (Wang and Murphy 1996).

The increasing demand for flavonoid consumption drives the search for efficient large-scale production platforms. A commonly chosen production alternative of natural products is through total or partial chemical synthesis. In general, the total synthesis of many complicated natural products is often not feasible due to the lack of mechanistic information. When reaction mechanisms are available, the synthesis of complicated structures requires several steps. The multiple chemical steps are feasibly conducted in bench-scale experiments; however, one-pot synthesis route is an important parameter that needs to be elucidated in order to implement efficient large-scale production. Requirement for high energy and toxic chemicals in the process of converting reactants to products also presents another disadvantage of chemical synthesis. Moreover, several chemical modifications such as glycosylation and acylation which are commonly associated to plant products are currently unattainable through chemical synthesis.

While many instances have demonstrated the feasibility of efficient production of plant-derived natural products through plant cell cultures, recent efforts in flavonoid production have focused on heterologous synthesis using well-characterized microbial hosts, such as yeast *Saccharomyces cerevisiae* and gram-negative bacterium *Escherichia coli* (Fig. 7.1). The microbial factories pose several advantages such as rapid growth, ease of cultivation, and convenient genetic manipulations. Due to these features, high-level production could be maintained. Moreover, the microbial-based production increases product selectivity and reduces the usage of toxic chemicals while conserves energy usage. In this section, advances in the area of microbial production of various flavonoids are presented.

7.1.1 Stilbenes

Resveratrol (**STB1**), a stilbene which is popularly associated with the health benefits of red wine, has recently attracted significant attention. This flavonoid has been shown to extend the lifespan of the simple eukaryote *S. cerevisiae* (Howitz et al. 2003), *Drosophila melanogaster* (Wood et al. 2004) and *Caenorabitis elegans* (Viswanathan et al. 2005) in a Sir2-dependent manner. Similarly, resveratrol exerted SIRT1-dependent activities which improved cellular functions and health. Based on these results, resveratrol (**STB1**) could potentially be used to promote human longevity, since this compound also improved health and survival of mice on a high-caloric diet (Baur et al. 2006).

To increase the availability of this important flavonoid, microorganisms have been engineered to serve as alternative production platforms for resveratrol (**STB1**) production. Commonly utilized heterologous production platforms such as *S. cerevisiae* and *E. coli* do not naturally produce flavonoids. Therefore, the engineering of flavonoid biosynthesis in these microorganisms requires grafting of the necessary plant biosynthetic pathways. The first endeavor of microbial flavonoid production was the metabolic engineering of resveratrol biosynthesis in *S. cerevisiae*. To enable

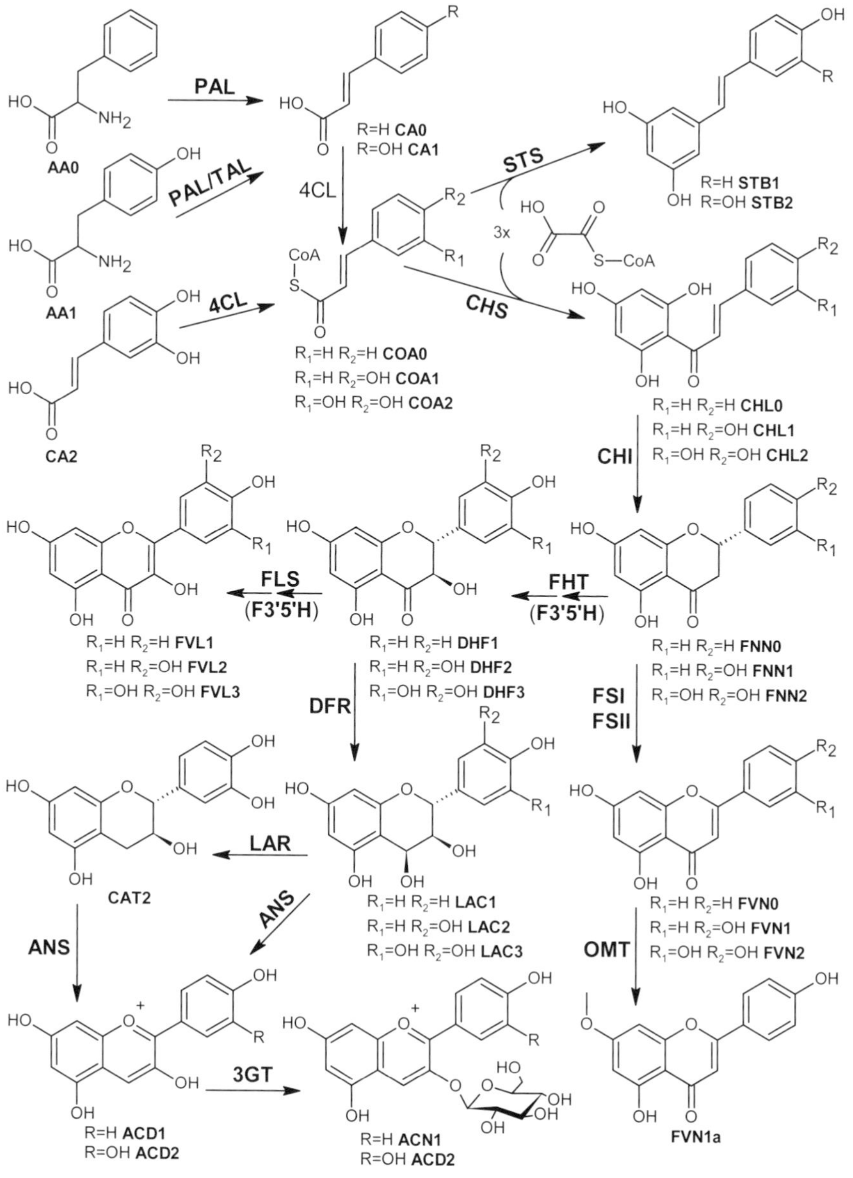

Fig. 7.1 Biotransformation of flavonoids in genetically engineered microorganisms

resveratrol (**STB1**) production from the *p*-coumaric acid (**CA1**) precursor, the structural gene of 4-coumarate: CoA-ligase (4CL) from poplar and of grape resveratrol synthase (STS) was inserted into *S. cerevisiae* strain FY23. Expression of the plant *4CL* and *STS* in FY23 was regulated under the control of the yeast alcohol dehydrogenase II promoter and the yeast enolase promoter, respectively. The resulting strain was capable of producing resveratrol (**STB1**) for the first time, up to 1.5 μg/L (Becker et al. 2003). Metabolic engineering of resveratrol biosynthesis was also recently demonstrated in *E. coli* (Watts et al. 2006). In this case, *Arabidopsis thaliana 4CL1* and peanut *STS* were introduced into *E. coli* strain BW27784. Expression of the plant enzymes under the *E. coli lac* promoter resulted in the synthesis of resveratrol (**STB1**) from *p*-coumaric acid (**CA1**), and piceatannol (**STB2**) from caffeic acid (**CA2**). Overall, these results have demonstrated the feasibility of synthesizing resveratrol compounds from genetically tractable microorganisms.

7.1.2 Flavanones

Flavanones (**FNN**) are a class of flavonoid molecules that serve as the committed precursors to the vast majority of flavonoid molecules such as flavones (**FVN**), flavonols (**FVL**), catechins (**CAT**), anthocyanins (**ACN**), and isoflavones (**ISO**). Due to the importance of these precursor molecules, extensive metabolic engineering endeavors have been pursued.

Engineering flavanone biosynthesis was first demonstrated in *E. coli*, through the episomal expression of phenylalanine ammonia lyase (PAL) from yeast *Rhodotorula rubra*, *Streptomyces coelicolor* 4CL, and licorice chalcone synthase (CHS), under the control of the strong *T7* phage derived promoter (Hwang et al. 2003). In this work, the effect of simultaneous expression efficiency of the plant enzymes towards flavonoid production was investigated. In particular, three types of the three-gene pathway ensemble were constructed. In the first construct, the transcription of all three genes in the pathway was controlled by one *T7* (monocistronic expression) and ribosome binding site (rbs) sequence, resembling the structure of an operon, a genetic architecture commonly found in bacteria. In the second construct, the expression of all three genes was controlled by one T7 promoter sequence; however, a rbs sequence was placed in front of each gene in the pathway. In the third ensemble, the transcription of each gene was individually controlled by a T7 and rbs sequence. By feeding the amino acid phenylalanine (**AA0**) or tyrosine (**AA1**), it was demonstrated that flavanone production was highest (approximately 400 μg/L naringenin (**FNN1**) from tyrosine (**AA1**)) using the *E. coli* strain harboring the individually expressed heterologous genes (third strain). In contrast, when the heterologous pathway was expressed through an operon system, flavanone productions were the lowest.

Flavanone biosynthesis was also engineered in *E. coli* by another group (Watts et al. 2004). In this case, tyrosine ammonia lyase (TAL), 4CL, and CHS were introduced in *E. coli* BW27784, and their expression was regulated by a *lac* promoter. Finally, baker's yeast *S. cerevisiae* was also employed for the biosynthesis of flavanones, mainly because of the ability of this organism, as a eukaryote, to functionally express P450 monooxygenases from plants that are required for flavonoid hydroxylation (Yan et al. 2005b). The construction of the heterologous pathway in yeast was similar to that of *E. coli*. Particularly, each biosynthetic gene sequence was placed under a galactose inducible promoter sequence (*GAL1*). The yeast production platforms were capable of flavanone production up to 28 mg/L from the *p*-coumaric acid (**CA1**) precursor (Yan et al. 2005b).

Based on these preliminary results, it appeared that higher production of fla-vanones could be accomplished using *S. cerevisiae*. This phenomenon could be ex-plained by the higher availability of malonyl-CoA which serves as the precursor for the formation of the flavanone backbone. Malonyl-CoA is a metabolite derived from the native metabolism of the microbial hosts. In the case for *E. coli*, it has been shown that these metabolites only exists in minute quantities (Takamura and Nomura 1988), hence limiting the synthesis of flavanones. We are currently pursuing various metabolic engineering strategies for manipulating the *E. coli* central metabolic path-ways for improving malonyl-CoA availability in this organism.

7.1.3 Isoflavones

Their unique core chemical structure distinguishes isoflavones (ISO) from the rest of the flavonoid subgroups. Isoflavones are also often referred to as "phytoestrogens", due to their resemblance to the human hormone 17β-estradiol, and their ability to exert estrogen-like activities. An extensive collection of publications based on epi-demiological, laboratory animal evidence, and mechanistic data have concluded the potential clinical importance of isoflavonoids, especially isoflavones, to promote health and prevent or delay the onset of certain chronic diseases. Currently, soy isoflavones are in chemoprevention phase I, II, III clinical trials sponsored by the National Cancer Institute, (http://www.cancer.gov/prevention/agents/Soy_Isofla-vones.html). Dietary supplements containing isoflavones are also popular for assist-ing in the treatment or prevention of various diseases, from post-menopausal disor-ders to cardiovascular diseases.

Isoflavones are only synthesized in legumes, such as soy. In plants, isoflavones hold various important functions which govern plant survival and reproduction (Dixon and Sumner 2003). The synthesis of the two major isoflavone structures, genistein (**ISO1**) and daidzein (**ISO1b**) (see Figs 7.13–7.14) is mediated by the membrane-bound protein, cytochrome-P450 isoflavone synthase (IFS). The mem-brane-bound protein performs the oxidative attack and novel aryl ring migration of the flavanone substrates. Belonging to the type II eukaryotic cytochrome-P450, the catalysis of the microsomal IFS requires electrons which are transferred through the P450-reductase partner proteins (CPR). The expression and catalysis of eukaryotic cytochrome-P450 enzymes can not be supported in *E. coli* due to the lack of P450-redox partner proteins and incompatibility of membrane insertion signal particles. Therefore, heterogolous expression of plant IFS has been done using yeast as a eu-karyotic host (Jung et al. 2000). Recently, Ralston et al. (2005) reported the partial reconstruction of isoflavonoid biosynthesis in *S. cerevisiae* by testing different isozymes of CHI and expressing them with IFS. Cultures fed with different chal-cones (**CHL**) were able to produce flavanones and isoflavones when IFS was com-bined with chalcone isomerase (CHI). Even though yeast as a recombinant platform demonstrated the feasibility of increasing isoflavonoid availability, the strategy still suffers from two disadvantages. The first is the low production yield, which could be attributed to the low turn-over rate of the IFS enzymes in yeast. The second draw-back is the necessity of chalcone supplementation into the yeast cultures as a precur-sor to isoflavones. In this case, since chalcones needed to be derived from chemical synthesis, one-pot isoflavone production could not be achieved. Indeed, this can be remedied through the engineering of the early pathway as has been demonstrated for other functional flavonoid engineering endeavors in yeast (Leonard et al. 2005).

7.1.4 Flavones

Flavones (**FVN**) are a ubiquitous flavonoid class within the plant kingdom that are involved in important functions for plant survival and growth. This class of flavonoids is also of medicinal importance to promote human health for it was shown to induce apoptosis of HER2/neu breast cancer cells (Way et al. 2004), retard the growth of melanoma (Caltagirone et al. 2000), and inhibit the replication of HIV-1 (Critchfield et al. 1997). However, despite its important health promoting properties, flavones only occur in a small fraction of human diets, as the primary sources are in herbs such as parsley, thyme, and chamomile.

   In order to increase flavone availability for nutraceutical or medicinal applications, recombinant strains of E. coli and baker's yeast have been engineered to produce flavones. In most plants, flavones are synthesized from flavanones through the membrane-bound P450 flavone synthase II (FSII) (Martens and Forkmann 1999). To engineer flavone biosynthesis in E. coli through heterologous pathway expression, a recombinant flavone biosynthetic circuit was grafted in E. coli BL21 strain (Leonard et al. 2006a). The pathway was build by simultaneous expression of the plant flavone biosynthetic genes, in which the expression of the genes derived from heterologous plant origins was individually controlled by the T7 phage promoter. Specifically, the artificial gene ensemble consisted of parsley 4CL, petunia CHS and CHI, and a FSII analog, the parsley FSI (Leonard et al. 2006a). Unlike FSII, which requires NADPH as an electron donor, FSI is a soluble dioxygenase that requires 2-oxoglutarate, ascorbate, and iron for catalysis (Martens et al. 2001). Moreover, unlike the widespread existence of FSII, the existence of FSI is only limited to the *Apiceae* (Martens et al. 2001).

   Utilization of the soluble FSI bypasses the requirement of functionalization of the P450 FSII in E. coli through extensive protein and cellular engineering. When the recombinant E. coli was cultured with the supplementation of p-coumaric acid (**CA1**), the flavone apigenin (**FVN1**) was produced (Martens et al. 2001). Attempts to produce additional flavones were met by limited success. Both caffeic acid (**CA2**) and ferulic (**CA2a**) are not native compounds to parsley but were previously shown to be suitable substrates for one of parsley's 4CL isozymes (at 28% and 66% efficiency, respectively, compared to p-coumaric acid (Lozoya et al. 1988). Incorporating this enzyme into the flavonoid cluster in E. coli, the tetrahydroxylated flavone, luteolin (**FVN2**), was produced from caffeic acid, albeit at a very small quantity (Martens et al. 2001). However, supplemental feeding of ferulic acid (**CA2a**) did not result in the synthesis of the respective flavanone, homoeriodictyol, or flavone, chrysoriol, which indicated that the feruloyl-CoA product of 4CL is a poor substrate for the downstream flavonoid enzymes in the recombinant assembly.

   Biotransformation of the flavone producing E. coli strain using p-coumaric acid (**CA1**) and caffeic acid (**CA2**) precursors should result in the synthesis of the flavanone intermediate metabolites. However, neither naringenin (**FNN1**) nor eriodictyol (**FNN2**) were detected in the medium because of the high activity of the soluble synthase that efficiently converted the flavanone substrates into flavones (Martens et al. 2001). In plants, many flavonoids, including flavones undergo further modifications, such as methylation and glycosylation. To generate methylated flavones from *E. coli*, the gene encoding 7-*O*-methyltransferase (OMT) from peppermint was cloned into the flavanone producing *E. coli* recombinant strain. Upon supplemental feeding with *p*-coumaric acid (**CA1**), the methylated flavone, genkwanin (**FVN1a**) was produced in addition to apigenin (**FVN1**) (Willits et al. 2004). However, due to the small quantity of luteolin (**FVN2**), no methylated luteolin could be detected.

In a recent study, baker's yeast was used as a metabolic engineering tool for the high production of various plant-specific flavones as well as a model to investigate the two distinct biosynthetic routes that exist *in planta* (Leonard et al. 2005). Specifically, the plant flavanone biosynthetic pathway which contained the *Arabidopsis thaliana* cinnamate 4-hydroxylase (C4H) P450 enzyme was coexpressed with either the soluble FSI or the membrane-bound P450 FSII. Even though P450 reductase proteins are present in yeast, the plant P450-reductase derived from *Catharanthus roseus* was coexpressed with FSII to improve the P450 activities of the heterologous plant enzymes. The introduction of the P450 C4H in to the flavone pathway allowed the utilization of cinnamic acid (**CA0**) as the precursor molecule to synthesize a wide flavonoid variety with and without B-ring hydroxylation, such as pinocembrin (**FNN0**), naringenin (**FNN1**), chrysin (**FVN0**) and apigenin (**FVN1**). Moreover, supplemental feeding of caffeic acid (**CA2**) resulted in the synthesis of flavonoids with two B-ring hydroxyl groups, namely eriodictyol (**FNN2**) and luteolin (**FVN2**). Feeding with *p*-coumaric acid (**CA1**), the flavone production levels from the soluble and the membrane-bound synthase were compared. The results demonstrated that the flavanone naringenin (**FNN1**) was metabolized more efficiently by the soluble dioxygenase and resulted in a larger apigenin (**FVN1**) production through this pathway. Therefore, it appeared that even though yeast is a natural host for the membrane-bound P450 enzymes, the catalysis performed by the soluble enzyme analog is more robust.

7.1.5 Flavonols

Flavonols (**FVL**) is another class of flavonoids that diverges from the flavanone branch pathway. Flavanone 3β-hydroxylase (FHT) is the leading enzyme to convert flavanones into dihydroflavonols. Dihydroflavonols then serve as the substrates for another 2-oxoglutarate dependent enzyme, flavonol synthase (FLS) to create flavonols. Similar to flavones, flavonols are also of pharmacological importance as antimutagenic agents or as effective inhibitors of angiogenesis (Formica and Regelson 1995; Lamson and Brignall 2000). In order to obtain substantial production, *E. coli* was engineered for flavonol biosynthesis through the episomal expression of flavonol biosynthetic enzymes. Extending the metabolic engineering endeavor, the expression of apple FHT and *A.thaliana* FLS were engineered in the flavanone producing *E. coli* strain to result in the synthesis of naringenin (**FNN1**), dihydrokaempferol (**DHF1**), and kaempferol (**FVL1**) from *p*-coumaric acid (**CA1**) (Leonard et al. 2006a). However, hydroxylated flavonoids (eriodictyol (**FNN2**), dihydroquercetin (**DHF2**), and quercetin (**FVL2**)) could not be detected in the culture medium, when the flavonol producing *E. coli* strain was supplemented with caffeic acid (**CA2**) due to the low affinity of 4CL toward the dihydroxylated phenylpropanoid, as discussed earlier.

7.1.6 Leucoanthocyanidins and Flavan-4-ols

Dihydroflavonol 4-reductase (DFR) is the enzyme located in the metabolic node that diverts the conversion of precursor flavonoids into anthocyanins, proanthocyanidins, and phlobaphenes (Fig. 7.2). DFR catalyzes the stereospecific reduction of (2*R*,3*R*)-dihydroflavonols to the respective colorless (2*R*,3*R*,4*S*)-leucoanthocyanidins. Besides its dihydroflavonol reductase activity, some DFRs have recently been shown to catalyze the reduction of another group of flavonoids, the flavanones, to flavan-4-ols (FNR activity) (Fischer et al. 1988, 2003). The three dihydroflavonol precursors of the most common anthocyanins and proanthocyanidins are dihydrokaempferol (**DHF1**), dihydroquercetin (**DHF2**) and dihydromyricetin (**DHF3**). These dihydroflavonols are very similar in structure, differing only in the number of hydroxyl groups on the B ring, which is not the site of DFR enzymatic reaction. Therefore, it is not surprising that DFRs from many species can utilize all three substrates (Forkmann and Heller 1999). However, DFR from certain plants demonstrate very distinct substrate specificity. For example, in some genera, the production of leucopelargonidin (**LAC1**), which is the anthocyanin aglycon derived from **DHF1**, is prevented due to the inability of the native DFR enzyme to accept **DHF1** as a substrate (Martens et al. 2002). Therefore, the pelargonidin (**ACD1**) orange-base color is not synthesized in these plants. In the recent past, there has been some effort in elucidating the underlying primary structural differences that dictate this substrate specificity (Johnson et al. 2001).

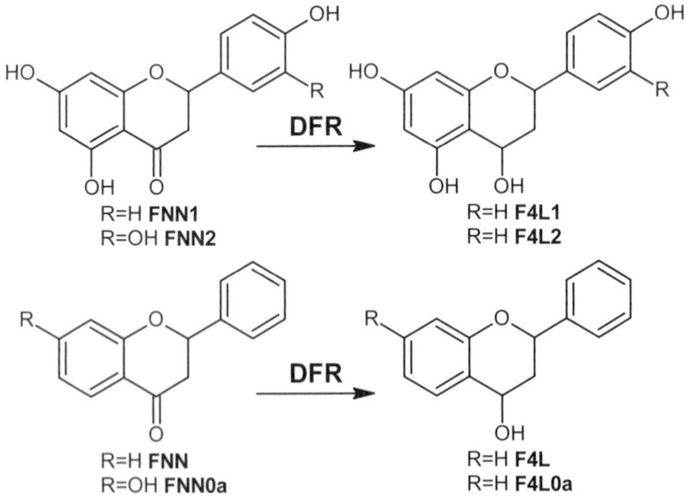

Fig. 7.2 Conversion of natural and unnatural flavanone into flavan-4-ols by DFR

DFR from petunia, a plant that does not accumulate pelargonidin type antho-cyanins does not reduce **DHF1** efficiently (Beld et al.1989). Based on amino acid sequence alignment of DFR from a few plants, the existence of a variable domain in the middle of the well-conserved regions was revealed (Beld et al. 1989; Johnson et al. 2001). In an effort to identify amino acids in this unique DFR protein region that could control substrate specificity, Petunia *dfr* was truncated at several positions in the central domain (at amino acid position 126, 170, and 235) and fused with nucleo-tide fragments of gerbera DFR enzyme that accepts **DHF1** (Johnson et al. 2001). The chimeric constructs were then used to transform a white colored petunia mutant line W80 which lacked DFR activities and accumulated all three DFR dihydroflavonol substrates. Selection of petunia mutant transformants with orange-red coloration, followed by thin layer chromatography (TLC) analysis showed the formation of pelargonidin (**ACD1**). The result indicated the existence of amino acids that were located in the gerbera DFR, but not in the petunia DFR that allowed for **DHF1** to serve as a substrate. To further identify the specific amino acids which were respon-sible for the alteration of substrate specificity, some hydrophilic residues of the ger-bera DFR were exchanged with hydrophobic amino acids that bear similar structures. Pigment analysis through TLC of the petunia flowers that were transformed using the mutant gerbera *dfr* variants indicated that substitution of the asparagine residue at position 134 with leucine abrogated the synthesis of cyanidin (**ACD2**) and del-phinidin (**ACD3**). Moreover, substitution of glutamic acid at position 145 with leu-cine abolished the DFR activities since no pigment synthesis was observed. No pig-ment phenotype change was observed in several other petunia transformants when the introduced gerbera DFR was mutated at amino acid positions 132, 138, 139, and 152. The mutagenesis studies indicated that a few of the amino acids in the central domain control the specificity of gerbera DFR enzymes. In this case, the amino acid exchange that caused the deceased specificity towards **DHF2** and **DHF3**, and only accepting **DHF1** were identified.

In order to further elucidate the DFR substrate specificity, a number of DFR en-zymes derived from various randomly chosen plants that accumulated cyanidin (**ACD2**) and delphinidin (**ACD3**) or the ones that also accumulated pelargonidin (**ACD1**) derivatives were expressed in *E. coli*. *In vitro* biochemical characterization revealed that all of the plant-derived DFR enzymes catalyzed the reduction of **DHF2** and **DHF3** with similar efficiency. The reduction of **DHF1** was selective, and only performed by recombinant enzymes derived from *Anthurium andraeanum*, *Ipomoea nil*, *Rosa hybrida*, *Fragaria ananassa*, and *A. thaliana*. The highest **DHF1** reduction activities were obtained from the *A. andraeanum* and *F. ananassa* DFRs, followed by *I. nil* and *R. hybrida* DFRs. It is interesting to note that *A. thaliana* DFR could reduce **DHF1** to leucopelargonidin (**LAC1**) albeit in small amount even though *A. thaliana* does not accumulate pelargonidin derivatives. This result supports a recent investigation of *A. thaliana* DFR which concluded that the enzyme is capable of utilizing **DHF1** only when the F3'H enzyme is inactive (Dong et al. 2001). Alto-gether, these results strongly suggest the presence of metabolic channeling in *A. thaliana*, something that has previously been demonstrated for other enzymes in the flavonoid biosynthetic pathway (Burbulis and Winkel-Shirley 1999). The bio-chemical assays also demonstrated the ability of all of the plant DFR enzymes to reduce flavanones with similar efficiency. Eriodictyol (**FNN2**) served as a universal substrate, as the synthesis of the flavan-4-ol, luteoferol (**F4L2**), was observed. However, the reduction of naringenin (**FNN1**), a flavanone with one B-ring hydroxyl group, was selective. The synthesis of apiferol (**F4L1**) from naringenin was only observed in assays using DFR from *A. andraenum, F. ananassa, I. nil*, and *R. hybrida*. Overall, the specific activities of DFR towards the flavanone substrates were lower when compared to the dihydroflavonol substrates. It is possible that different types of enzymes are present and serve as the primary catalysts for the synthesis of the rare flavan-4-ol molecules.

By correlating biochemical data with protein primary structures, it was shown that the two recombinant DFR enzymes with the highest specific activities towards **DHF1**, which were derived from *A. andraeanum* and *F. ananassa,* possessed a serine and alanine respectively at position 133, instead of the consensus amino acid asparagine. This result further supports the observation made by Johnson et al. (2001) that the residue at position 134 in gerbera DFR (this residue aligned with the 133th residue in the *A. andraeanum* DFR sequence) played a key role in dictating substrate specificity. It is unlikely that steric hindrances due to the 133th amino acid side chain are involved in the efficiency of **DHF1** reduction because the unique alanine and serine residues found in the *A. andraeanum* and *F. ananassa* DFR that efficiently reduced **DHF1** do not carry bulky side chains. Supporting this observation, two recombinant *Medicago* DFR isoforms, which possess either an aspartic acid or asparagine at position 133, accepted **DHF1** *in vitro* with different efficiency (Xie et al. 2004). Aspartic acid and asparagine are both similar in structure, carrying a hydroxyl and amino side-chain group respectively. Therefore, bulkyness of the amino acid at position 133 is unlikely to be the substrate specificity-determining factor. Instead, the charge of the corresponding side chain appears to be more important. The flexibility of DFR was also examined through the biotransformation of unnatural flavanones by *E. coli* expressing *A. andraeanum* DFR. When unnatural flavanones were supplemented into the recombinant *E. coli* cultures, the conversions of unsubstituted flavanone (**FNN**) and 7-hydroxyflavanone (**FNN0a**) to the corresponding flavan-4-ols (**F4L** and **F4L0a**) were obtained. However, the reduction products of hesperetin (**FNN2a**) or 5,7-dimethoxyflavanone (**FNN0b**) were not observed. The ability to reduce certain unnatural substrates could be used in conjunction with the crystal structure of DFR (which is currently not available) to study the reaction mechanism of this important enzyme.

7.1.7 Anthocyanin

Anthocyanins (**ACN**) are colorful phytochemicals that contribute to the red and blue colorations in plants. In plants, anthocyanins are found attached to several other molecules, such as sugar groups and other flavonoids, an arrangement that helps improve their stability and coloration (Mol et al. 1998; Harborne and Williams 2000). For humans, anthocyanins have important pharmacological properties, as they serve among others, as antioxidants (Weisel et al. 2006), and cancer-preventive metabolites (Lala et al. 2006). Anthocyanins are available to humans through diets rich in colorful grains, fruits and vegetables, and also red wine.

Several attempts have been made to produce and extract anthocyanins from plants with the purpose of utilizing them as colorants. The most abundant and historically oldest anthocyanin extract used as natural colorant is derived from grape pomace (Wrolstad 2000). Others, either available commercially or used locally as natural extracts for coloring foodstuffs include flower petals (Kamei et al. 1995), grape rinds and red rice (Koide et al. 1996), red soybeans and red beans (Koide et al. 1997), *Vaccinium* species (Bomser et al. 1996), purple corn (Hagiwara et al. 2001), and different cherry extracts (Harris et al. 2001; Kang et al. 2003; Katsube et al. 2003). However, in all cases, one of the biggest obstacles in commercialization is the "browning effect". This refers to the formation of a brown color in plant anthocyanin extracts as a result of a two-step process. First, anthocyanins are oxidized by plant polyphenol oxidases present in the plant extract (Oszmianski and Lee 1990; Mclellan et al. 1995; Tsai et al. 2004). Second, the oxidized anthocyanins undergo condensation and form brown pigments, which are usually undesired by the food industry. The red cabbage colorant is the only one that does not undergo browning and as a result it is the most commonly used anthocyanin mixture.

As an alternative to plant extracts, bioreactor-based systems of plant cells for mass production of flavonoids, anthocyanins in particular, have been described for a few species (Zhong et al. 1991; Kobayashi et al. 1993). However, economic feasibility has not been established yet, in part because of engineering challenges in plant cell cultures. One such challenge is that plant cells tend to form aggregates that influence anthocyanin culture productivity (Hanagata et al. 1993). Cells within aggregates are not adequately exposed to lighting required for flavonoid biosynthesis. For example, formation of PAL, a key enzyme in the biosynthetic pathway is promoted primarily by UV wavelengths, particularly those of the UV-B region (Wellmann 1975). Other enzymes in the pathway, particularly those of the anthocyanin biosynthetic branch, appear to be regulated in part by UV and in part by phytochrome-activating wavelengths (700–800 nm) (Meyer et al. 2002). In that respect, irradiance becomes a limiting factor to productivity. This is because cells at the interior of an aggregate have limited or no exposure to light (Hall and Yeoman 1986). Also, the average light dosage is reduced or insufficient within a dense cell culture since the cell wall composition selectively restricts certain wavelengths (Smith and Spomer 1995). On the other hand, flavonoid production in recombinant microorganisms is advantageous because the cloned pathway(s) are under microbial promoters and therefore the production is independent of light or other regulatory elements (such as the MYB transcription factors) required by plants. In addition, *E. coli* and *S. cerevisiae* cultures can achieve higher yields than plant cell cultures because they do not form aggregates and have better duplication times. In addition, no plant peroxidases are present in bacteria and yeast and therefore the "browning effect" problem is significantly reduced. A simplified extraction procedure is another advantage of using microbial production platforms over plant crops or cultures. Since anthocyanins are not naturally produced in microbial hosts, a much less complicated matrix of products is generated through the heterologous expression of pathways that lead to specific product targets. This minimizes the downstream processing required for purification of the target molecules.

Recently, *E. coli* was engineered to produce anthocyanins (Yan et al. 2005a). To achieve this goal, the flavanone pathway was bypassed by supplemental feeding of flavanones into *E. coli* JM109 strain carrying a gene assembly which consisted of apple FHT, *A. andraeanum* DFR, anthocyanidin synthase (ANS) from apple, and a UDP-glucose:flavonoid 3-*O*-glucosyltransferase (3GT) from petunia. Unlike the production of other flavonoids, the synthesis of anthocyanin glucoside from the fermentation of *E. coli*, supplemented with flavanones was very low. Characterization of *A. thaliana* ANS activities *in vitro* indicated that the major product from the conversion of (2*R*,3*R*,4*S*)-leucoanthocyanidins were flavonols, with only minor anthocyanin aglycons. Recently, however, the role of ANS in the anthocyanin branch pathway was clarified through *in vitro* characterization of recombinant gerbera ANS heterologously expressed in yeast. When (+)-catechin (**CAT2a**) was added as the reaction substrate, the majority of the ANS reaction product was the 4,4-dimer of oxidized (+)-catechin, and a trace amount of cyanidin (**ACD2**) and the flavonol quercetin (**FVL2**). However, the addition of recombinant *Fragaria* 3GT derived from expression in *E. coli* resulted in conversion of (+)-catechin (**CAT2a**) into cyanidin-3-*O*-glucoside (**ACN2**) as the major product and cyanidin and dimeric catechin as the minor products. Based on these results, (+)-catechin (**CAT2a**) may be a better substrate for ANS, which could explain the minute productions of anthocyanins from the engineered *E. coli*.

7.1.8 Hydroxylated Flavonoids from *E. coli* Expressing Plant P450s

Cytochrome P450 enzymes catalyze the regiospecific and stereospecific oxidation of nonactivated hydrocarbons at moderate temperatures. In plants, the P450 enzymes (type II) are bound to the membrane of the endoplasmic reticulum (ER) and require P450 reductase enzymes to shuttle electrons derived from NADPH into the heme core of the P450s. Electrostatic fields generated by the P450s and the reductase proteins allow interactions between the two proteins to mediate electron transfer. An advantageous feature of utilizing yeast as a flavonoid production platform is the ability to naturally support the functional expression of eukaryotic P450 enzymes. As a eukaryote, yeasts contain ER and P450 redox partner proteins and are readily adaptable to accept P450s from higher organisms.

E. coli is often the preferred heterologous production platform, not only because of its robust growth and ease of cultivation, but also because it is the most genetically tractable microbial host. Extensive characterization of *E. coli* has resulted in a wealth of available genetic engineering tools that are useful for metabolic engineering purposes. However, functional expression of many eukaryotic proteins is hindered in *E. coli* due to, for example, codon incompatibility or post-translational modifications essential for enzymatic activity. Such is the case for the functional expression in *E. coli* of the cytochrome P450 flavonoid biosynthetic enzymes. The requirement of membrane attachment and an electron transfer system prohibits the heterologous functional expression of flavonoid P450 enzymes in *E. coli* because of the lack of ER and P450-redox partner proteins. Understanding the features of P450s is crucial to obtain their successful enzymatic expression in *E. coli*. Membrane-associated P450s are synthesized with an N-terminal leader sequence rich in hydrophobic residues that serve as recognition signal for translational insertion into the ER membrane (Sakaguchi et al. 1984). Adjacent to the hydrophobic-rich residue is a short stretch of positively charged amino acid residues that act as a stop signal of the translocation across the membrane (Williams et al. 2000a). Moreover, a short proline-rich region that follows the signal-anchor sequence serves as the hinge between the N-terminal anchor and the cytoplasmic catalytic domain since proline is a well known α-helix breaker (Yamazaki et al. 1993). In the past, mutations in the proline cluster did not affect protein expression and membrane association was retained. However, carbon monoxide-binding of the mutated proteins failed to display absorbance at 450 nm, a characteristic of P450-type proteins, which indicated the lack of proper heme incorporation into the expressed proteins due to improper folding (Yamazaki et al. 1993).

Attempts to express membrane-bound P450 proteins in *E. coli* have been primarily geared for crystallography purposes. So far, there are only three eukaryotic microsomal P450 enzymes that have been crystallized when using *E. coli* as a recombinant host (Williams et al. 2000b, 2003, 2004). Removal of the transmembrane domain of various microsomal P450 has generally produced enzymes with low solubility and required detergents for solubilization. This indicates that there are regions within the protein body other than the membrane anchor module that are responsible for membrane association. Removal of the membrane spanning N-terminus of the rabbit microsomal 2C5 caused release of the proteins from membranes following expression in *E. coli* in a high ionic environment and without the need for detergent. However, aggregates were formed by the dissociated P450s which was not conducive to crystallization. To minimize/abolish monofacial membrane interactions that caused aggregation, several amino acid substitutions were made within the protein body. The resulting modified microsomal P450 was soluble in a high-salt buffer and formed monomer unit which was then successfully crystallized. In the case of the truncated 2C5, the catalytic activities could be reconstituted *in vitro* after the addition with rabbit NADPH P450 reductase, which indicated that the lack of substantial membrane attachment did not affect the catalysis of the proteins.

From the study on the expression of plant membrane-bound P450 limonene hydroxylase in *E. coli*, removal of the membrane spanning region did not result in the detection of the P450 absorbtion peak at 450 nm (Haudenschild et al. 2000). A similar result was also observed with the native, untruncated limonene hydroxylase. Furthermore, SDS-PAGE immunoblot analysis of the membrane fractions from *E. coli* expressing the truncated hydroxylase only detected a small amount of protein recognized by the rabbit polyclonal antibodies. This assay did not detect any protein in the membrane fractions of *E. coli* expressing the native hydroxylase which indicated the absence of functional protein expression. The flavonoid P450 enzyme C4H has also been studied for expression in *E. coli*. In this case, it was found that immunoblots failed to detect the full-length protein as well as the construct with complete removal of the membrane anchor module after expression in various *E. coli* strains indicating the lack of functional expression. It is interesting to observe the phenomenon of nonfunctional expression occurred in both cases of intact P450 and the N-terminally truncated proteins. The variations in sensitivity of N-terminal removal for strategy for *E. coli* expression indicate a substantial structural divergence among microsomal P450 enzymes. Overall, studies of plant P450s revealed that the N-terminal hydrophobic region is required for the expression of the protein in *E. coli*. However, the native module seemed to lead to the synthesis of inactive, misfolded proteins that fail to incorporate heme into the active site. Moreover, high expression of the misfolded proteins exerted cellular damages to the *E. coli* host which resulted in cell lysis and reduced growth after protein induction (Hotze et al. 1995).

Synthesis of misfolded proteins often occurs when RNA polymerase fails to recognize translation codons of the RNA transcripts. This phenomenon is especially prevalent when the RNA polymerase of a species organism is responsible for the translation of foreign proteins which originate from a different source organism. In the case of plant P450s, the N-terminal hydrophobic sequence contains signal specific for the recognition of endoplasmic reticulum membranes. Therefore, since endoplasmic reticulum does not exist in *E. coli*, it is not surprising that the N-terminal codons of the plant P450 are not compatible for expression in this microorganism. Changes of several N-terminal codons of some plant P450s have been reported to remedy the expression problems. In the case of successful C4H expression in *E. coli*, the protein engineering method involved the combination of N-terminal truncation and codon replacements. Specifically, the first six codons encoding the hydrophobic membrane-spanning residues were deleted. The seventh codon, which encoded glutamic acid, was replaced by the *E. coli* ATG codon corresponding to methionine in order to allow translation initiation. The second codon of the truncated P450 encoding lysine residue (the eighth codon of the native P450) was changed into alanine because alanine is the preferred second codon for the *E. coli* LacZ gene (Looman et al. 1987). Moreover, the sixth codon of the truncated construct (the twelfth codon of the original protein), which encoded for glycine, was replaced with alanine to eliminate the amino acid that could serve as a helix breaker (Hotze et al. 1995). Expression of the modified C4H in *E. coli* strain M15, DH5α and JM109 did not result in the detection of any protein. However, expression of the modified construct in the minicell producing *E. coli* strain DS410 resulted in the synthesis of the modified P450 that could be isolated without solubilization steps or the addition of lipids. Even though the recombinant proteins isolated from DS410 membranes exhibited catalytic activity *in vitro* that was 4-7 folds lower than in microsomes from induced plant cell, the reaction products could be detected through radiochromatography (Hotze et al. 1995).

Successful expression of the P450 limonene hydroxylase was also achieved through modifications of the N-terminal membrane-anchor signal (Haudenschild et al. 2000). In this case, similar to N-terminal truncation of C4H, the successful expression involved removal of six amino acid codons from the P450 hydroxylases. Following the codon removal, the truncated gene constructs were fused with a modified leader sequence of eight amino acids derived from CYP17A, the bovine P450 17α-hydroxylase (Haudenschild et al. 2000). This construct seemed to provide good expression in *E. coli* strain JM109, but not in XL1Blue and the catalysis of the protein products could be restored *in vitro* after the addition of P450 reductase. CYP17A was functionalized for *E. coli* expression through extensive modifications of the N-terminal codons (Barnes et al. 1991). Specifically, the second codon which encoded for tryptophan was replaced with the *E. coli* codon for alanine since this codon is preferred for the expression of the *LacZ* gene (Looman et al. 1987) as mentioned previously. Nucleotide sequences for the fourth and fifth codons which both encode for leucine were replaced with another alanine encoding codon (silent mutations) that is rich in adenosine and thymine nucleotides since this region of *E. coli* mRNAs were shown to be rich in adenosine and uridine (Stormo et al. 1982). Two silent mutations were also made at the last nucleotides of the sixth and seventh codon by replacing with adenosine and thymine, respectively to minimize mRNA secondary structure formation that can halt RNA polymerase processing (Schauder and McCarthy 1989).

Another route to synthesize hydroxylated flavonoids is through the oxidation process catalyzed by the flavonoid 3′-hydroxylase (F3′H) and flavonoid 3′,5′-hydroxylase (F3′5′H) (see Fig. 7.3 for positional numbering). Metabolic engineering hydroxylated flavonol biosynthesis in *E. coli* was achieved through protein engineering F3′5′H and grafting of the flavonol biosynthetic pathway (Leonard et al. 2006b). F3′5′H from *C. roseus* was chosen for the recombinant pathway because this enzyme exhibits broad substrate specificity and introduces both single and double hydroxyl groups into the flavonoid substrates. The synthesis of hydroxylated flavonoids required both the expression of the F3′5′H and the ability of the enzyme to catalyze the oxidation reaction *in vivo*. Specifically, four N-terminal codons were removed and the nucleotides of the fifth codon were replaced to ATG. Moreover, the second codon of the shortened F3′5′H which encodes for leucine was changed into alanine.

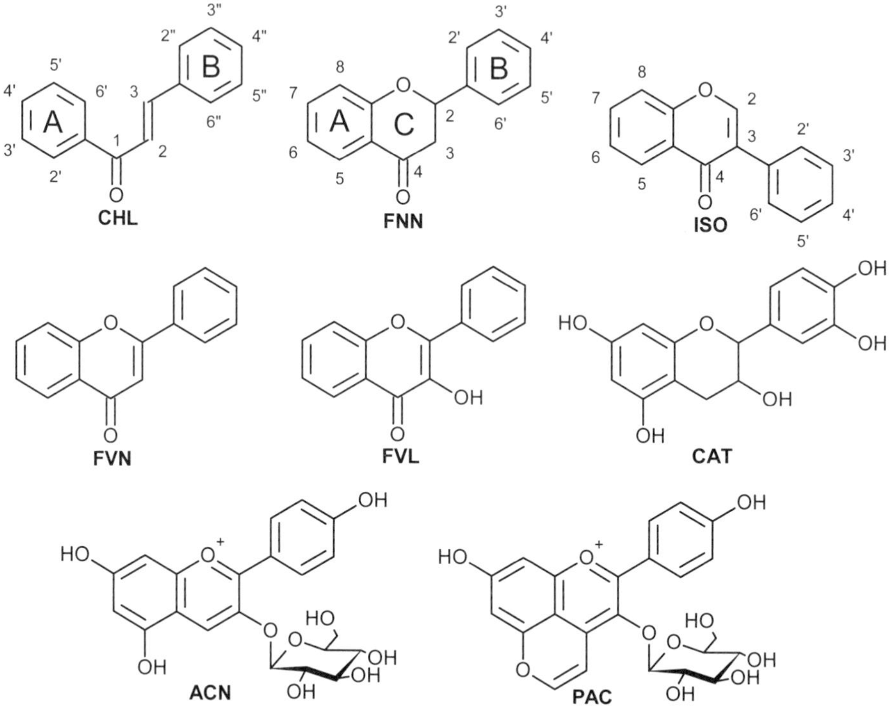

Fig. 7.3 Structures, numbering and lettering of chalcones, flavonoids and isoflavonoids

To obtain enzymatic activities *in vivo*, the lack of P450-reductase enzymes in *E. coli* had to be compensated. For that reason, a shortened P450-reductase (CPR) gene derived from *Catharanthus* was introduced along with the modified *F3′5′H* as a functional translational fusion. *CPR* was fused with the modified *F3′5′H* through a short sequence that did not favor the formation of secondary structures which can interfere with the interactions between the two proteins. When the chimeric F3′5′H was co-expressed with the flavonol biosynthetic pathway, in the presence of *p*-coumaric acid (**CA1**), a small amount of the dihydroxylated flavonol, quercetin (**FVL2**), could be recovered from the culture media which indicated that the chimeric P450 was expressed and was catalytically active in *E. coli*. Supplemental feeding of the flavanone naringenin (**FNN2**) to the recombinant *E. coli* strain resulted in the synthesis of dihydrokaempferol (**DHF1**) and kaempferol (**FVL1**), as well as the di- and tri- hyrdoxylated flavonoid dihydroquercetin (**DHF2**), quercetin (**FVL2**), and myricetin (**FVL3**). Even though protein engineering of the F3′5′H resulted in the functional expression of the enzyme in *E. coli*, the low amount of the hydroxylated products indicated that the engineered enzyme did not attain optimum activities (Leonard et al. 2006b). Moreover, it was also reported that the some cell lysis occurred after induction of protein expression which present evidence that the expression of the proteins was toxic to the cells. In the metabolic engineering strategy, the toxicity of the P450 proteins was minimized through the placement of the gene into a low copy-number expression plasmid to reduce protein production. Another important parameter for successful P450 expression in *E. coli* is the use of low to medium strength tightly regulated promoters. The expression of P450s in *E. coli* under the regulation of strong promoters appeared to favor the formation of inclusion bodies. The use of tightly regulated promoters is also especially important for toxic protein expression to prevent early protein synthesis, which occurs when using "leaky" promoters. In general, this study demonstrated that the synthesis of hydroxylated products need not rely on more complicated organism, such as yeast or insect cells. *E. coli*, a commonly used and robust biocatalyst could be engineered for hydroxylated product synthesis through the functionalization of microsomal enzymes.

7.2 Microbial Metabolism of Flavonoids

The previous section discussed the metabolic engineering of microorganisms as potential platforms for the production of flavonoids. The following section focuses on the natural biotransformations of flavonoids by non-recombinant microorganisms. A variety of reactions are reviewed including oxidations, reductions, ring cleavage, deglycosylations, and conjugations. The interest of researchers to study the reactions reviewed here stems from several concepts. These include understanding the role of gastrointestinal bacteria on the fate of flavonoids following consumption; the potential of microorganisms to convert flavonoid feed stocks into novel or rarer products; and how the chemical constituents of consumer produces are altered by preparative processes involving microorganisms. The focus of this review is on the microbial transformations of flavonoids, whether they are directly linked to enzymatic catalysis or spontaneous reactions associated with an organism's metabolites. The following section is divided into topics pertaining to biotransformation of chalcones (**CHL**), flavanones (**FNN**), isoflavones (**ISO**), flavonols (**FVL**), flavones (**FVN**), catechins (**CAT**), anthocyanins (**ACN**) and the formation of pyranoanthocyanins (**PAC**).

7.2.1 Chalcones

The reported fungi biotransformations of chalcones commonly involve the cyclization reaction of the chalcone into a flavanone as well as modification of the chalcone to afford novel chalcones (Figs. 7.4–7.6). The fungus, *Pichia membranifaciens* transformed xanthohumol (**CHL1b**) into the flavanone, (2*S*)-2″-(2‴-hydroxyisopropyl)-dihydrofurano[2″,3″:7,8]-4′-hydroxy-5-methoxyflavanone (**FNN1a**). The syntheses of two altered chalcones, (*E*)-2″-(2‴-hydroxyisopropyl)-dihydrofurano-[2″,3″:3′,4′]-2′,4-dihydroxy-6′-methoxychalcone (**CHL1c**) and (*E*)-2″-(2‴-hydroxyisopropyl)-dihydrofurano-[2″,3″:2′,3′]-4′-hydroxy-5-methoxy-chalcone (**CHL1d**) (Herath et al. 2003) were also reported. The conversion of xanthohumol (**CHL1b**), a prenylated chalcone isolated from hops (*Humulus lupulus*), by *Cunninghamella elegans* var. *elegans* 6992 afforded a new glycosylated derivative, 5-methoxy-8-prenylnaringenin 7-*O*-β-glucopyranoside (**FNN1b**) (Kim and Lee 2006). Also, *Penicillium chrysogenum* 6933 glycosylated xanthohumol (**CHL1b**) into two novel chalcones, xanthohumol 4′-*O*-β-glucopyranoside (**CHL1e**) and xanthohumol 4,4′-*O*-β-diglucopyranoside (**CHL1f**) while *Rhizopus oryzae* KCTC 6946 produced a known compound, isoxanthohumol (**FNN1c**).

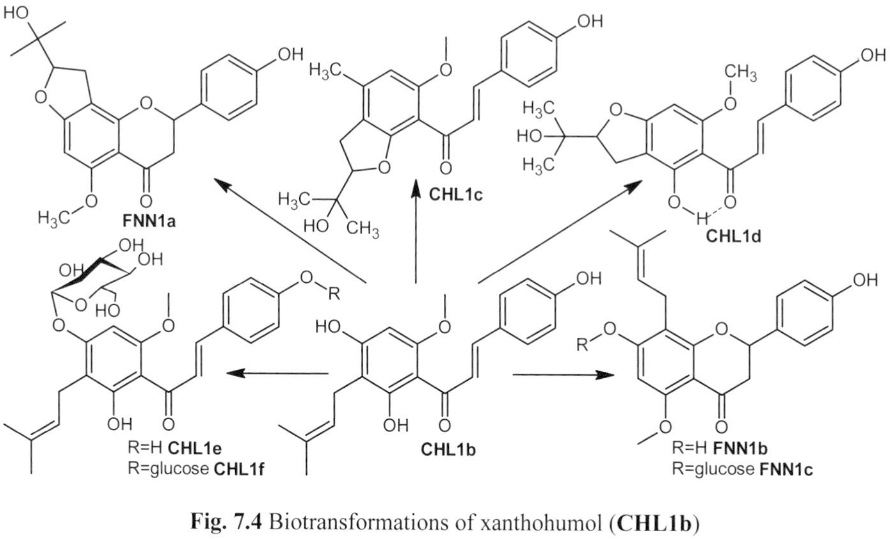

Fig. 7.4 Biotransformations of xanthohumol (**CHL1b**)

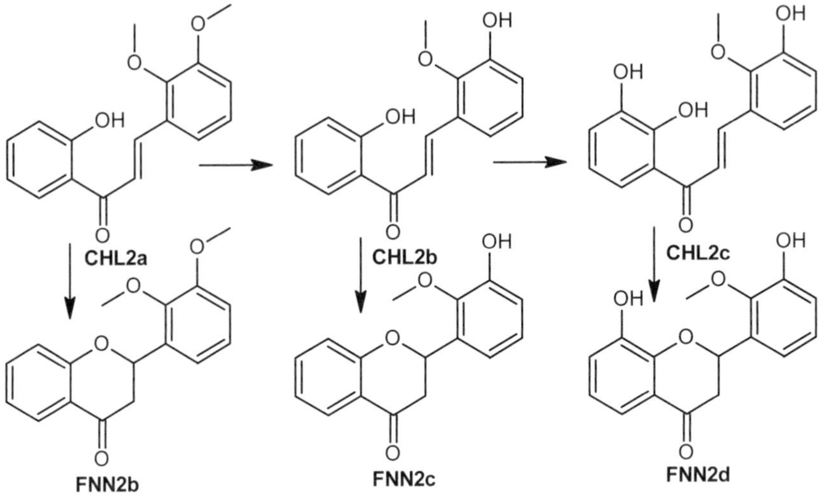

Fig. 7.5 Biotransformation of chalcone by *Aspergillus alliaceus*

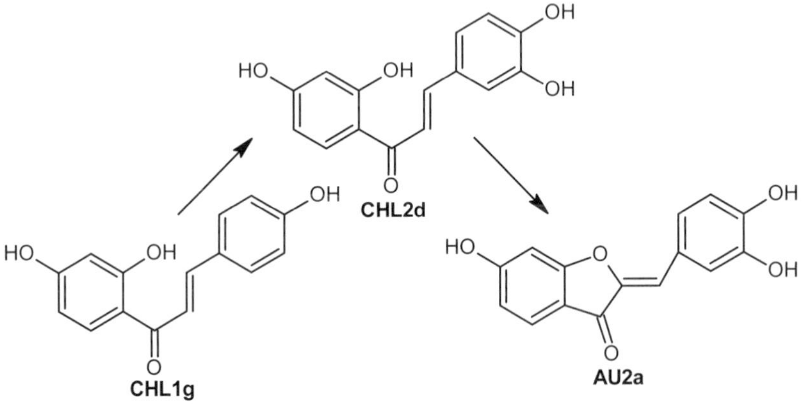

Fig. 7.6 Biotransformation of isoliquiritigenin (**CHL1g**) by *A. alliaceus*

Another fungus, *Aspergillus alliaceus*, was examined for its potential to catalyze the cyclization of chalcones into flavonoids and other products, a process analogous to pathways that occur in plants. *A. alliaceus* UI 315 transformed 2′-hydroxy-2″,3″-dimethoxychalcone (**CHL2a**) into 2′,3″-dihydroxy-2″-methoxychalcone (**CHL2a**) and 2′,3′,3″-trihydroxy-2″-methoxychalcone (**CHL2a**). The three molecules served as intermediates to form their respective flavanones, 2′,3′-dimethoxyflavanone (**FNN2b**), 3′-hydroxy-2′-methoxyflavanone (**FNN2c**), and 3′,8-dihydroxy-2′-methoxyflavanone (**FNN2d**) (Sanchez-Gonzalez and Rosazza 2004).

Investigating the role of cytochromic P450 enzymes using the P450 inhibitors SKF525A, metyrapone, and phenylthiocarbamide showed that up to three P450 enzymes may be responsible for catalyzing the cyclization, *O*-demethylation and hydroxylation of chalcones. The resulting flavanone products were racemic, unlike the same products that are stereoselectively cyclized in plants. The nonstereospecific products were speculated to be a result of a radical-based, intramolecular cyclization of chalcones to flavanones (Sanchez-Gonzalez and Rosazza 2004). In plants, CHI is responsible for the stereospecific intramolecular reaction in which a phenolate ion cyclizes by 1,4-Michael addition with the α,β-unsaturated carbonyl functionality to form (2S)-flavanones (Jez et al. 2000).

Recently, *A. alliaceus* biotransformed differently substituted chalcones into aurones (Das and Rosazza 2006). Isoliquiritigenin (**CHL1g**) was first hydroxylated at 3′-position to form butein (**CHL2d**), which was further cyclized to the aurone product sulfuretin (**AU2a**). The use of cytochromic P450 inhibitors gave evidence that initial C-3′ hydroxylation of isoliquiritigenin (**CHL1g**) to butein (**CHL2d**) was catalyzed by a CYP450 enzyme system. Indeed, a partially purified *A. alliaceus* catechol oxidase cyclized butein (**CHL2d**) to sulfuretin (**AU2a**). The biosynthesis of aurones in plants was suggested to be a two-step process (Nakayama et al. 2001) and the synthesis of aurones by *A. alliaceus* could likely follow a similar process. In the first step, cytochrome P450 hydroxylates isoliquiritgenin (**CHL1e**) at the 3′-position, giving butein (**CHL2d**), while a catechol oxidase catalyzes the intramolecular cyclization to form the aurone (**AU2a**) and likely involves an *o*-quinone intermediate.

7.2.2 Flavanones

Formed from chalcones, flavanones (**FNN**) are gateway compounds for the biosynthesis of a plethora of flavonoids in plants and dominate the flavonoid content in citrus fruits (Peterson et al. 2006). Natural and synthetic flavanones have attracted considerable attention for their interesting biological activity in mammals including antioxidant, antibacterial, anti-inflammatory and nueroprotective activities (Havsteen 1983; Khan and Hasan 2003; Cao et al. 2006; Cho 2006; Moorthy et al. 2006). Modification of flavanones by microorganisms increases their diversity with the potential of instilling improved or new biological properties (Figs. 7.7–7.12).

Halogenation of flavonoids was observed when *Caldariomyces fumago* (ATCC 11925) was fermented in the presence of flavanones but not flavones (Yaipakdee and Robertson 2001). *C. fumago* halogenated both naringenin (**FNN1**) and hesperetin (**FNN2a**) at the C-6 and/or C-8 positions (**FNN1d,e,f,g** and **FNN2e,f**) when KCl or KBr was added to the cultures. Assays with the well know chloroperoxide enzyme from *C. fumago* resulted in the production of halogenated flavanones, lending support that it is the enzyme responsible for the biotransformation. Halogenated flavonoids could possess potent biological activities as exhibited with halogenated chalcones (Bois et al. 1998).

Sulfation reactions of flavonoids has notably been observed as products of mammalian phase II metabolism (Ruefer et al. 2005; Williamson et al. 2005). With respect to flavanones, fermentation of naringenin (**FNN1**) with the fungus *Cunninghamella elegans* (NRRL 1392) yielded naringenin 7-*O*-sulfate (**FNN1h**) (Ibrahim 2000). Although parallel to mammalian metabolism, sulfation is a rare microbial transformation.

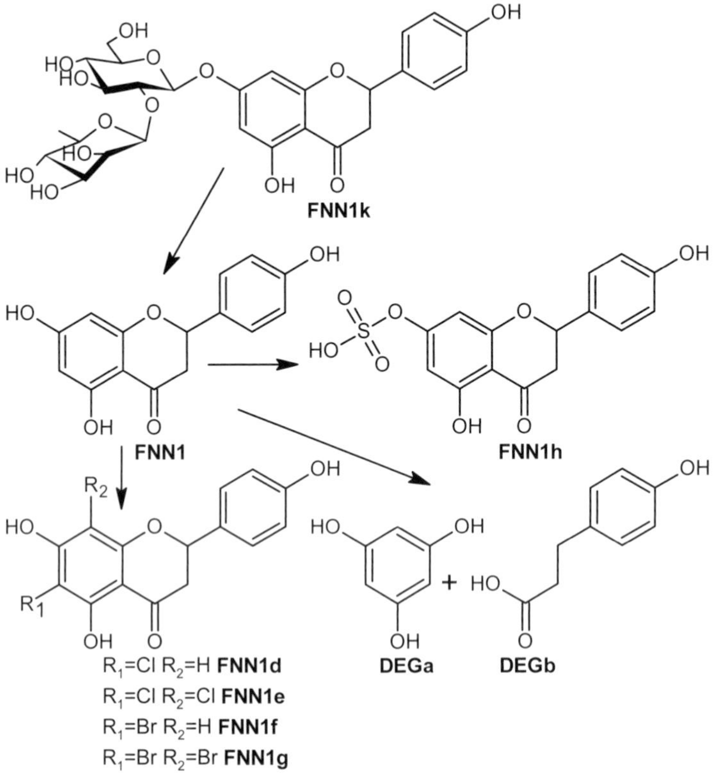

Fig. 7.7 Biotransformation of naringenin (**FNN1**)

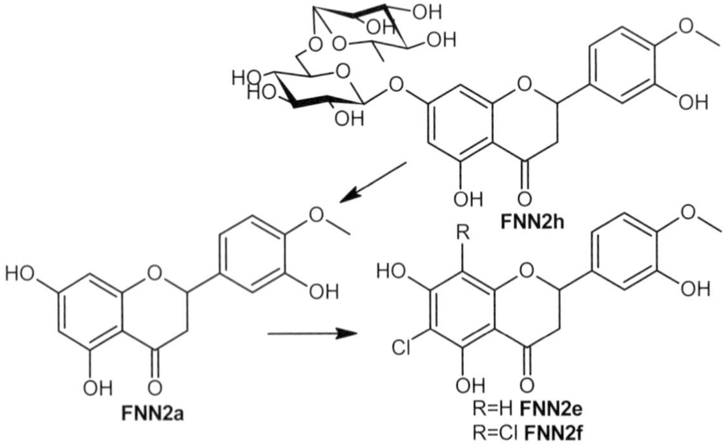

Fig. 7.8 Biotransformations of hesperetin (**FNN2a**)

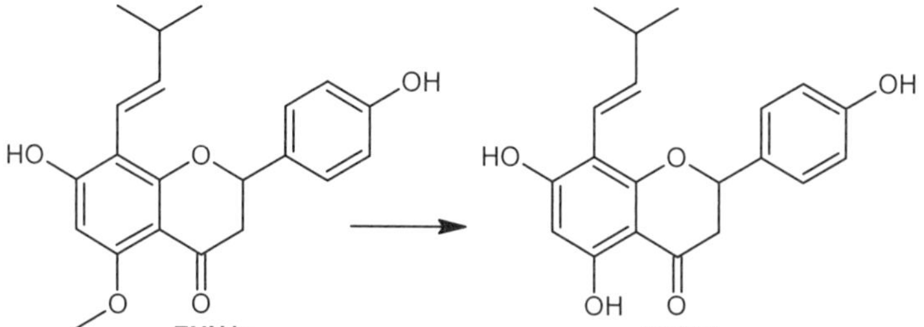

Fig. 7.9 Biotransformation of isoanthohumol (**FNN1c**) into 8-prenylnaringenin (**FNN1i**)

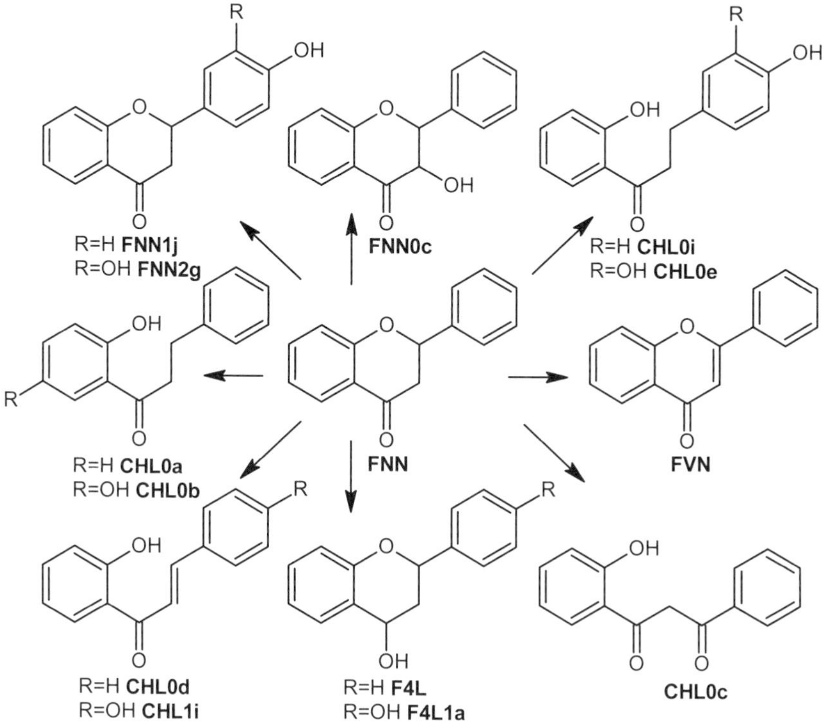

Fig. 7.10 Biotransformation of flavanone (**FNN**)

In the past few years, hops has gained attention as the source of prenylflavonoids. Among other flavonoids, hops contain two prenylchalcones, xanthohumol (**CHL1b**) and desmethylanthohumol and three prenylflavanones, isoxanthohumol (**FNN1c**), 8-prenylnarigenin (**FNN1i**) and 6-prenylnaringenin. Xanthohumol (**CHL1b**) is a strong cancer chemopreventive agent (Gerhauser et al. 2002; Gerhauser 2005), and 8-prenylnaringenin (**FNN1i**) is one of the most potent phytoestrogens identified so far (Milligan et al. 1999, 2002). There is considerable interest whether human exposure to phytoestrogens has either health risks or benefits (Barnes 2003; Magee and Rowland 2004).

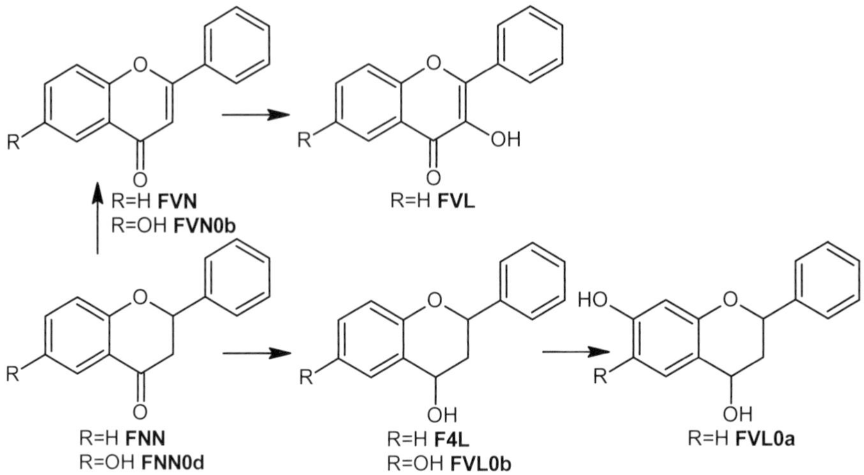

Fig. 7.11 Reduction and hydroxylation of flavanones by *A. niger*

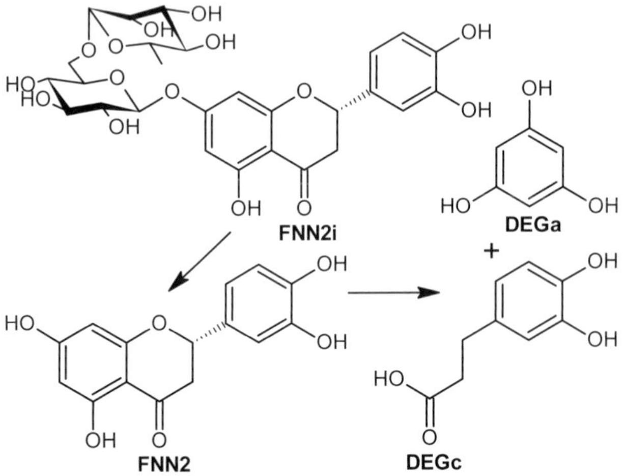

Fig. 7.12 Metabolism of eriocitrin (**FNN2i**)

Isoanthohumol (**FNN1c**) is the prevalent prenylflavonoid found in beer in concentrations as high as 4mg/L in strong ales (Stevens et al. 1999; Rong et al. 2000). A study determined that human intestinal microflora is capable of *O*-demethylating isoanthohumol (**FNN1c**) into 8-prenylnaringenin (**FNN1i**). Indeed, *Eubacterium limosum* isolated from the microflora was capable of *O*-demethylation of isoanthohumol (**FNN1c**) into 8-prenylnaringenin (**FNN1i**) with 90% efficiency (Possemiers et al. 2005). This study raises the question whether moderate beer consumption elevates the level of 8-prenylnaringenin (**FNN1i**) *in vivo* and its affect on human health.

The intestinal conversion of isoanthohumol (**FNN1c**) to 8-prenylnaringenin (**FNN1i**) was further investigated *in vitro* using the Simulator of the Human Intestinal Microbial Ecosystem (SHIME), a succession of five reactors representing the different parts of the human gastrointestinal tract (Molly et al. 1993). The results found that isoanthohumol (**FNN1c**) could reach the large intestine intact followed by up to 85% conversion to 8-prenylnaringenin (**FNN1i**) within the transverse and descending colon (Possemiers et al. 2006).

Microorganisms are not limited to biotransforming dietary flavanones that occur naturally in plants. Chemically synthesized unsubstituted (±)-flavanone has been transformed by a number of microorganisms yielding an array of products. After screening 20 genera (80 species), *Absidia blackesleeana* NRRL 1306, *Aspergillus niger* X172 and NRRL 599, *Penicillium chrysogenum* 10002 K, and *Streptomyces fulvissimus* NRRL B1453 were identified as being efficient modifiers of unsubstituted flavanone into numerous metabolites at various positions by hydroxylation, C-ring dehydrogenation, and C-ring cleavage (Abdelrahim and Abulhajj 1990). Ten metabolites of (±)-flavanone were isolated and identified as 4′-hydroxyflavanone (**FNN1i**), 3′,4′-dihydroxyflavanone (**FNN2g**), 3-hydroxyflavanone (**FNN1c**), flavone (**FVN**), 2′-hydroxydihydrochalcone (**CHL0a**), 2′,4″-dihydroxydihydrochalcone (**CHL1h**), 2′,3″,4″-trihydroxydihydrochalcone (**CHL2e**), 2′,5′-dihydroxydihydrochalcone (**CHL0b**), 4′-hydroxyflavan-4α-ol (**F4L1a**), and 2′-hydroxydibenzoylmethane (**CHL0c**). Evidence also indicated that some reactions occur in a stereospecific manner. More recently, *A. niger* reduced the carbonyl group at the C4 position or dehydrogenated at the C2 and C3 positions of (±)-flavanone and 6-hydroxyflavanone (**FNN0d**) to give their respective flavones (**FVN** and **FVN0b**) and flavan-4-ols (**F4L** and **F4L0b**) (Kostrzewa-Suslow et al. 2006). Additionally, flavanone reduction of C4 together with hydroxylation at C7 gave 7-hydroxyflavan-4-ol (**F4L0a**) while dehydrogenation at C2, C3 along with hydroxylation at C3 gave flavonol (**FVL**).

C-Ring opening of (±)-flavanone was also achieved by the fungus *Gibberella fujikuroi* to give 2′-hydroxychalcone (**CHL0d**) and 2′,4″-dihydroxychalcone (**CHL1i**) (Udupa et al. 1969). Additional products from *G. fujikuroi* metabolizing flavanone included (−)-flavan-4α-ol (**F4L**), (±)-4′-hydroxyflavanone (**FNN1i**) and (−)-4′-hydroxyflavan-4α-ol (**F4L1a**). Results of these studies showed the fungus preference to hydroxylate the B-ring.

Intestinal bacteria played a key role in the degradation of flavanone glycosides. Both naringin (**FNN1k**) and hesperidin (hesperetin-7-*O*-rhamnoglucoside) (**FNN2h**) were hydrolyzed at the glycosidic bond liberating the aglycons by a bovine rumen isolate, a *Butyrivibrio* sp. C_3. When naringin (**FNN1k**) was anaerobically incubated, the products were naringenin (**FNN1**), phloroglucinol (**DEGa**), and *p*-hydroxyphenylpropionic acid (**DEGb**) (Krishnamurty et al. 1970). Both phloroglucinol (**DEGb**), and *p*-hydroxyphenylpropionic acid (**DEGb**) presumably resulted from the C-ring cleavage of the precursor naringenin (**FNN1**).

Eriocitrin (eriodictyol 7-*O*-rutinoside) (**FNN2i**), a common flavanone in lemon-fruit, was hydrolyzed into eriodictyol (**FNN2**), its aglycon, by the intestinal bacteria *Bacteroides distasonis* JCM 5825, *Bacteroides uniformis* JCM 5828 and *Enterobacter cloacae* IAM 12349. Furthermore, eriodictyol (**FNN2**) was converted to 3,4-

dihydroxyhydrocinnamic acid (**DEGc**) and phloroglucinol (**DEGa**) by *Clostridium butyricum* IFO 14252 but not by the other strains (Miyake et al. 1997). The metabolism of eriodictyol (**FNN2**) by *C. butyricum* showed that the bacteria cleaved the flavonoid ring of eriodictyol (**FNN2**). Incubation of narigin (**FNN1k**) and hesperidin (**FNN2H**) with human fecal microflora resulted in their degradation into their respective aglycons (**DEGa, DEGc**) as well as smaller phenolic compounds that were not positively identified (Justesen et al. 2000). Biodegradation by gut microflora may decrease the overall bioavailability of rapidly degraded flavonoids because they are less likely to be absorbed if they are rapidly degraded.

7.2.3 Isoflavones

The most abundant isoflavones in soy beans are genistin (**ISO1a**) and daidzin (**ISO1c**) (Hosny and Rosazza 1999) which are the 7-(6-*O*-acetyl)-glucosides of the aglycons genistein (**ISO1**) and daidzein (**ISO1b**), respectively. Available in abundant quantities, genistein (**ISO1**) and daidzein (**ISO1b**) are ideal candidates for use as chemical feed stocks for the microbial synthesis of rarer isoflavones. Microbes are able to perform a number of transformations of isoflavones (Figs. 7.13–7.15). For instance, the strain *Strepomyces griseus* (ATCC 13273) cultivated with media containing soybean-meal resulted in halogenated isoflavonoids. Along with genistein (**ISO1**), 8-chlorogenistein (**ISO1d**) and 6,8-dichlorogenistein (**ISO1e**) were isolated and characterized. Incubation of *S. griseus* with genistein (**ISO1**) also produced the same chlorinated genisteins (**ISO1d,e**) showing that microbial activity is responsible for the halogenation (Anyanwutaku et al. 1992).

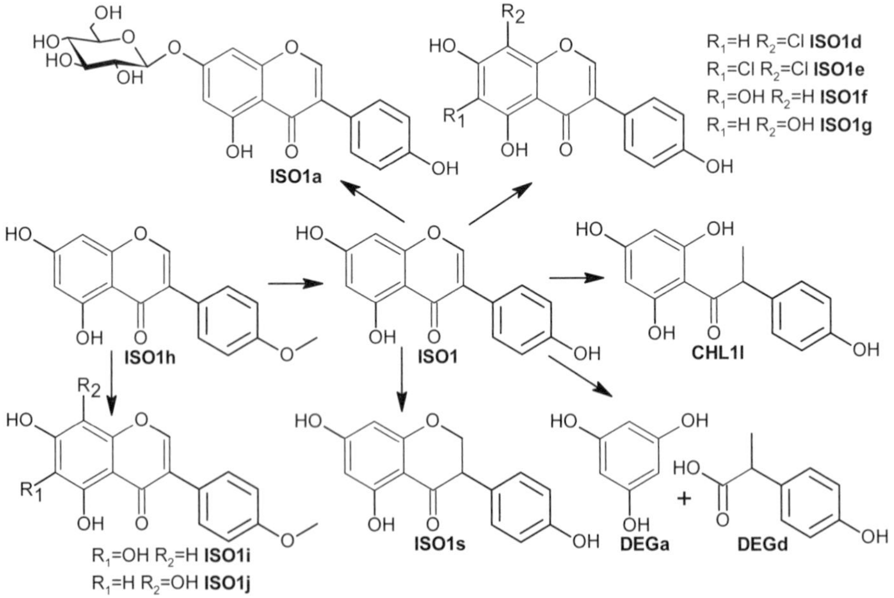

Fig. 7.13 Biotransformation of genistein (**ISO1**)

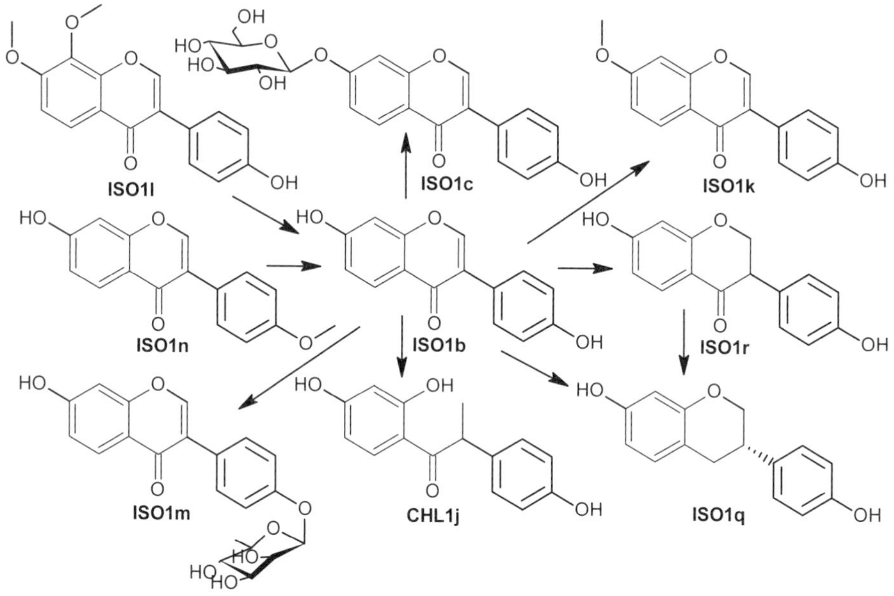

Fig. 7.14 Biotransformation of daidzein (**ISO1b**)

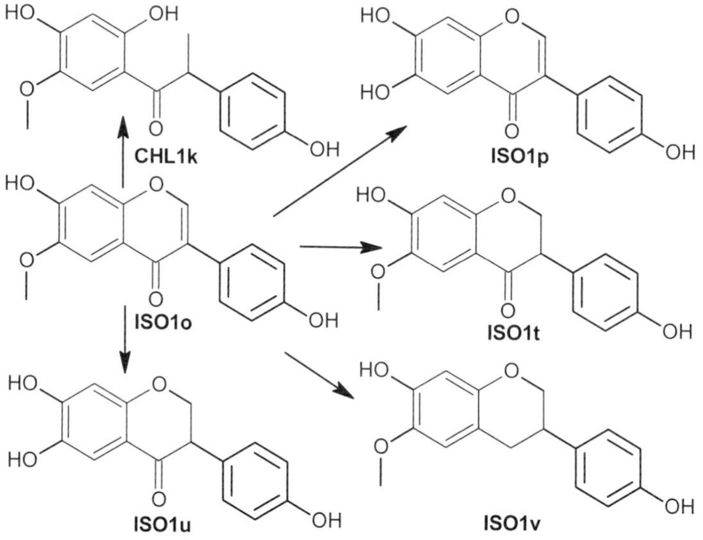

Fig. 7.15 Biotransformations of glycitein (**ISO1o**)

Two strains identified as *Micrococcus* or *Arthrobacter* species were isolated from fermenting soybeans used to make tempe, a traditional Indonesian food. Both strains hydroxylated genistein (**ISO1**) at the 6- or 8-position to give 6-hydroxygenistein (**ISO1f**) and 8-hydroxygenistein (**ISO1g**) (Klus and Barz 1998).

Biochanin A (**ISO1h**), the 4'-methoxy form of genistein (**ISO1**), was also hydroxylated by both strains to form 8-hydroxybiochanin A (**ISO1i**) but only the *Micrococcus* bacteria produced 6-hydroxybiochanin A (**ISO1j**). The formation of these new isoflavones lends importance to the study of flavonoid containing food prepared using microbial fermentation for possible nutritional value and biological activity.

Nocardia species NRRL 5646 and *Mortierella isabellina* ATCC 38063 hydroxylated and methylated daidzein (**ISO1**) to isoformononetin (7-methoxy-4'-hydroxyisoflavone) (**ISO1k**) and 7,8-dimethoxy-4'-hydroxyisoflavone (**ISO1l**) (Maatooq and Rosazza 2005). In addition, the fungus *M. isabellina* produced a new compound from daidzein (**ISO1**) identified as daidzein-4'-O-rhamnopyranoside (**ISO1m**).

O-Demethylation is another significant process that may increase the bioactivity of isoflavones (Arora et al. 1998; Kuiper et al. 1998). *Eubacterium limosum*, an acetogenic bacterium, commonly found in human intestinal contents (Wang et al. 1996), is known to *O*-demethylate the methoxyl derivatives of benzoic acid (Genthner et al. 1981; Deweerd et al. 1988). Incubation of *E. limosum* (ATCC 8486) with three methoxylated isoflavonoids, biochanin A (**ISO1h**), formononetin (**ISO1n**), and glycitein (**ISO1o**), resulted in the detection of genistein (**ISO1**), daidzein (**ISO1b**), and 4',6,7-trihydroxyisoflavone (**ISO1p**), respectively (Hur and Rafii 2000).

Metabolism of daidzein (**ISO1b**) is not well characterized but equol (**ISO1q**), dihydrodaidzein (**ISO1r**), and *O*-desmethylanglesin (**CHL1j**) are three metabolites that have been investigated recently for their health benefits (Heinonen et al. 1999, 2003; Hur et al. 2000). Equol (**ISO1q**), [7-hydroxy-3-(4'hydroxyphenol)-chroman], is a nonsteriodal estrogen discovered during the early 1980s in the urine of adults consuming soy foods (Axelson et al. 1982). It has a high affinity for estrogen receptors, ERα and ERβ (Setchell et al. 2005). The health benefits of producing equol (**ISO1q**) remain unclear but some data suggests it is useful in relation to markers of breast cancer risk (Adlercreutz et al. 1987; Cassidy et al. 1993; Ingram et al. 1997; Duncan et al. 2000). The amounts and ability to produce certain isoflavone derivatives varies among individuals. Studies have shown that about 30–50% of individuals of the studied populations are able to produce equol (**ISO1q**) from daidzein (**ISO1b**) (Hutchins et al. 1995; Kelly et al. 1995; Lampe et al. 1998; Arai et al. 2000; Akaza et al. 2002) and approximately 80–90% are capable of producing *O*-desmethylangolesin (**CHL1j**) from daidzein (**ISO1b**) (Kelly et al. 1995; Arai et al. 2000; Frankenfeld et al. 2004b). The interindividual differences of daidzein (**ISO1b**) metabolism maybe unique to humans. The majority of animal studies, including monkeys, chimpanzees, mice, rats, hamsters, cows, pigs, sheep and dogs, suggest that all have the ability to produce equol (**ISO1q**) (Monfort et al. 1984; Adlercreutz et al. 1986; Juniewicz et al. 1988; Lundh 1995; Brown and Setchell 2001; Lamartiniere et al. 2002; Ohta et al. 2002; Blair et al. 2003).

Production of equol is thought to be linked to the gut microflora (Setchell et al. 1984). Equol (**ISO1q**) was not detected in urine from germ-free rats fed with a soy-isoflavone containing diet but when the germ-free rats where inoculated with intestinal microflora from an equol producing human, equol (**ISO1q**) was detected in the urine (Bowey et al. 2003). The lack of equol (**ISO1q**) in plasma of infants fed formula also highlights the role of active microflora for its formation (Setchell et al. 1997, 1998). *In vitro* cultures of human feces with daidzein (**ISO1b**) produced equol (**ISO1q**) and *O*-desmethylanglesin (**CHL1j**) and the conversion was inhibited by some antibiotics (Atkinson et al. 2004). Conversion of daidzein (**ISO1B**) by microflora was also shown to exclusively produce the *S*-equol enantiomer (**ISO1q**). The *S*-equol enantiomer (**ISO1q**) was also determined to have a high affinity for estrogen receptor *β*, whereas *R*-equol was relatively inactive (Setchell et al. 2005).

It is not entirely clear as to which microbes are responsible for the conversion of daidzein (**ISO1b**) into equol (**ISO1q**) as several candidates have been suggested. The conversion of daidzin (**ISO1c**) to daidzein (**ISO1b**) and equol (**ISO1q**) was detected in fermented soymilk by a few *Bifidobacterium* species including *B. pseudolongum*, *B. longum*-a and *B. animalis* (Tsangalis et al. 2002, 2003). Bacteria isolates from equol-producing individuals, including *Bacteroides ovatus* ssp., *Ruminococcus productus* ssp., and *Streptococcus intermedius* ssp, were also reported to metabolize pure daidzein (**ISO1b**) to equol (**ISO1q**) *in vitro* (Ueno et al. 2002). A mixed culture containing four different bacteria species, including a tentatively identified *Veillonella* sp. strain EP, produced equol (**ISO1q**) (Decroos et al. 2005). Three of the strains that could be isolated separately did not produce equol alone and the tentatively identified *Veillonella* sp. strain EP could not be isolated as a single strain.

Dihydrodaidzein (**ISO1r**) is a potential intermediate for the conversion of daidzein (**ISO1b**) to equol (**ISO1q**) (Setchell et al. 1984; Chang and Nair 1995) and therefore, the metabolic pathway may even involve multiple bacterial strains. Both *Escherichia coli* HGH21 and a Gram-positive *Clostridium* sp. strain HGH6 converted daidzein (**ISO1b**) and genistein (**ISO1**) to dihydrodaidzein (**ISO1r**) and dihydrogenistein (**ISO1s**), respectively (Hur et al. 2000). More recently, a rod-shaped and Gram-positive anaerobic bacterium, named Niu-O16, which was isolated from bovine rumen contents, had a higher conversion rate than the HGH6 strain (Wang et al. 2005b). Also, a newly isolated rod-shaped, gram-negative anaerobic bacterium from human feces, named Julong 732, was found to be capable enantioselective synthesis of *S*-equol (**ISO1q**) from a racemic mixture of dihydrodaidzein (**ISO1r**) (but not daidzein (**ISO1b**), tetrahydrodaidzein, dehydroequol or *R*-equol) under anaerobic conditions (Wang et al. 2005a).

Equol (**ISO1q**) and *O*-desmethylangolesin (**CHL1j**) are likely produced by different strains. *In vitro*, fecal bacteria from equol nonproducing individuals converted daidzein (**ISO1b**) to *O*-DMA (**CHL1j**) but not equol (**ISO1q**) (Atkinson et al. 2004; Decroos et al. 2005) and observational studies show that some equol producers did not produce *O*-DMA (**CHL1j**), and vice versa (Rowland et al. 2000; Frankenfeld et al. 2004a, 2004b). Germ-free rats inoculated with fecal flora from a good equol producing human produced equol (**ISO1q**) but not *O*-DMA (**CHL1j**) and rats with microflora from a poor equol producer made *O*-DMA (**CHL1j**) but not equol (**ISO1q**) (Bowey et al. 2003). A gram-positive intestinal anaerobic bacterium, strain HGH 136, identified as a *Clostrium* sp., cleaved the C-ring of daidzein (**ISO1b**) to produce *O*-demethylangolensin (**CHL1j**) (Hur et al. 2002).

Another important isoflavone is glycitein (**ISO1o**) which makes up less than 10% of the total isoflavone content in soybeans and soybean foods but comprises over 50% of the isoflavone mass in soy germ (Song et al. 1998). Glycitein (**ISO1o**) was shown to have a more estrogenic effect in a mouse uterine growth assay compared to genistein (**ISO1**), when fed in equal amounts(Song et al. 1999). Five tentative metabolites, dihydroglicitein (**ISO1t**), dihydro-6,7,4′-trihydroxyisoflavone (**ISO1u**), 5′-*O*-methyl-*O*-desmethylangolensin (**CHL1k**), daidzein (**ISO1b**) and 6-*O*-methylequol (**ISO1v**), were identified when glycitein (**ISO1o**) was anaerobically incubated with human feces (Simons et al. 2005b). A similar metabolic mechanism was proposed to explain the isoflavone metabolites found in urine samples of subjects on a soy-bar supplemented diet (Heinonen et al. 2003).

Human intestinal microflora plays an important role in the metabolism of isoflavones. Isoflavones occur mainly as glucosides in nature (Goldin 1990). A typical step in the metabolism of isoflavones is the cleavage of the glucosides to release the aglycons (Xu et al. 1995; Kim et al. 1998; Rowland et al. 1999; Scalbert and Williamson 2000). Some glucosidase activity may be assigned to non-bacterial enzymes in the intestinal tract (Day et al. 2000) but microflora are attributed to the majority of the activity (Turner et al. 2003). For example, bacterial species of *Bifidobacterium* fermenting soymilk, caused the hydrolysis of isoflavone malonyl-, acetyl- and *β*-glucosides and an increase of aglycons (Tsangalis et al. 2002). Similarly, *Bifidobacterium* sp. Int-57 completely hydrolyzed the isoflavone glycosides, genistin (**ISO1a**) and daidzin (**ISO1c**), in 18 hours (Jeon et al. 2002). Glucosidase activity may affect the bioavailability of isoflavones since isoflavone glucosides are poorly absorbed, while human and bacterial metabolism can alter the bioactivity since many of the metabolites are pharmacologically inactive (Joannou et al. 1995; Xu et al. 1995; Hur et al. 2000). Following the removal of the sugar group, the isoflavone aglycon may be further metabolized or absorbed and subjected to first pass hepatic metabolism (Xu et al. 1995; Kim et al. 1998; Scalbert and Williamson 2000).

Isoflavone excreted in urine only make up a small fraction of the total isoflavone intake. Therefore, isoflavones are either not absorbed by the gut, absorbed and released in bile followed by fecal excretion, or degraded by the liver or gut microflora (Scalbert and Williamson 2000). Only a low amount of free aglycon isoflavones was detected in feces (Watanabe and Adlercreutz 1998); thus direct elimination by fecal excretion seems unlikely. The total isoflavone content in fecal matter was determined to be about 1–2% of the total intake (Xu et al. 1994, 1995). Therefore, the majority of isoflavones consumed are apparently further degraded beyond deglycosylation. The microflora that aid in the digestion of food likely plays an important role in this process. Incubation of rat caecal and human fecal microflora with [^{3}H] and [^{14}C]genistein (**ISO1**) yielded intermediates dihydrogenistein (**ISO1s**) and 6′-hydroxy-*O*-desmethylangolensin (**CHL1l**) and end-products 2-(4-hydroxyphenyl) propanoic acid (**DEGd**) and presumably (unlabeled and therefore undetectable) phloroglucinol (**DEGa**) (Coldham et al. 2002). Subsequently, metabolism would explain the low bioavailability of biologically active isoflavones from food (Turner et al. 2003).

7.2.4 Flavones

Flavones (**FVN**) share similar chemical structures but differ from flavanols by their lack of a hydroxyl group at the C3 position. The most abundant flavones in plants are apigenin (**FVN1**) and luteolin (**FVN2**) and are usually found with attached sugars as

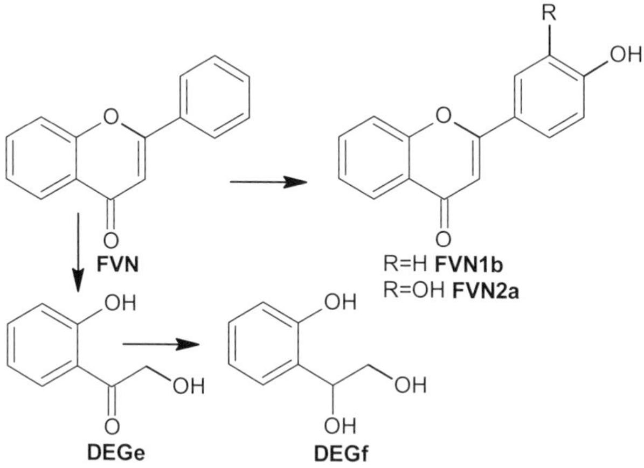

Fig. 7.16 Biotransformations of flavone (**FVN**)

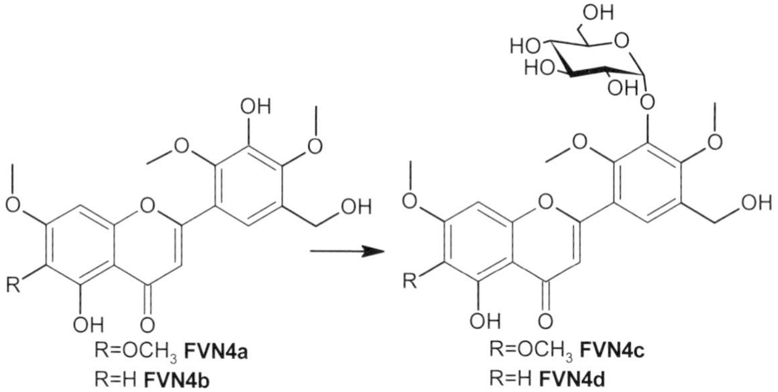

Fig. 7.17 Phase II reactions by *Cunninghamella elegans*

O-glucosides and even C-glucosides (Hollman and Arts 2000). Flavones serve a variety of roles in plants including UV protection and may promote human health. For example, diosmin, found in citrus fruits, is used in combination with hesperidin (**FNN2h**) under the trade name Daflon for the treatment of venous diseases.

Unsubstituted (±)-flavone has been biotransformed by a number of microorganisms yielding a wide array of metabolites (Figs. 7.16–7.20). Fermentations with several species of *Aspergillus*, *Cunninghamella*, *Helicostylum*, *Penicillium*, and *Liderina* and a *Streptomycete* showed the formation of 4′-hydroxyflavone (**FVN1b**) and 3′,4′-dihydroxyflavone (**FVN2a**) (Ibrahim and Abulhajj 1990). Fungal species of *Absidia*, *Gongronella*, *Rhizopus*, *Manascus*, and *Gymnascella* produced *o*-hydroxyphenylhydroxymethyl ketone (**DEGe**) by C-ring cleavage of (±)-flavone and *Rhizopus nigricans* further reduced the ketone to form 1-(o-hydroxyphenyl)-1,2-ethanediol (**DEGf**) (Ibrahim and Abulhajj 1990). A number of fungi tested also transformed (±)-flavone with higher yields compared to unsubstituted isoflavone.

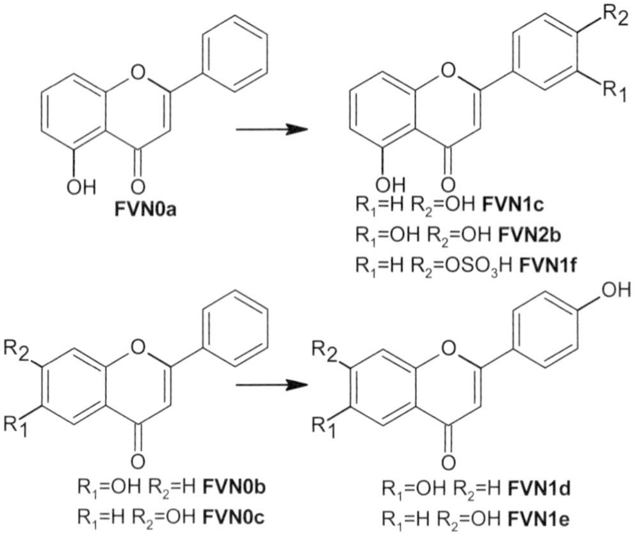

Fig. 7.18 Biotransformations of flavones by *S. fulvissimus*

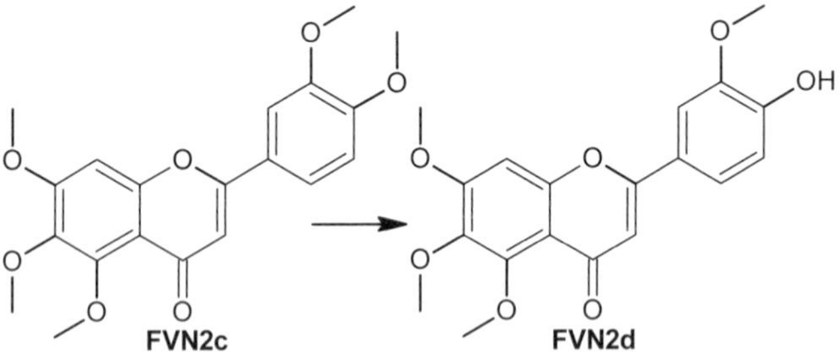

Fig. 7.19 Demethylation of sinesetin (**FVN2c**) by *Aspergillus niger*

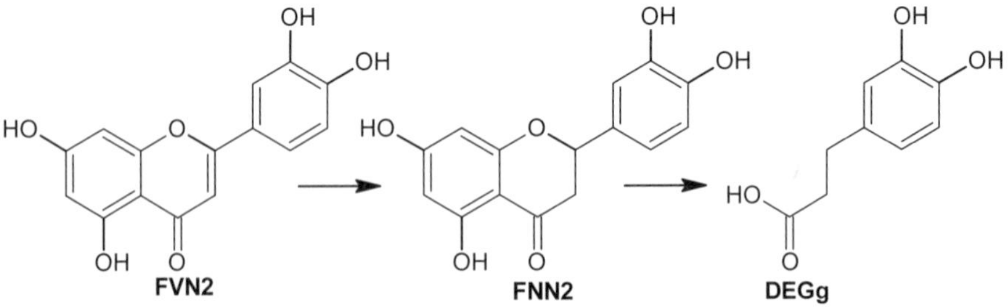

Fig. 7.20 Metabolism of luteolin (**FVN2**) by *Eubacterium ramulus*

Several studies reported phase II conjugates such as flavonoid glucosides and sulfates. Microbial transformation of psiadiaradin (**FVN4a**) and 6-desmethoxypsiadiarabin (**FVN4b**), two flavones from the bushy shrub *Psiadia ara-*

bica, by *Cunninghamella elegans* NRRL 1392 gave the corresponding 3'-*O*-α-D-glucoside conjugates (**FVN4c,d**) (Ibrahim et al. 1997). The psiadiaradin glucosides were also imbued with antioxidant activity towards human leukemia cell line HL-60, whereas the aglycons were inactive (Takamatsu et al. 2003).

Microorganisms are known to perform hydroxylations upon flavones. Aromatic hydroxylation of 5-hydroxyflavone (**FVN0a**), 6-hydroxyflavone (**FVN0b**) and 7-hydroxyflavone (**FVN0c**) by *Streptomyces fulvissimus* NRRL B1453 gave their corresponding 4'-hydroxylated products (**FVN1c,d,e**) and produced 3',4',5-trihydroxyflavone (**FVN2b**) (Ibrahim and Abulhajj 1990). Furthermore, *S. fulvissimus* converted 5-hydroxyflavone (**FVN0a**) into more polar metabolite identified as 5,4'-dihydroxyflavone-4'-sulfate (**FVN1f**) (Ibrahim and Abulhajj 1989). In general, sulfation is a rare microbial transformation parallel to phase II mammalian metabolism.

Sinesetin (**FVN2c**), found in *Citrus* fruit, was demethylated at the 4'-position to give ageconyflavone B (5,6,7,3'-tetramethoxy-4'-hydroxyflavone) (**FVN2d**) by *A. niger* and the metabolite showed antimutagenic activity against chemical mutagens (Okuno and Miyazawa 2006).

The basic structure of a flavonoid may contribute to its resistance or susceptibility to metabolism by bacteria. A study using gut microflora compared the degradation rates of a number of flavonoids and found flavones hydroxylated at both the 5- and 7-positions had significantly higher degradation rates than less substituted flavones, most of which are not found naturally (Simons et al. 2005a). Another example of flavone degraded was by *Eubacterium ramulus* strain wK1 which converted luteolin (**FVN2**) into eriodictyol (**FNN2**) as an intermediate followed by further degradation to 3-(3,4-dihydroxyphenyl)propionic acid (**DEGg**) (Braune et al. 2001).

7.2.5 Flavonols

Plants often modify the core structure of flavonoids by adjusting the number of hydroxyl or methoxy groups. In a similar fashion (Fig. 7.21–7.22), *Streptomyces griseus* (ATCC 13273) methylated the flavonol quercetin (**FVL2**) at the B-ring to afforded isorhamnetin (5,7,3'-trihydroxy-4'-methoxyflavonol) (**FVL2a**) and dillenetin (5,7-dihydroxy-3',4'-dimethoxyflavonol) (**FVL2b**), and hydroxylated at the A-ring or B-ring to give gossypetin (5,7,8,3',4'-pentahydroxyflavonol) (**FVL2c**) and myercetin (**FVL3**), respectively (Hosny et al. 2001). Gossypetin (**FVL2c**) was also further modified with the addition of a *O*-methyl group to the A-ring to form corniculatusin (3',4',5,7-tetrahydroxy-8-methoxyflavonol) (**FVL2d**). (**FVL2f**) and performed B-ring *O*-methylations giving mono and Additionally, *S. griseus* catalyzed A-ring hydroxylation of fisetin (**FVL2e**) to melanoxetin (3',4',7,8- tetrahydroxyflaonol) dimethoxy products, geraldol (3'-methoxyfisetin) (**FVL2g**), 4'-methoxyfisetin (**FVL2h**), and 3',4'-dimethoxy-fisetin (**FVL2i**) (Hosny et al. 2001). Apparently, *S. griseus* only methylated metabolites that contained catechol moieties (Ibrahim and Abulhajj 1990; Hosny et al. 2001). Purified *S*-adenosyl methionine-dependent, catechol-*O*-methyl transferase (COMT) from *S. griseus* (Dhar and Rosazza 2000), the first COMT found in bacteria, methylated only catechol groups. In mammals, *O*-methyltransferases exist that are capable of methylating flavonoid catechols and is a well known detoxification pathway (Czeczot et al. 1990). Microorganisms are an excellent means to mimic metabolic patterns observed in both plants and animals.

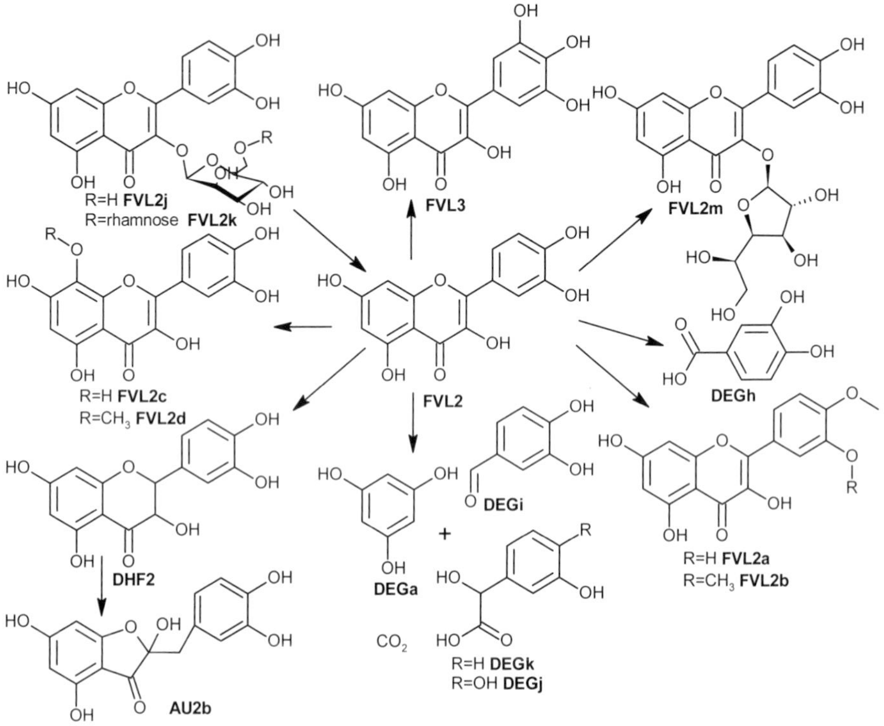

Fig. 7.21 Biotransformations of quercetin (**FVL2**)

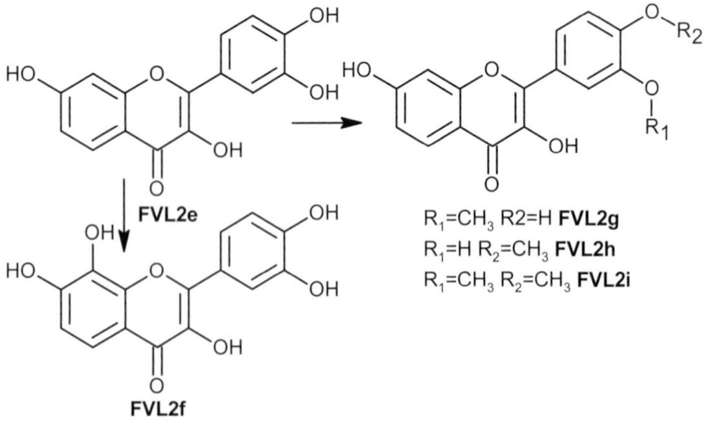

Fig. 7.22 Biotransformations of fisetin (**FVL2e**) by *Streptomyces griseus*

Quercetin (**FVL2**) is often found in plants as a glucoside. Glucosides are often cleaved by microorganism but *Bacillus cereus* catalyzed 3-*O*-glycosylation of quercetin, forming isoquercitrin (**FVL2m**) (Rao and Weisner 1981). The appearance of protocatechuic acid (**DEGg**) was likely the degradation product following an initial C-ring cleavage.

Rutin (**FVL2k**), a flavonol rutinoside, had a delayed absorption (t_{max} 9.3 h), based on the appearance of quercetin (**FVL2**) in plasma, compared with that of flavonol glucosides (t_{max} 0.7 h), suggesting that rutin (**FVL2k**) undergoes hydrolysis by colon microflora before adsorption (Hollman et al. 1997). Indeed, anaerobic degradation of the glycoside rutin by *Butyrivibrio* sp. C_3, an organism isolated from bovine rumen, gave quercetin (**FVL2**), glucose, and rhamnose by sugar hydrolysis (Krishnamurty et al. 1970). Quercetin (**FVL2**) was further metabolized to phloroglucinol (**DEGa**), CO_2, 3,4-dihydroxybenzaldehyde (**DEGi**), and 3,4-dihydroxyphenylacetic acid (**DEGj**).

Rutin (**FVL2k**) and quercitrin (quercetin-3-*O*-rhamnose) (**FVL2l**) were also hydrolyzed, releasing the aglycon, quercetin (**FVL2**), by human gut bacteria (Macdonald et al. 1983). A fecal bacteria suspension cleaved the C-ring of rutin (**FVL2k**) into 3,4-dihydroxyphenylacetic acid (**DEGj**) following the transient appearance of quercetin (**FVL2**) (Winter et al. 1989). A bacterial strain isolated from human feces, *Enterococcus casseliflavus*, hydrolyzed the sugar moiety of isoquercetrin (quercetin-3-*O*-glucoside) (**FVL2j**) but did not further metabolize the aglycon (**FVL2**) (Schneider et al. 1999). Another strain in the same study, *Eubacterium ramulus*, degraded isoquercetrin (**FVL2j**) to phloroglucinol (**DEGa**) and 3,4-dihydrophenylacetic acid (**DEGj**) by means of deglycolsylation and C-ring cleavage. After degrading rutin (**FVL2k**), human fecal flora was able to dehydroxylate the main metabolite 3,4-dihydrophenylacetic acid (**DEGj**) to 3-hydrophenylacetic acid (**DEGk**) and no methylated products were detected (Aura et al. 2002).

Quercetin (**FVL2**) was metabolized stepwise by *Eubacterium ramulus* strain wK1 into dihydroquercetin (taxifolin) (**DHF2**), then into alphitonin (**AU2a**) and finally into 3,4-dihydrophenylacetic acid (**DEGj**) (Braune et al. 2001). *E. ramulus* was also reported to grow on isoquercetrin (**FVL2j**) as the sole carbon source (Schneider et al. 1999). Quercetin (**FVL2**) and phloroglucinol (**DEGa**) were detected as intermediates and 3,4-dihydroxyphenylacetic acid (**DEGj**), butanoic acid and acetic acid were the final fermentation products.

A study aimed at finding a relationship between flavonoid structure and degradation rate by microflora found kaempferol and other flavonoids with 5,7,4'-hydroxyl groups had the highest degradation rate (Simons et al. 2005a). The rates of degradation of hesperetin (**FNN2a**) and naringenin (**FNN1**) by human microflora were not significantly different suggesting flavanone degradation was not affected by methylation.

7.2.6 Catechins

Catechins (**CAT**), classified by their flavan-3-ol structure (Figs. 7.23–7.25), are commonly found in many vegetables, herbs and teas and are one of the basic building blocks of proanthocyanidins. Green tea leaves have especially high polyphenol

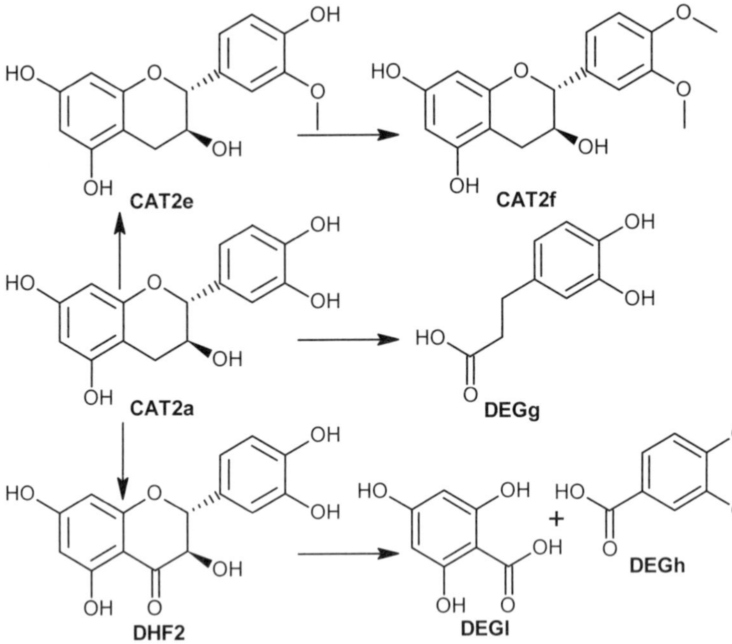

Fig. 7.23 Biotransformations of (+)-catechin (**CAT2a**)

content including catechins of up to 30% dry weight (Pan et al. 2003). Catechins are known for their antioxidant (Yoshino et al. 1994; Yen and Chen 1995), nueroprotective (Li et al. 2005), and anti-inflammatory activities (Ferrandiz and Alcaraz 1991; Middleton and Kandaswami 1992; Gil et al. 1994). Their ability to scavenge free radicals (Rice-Evans et al. 1995; Cotelle et al. 1996) is thought to provide protection against cancer (Jankun et al. 1997; Inoue et al. 1998; Cao and Cao 1999; Setiawan et al. 2001) and heart disease (Vinson and Dabbagh 1998).

Mammalian phase II type metabolism was mimicked by *Streptomyces griseus* (ATCC 13273) by methylating (+)-catechin (**CAT2a**) in a stepwise fashion first to 3′-*O*-methyl-(+)-catechin (**CAT2e**) and then to 3′,4′-*O*-dimethyl-(+)-catechin (**CAT2f**) (Hosny et al. 2001). The *S. griseus* catechol *O*-methyltransferase is a likely candidate responsible for the methylation of catechin (Dhar and Rosazza 2000).

As with previously mentioned flavonoids, catechins are also subject to metabolism by microorganism into smaller molecules such as phenolic acids. (+)-Catechin (CAT2a) was oxidized by the fungi Penicillium expansum into dihydroquercetin (DHF2) before further degradation and gave protocatechuic acid (DEGh) and phloroglucinolcarboxylic acid (DEGl) as the final fermentation products (Contreras et al. 2006). Both (+)-catechin (**CAT2a**) and (−)-epicatechin (**CAT2b**) were anaerobically fermented *in vitro* with pooled human fecal microbiota (Aura et al. 2006). Whereas (+)-catechin (**CAT2a**) degraded into 3,4-dihydroxyphenylpropionic acid (**DEGj**) as a primary product, the stereoisomer (−)-epicatechin did not. Yet, both isomers shared a common secondary metabolite 3-hydroxyphenylpropionic acid

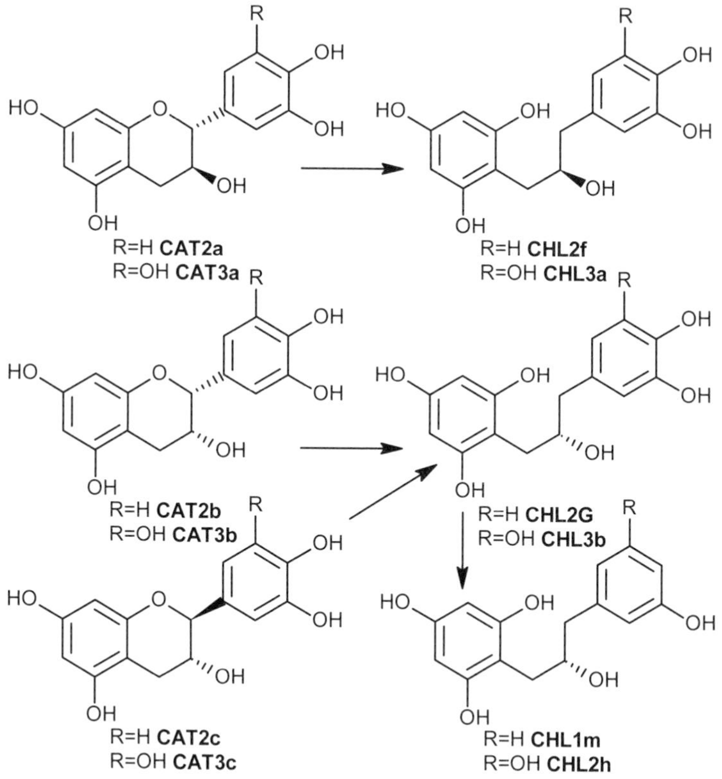

Fig. 7.24 Metabolisms of catechins by *Eubacterium* sp strain SDG-2

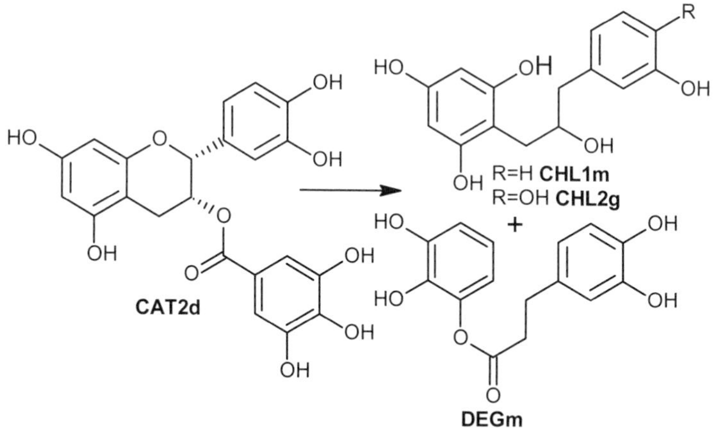

Fig. 7.25 Biotrasformation of (−)-epicatechin-3-*O*-gallate (**CAT2d**)

(**DEGk**). A study using human intestinal bacterium isolate *Eubacterium* sp. transformed various flavan-3-ol enantiomers (**CAT2a,b,c, CAT3a,b,c**) into 1,3-diphenylpropan-2-ol derivatives(**CHL2f,g, CHL3a,b**) by cleaving the C-ring of the flavan-3-ol (Wang et al. 2001). The *Eubacterium* sp strain SDG-2 also favored the dehydroxylation of the B-ring 4′-position of 1,3-diphenylpropan-2-ol derivatives (**CHL2g, CHL3b**) originating from (3*R*)-flavan-3-ols, such as (−)-catechin (**CAT2c**), (−)-epicatechin (**CAT2b**), (−)-epigallocatechin (**CAT3b**) and (−)-gallocatechin (**CAT3c**) to give additional 1,3-diphenylpropan-2-ol (**CHL1m, CHL2h**), but not the (3*S*)-flavan-3-ols such as (+)-catechin (**CAT2a**) and (+)-gallocatechin (**CAT3a**).

Investigation into the biotransformation of (−)-epicatechin-3-*O*-gallate (**CAT2d**) and related compounds was undertaken using rat and human fecal suspension resulting in the isolation of fifteen metabolites (Meselhy et al. 1997). Four compounds were new, namely, two isomers of 1-(3′-hydroxyphenyl)-3-(2″,4″,6″-trihydroxyphenyl)propan-2-ol (**CHL1m**), 1-(3′,4′-dihydroxyphenyl)-3-(2″,4″,6″-trihydroxyphenyl)propan-2-ol (**CHL2g**) and 2″,3″-dihydroxyphenoxyl-3-(3′,4′-dihydroxyphenyl)propionate (**DEGm**). (−)-Epicatechin (**CAT2b**), (−)-epigallo-catechin (**CAT3b**) and their gallates (**CAT2d, CAT3d**) were all extensively metabolized by human fecal microflora but the flavan-3-ol gallates (**CAT2d, CAT3d**) were resistant to degradation by rat fecal microflora, even after prolonged incubation, which suggests that intestinal microflora from different species may result in different metabolic activities.

7.2.7 Anthocyanins

Among the natural pigments in plants, anthocyanins are the largest water-soluble group, responsible for red, purple, or blue plant pigments found in most fruits, flower petals, and leaves (Mazza and Miniati 1993). Naturally occurring anthocyanins are composed of six anthocyanidin aglycons namely pelargonidin (**ACD1**), cyanidin (**ACD2**), peonidin (**ACD2a**), delphinidin (**ACD3**), malvidin (**ACD3a**) and petunidin (**ACD3b**) (Figs. 7.26–7.27). They are derivatives of a 2-phenylbenzopyrylium (flavylium) cation varying in the degree of hydroxylation and methylation and linked to sugar groups at positions 3 and/or 5. These naturally occurring compounds are industrially important for use as food colorants (Stintzing and Carle 2004) and as potential replacements for banned dyes because they have no apparent adverse effects on human health (Brouillard 1982; Stich et al. 1999; Boyd 2000). Anthocyanins make up the greatest portion of flavonoids consumed in human diets, mainly from berries and beverages, with an average daily intake of around 200 mg/day in the United States (Kuhnau 1976). Recently, however, much attention has been drawn to anthocyanin-derived plant products due to their range of biological activities including antioxidant (Noda et al. 2000; Kahkonen and Heinonen 2003), anti-inflammatory (Wang and Mazza 2002; Youdim et al. 2002), and anti-carcinogenic properties (Kamei et al. 1995; Kang et al. 2003; Katsube et al. 2003; Zhang et al. 2005). They may lower risk of coronary heart disease (Renaud and Delorgeril 1992) by inhibiting oxidation of human low-density lipoproteins (SatueGracia et al. 1997) or through vasoprotective activities (Lietti et al. 1976) and effects on arterial vasomotion (Colantuoni et al. 1991).

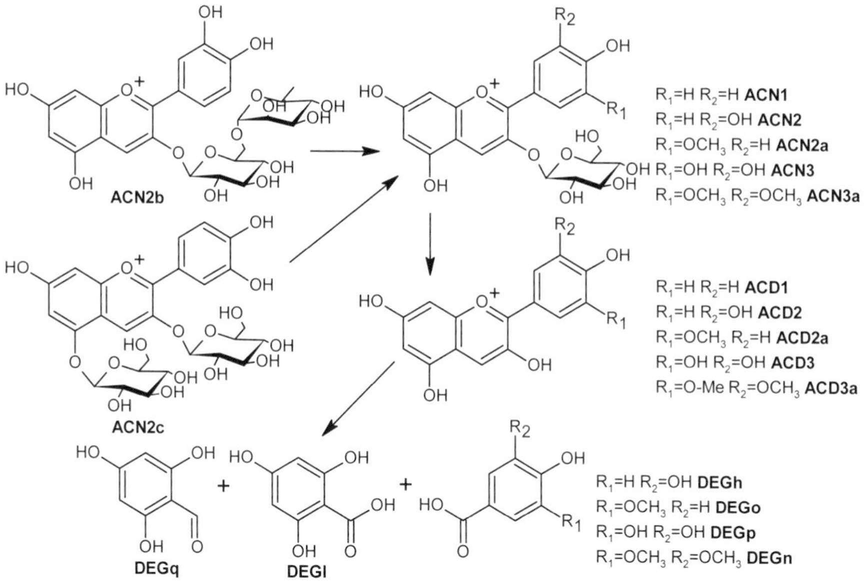

Fig. 7.26 Metabolism of anthocyanins by microflora

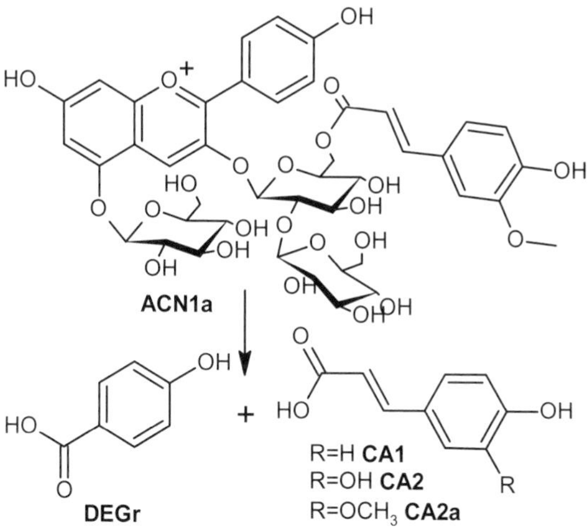

Fig. 7.27 Degradation of pelargonidin-3-*O*-sophoroside-5-*O*-glucoside acylated with ferulic acid (**ACN1a**) by human fecal flora

Knowledge of the bioavailability of anthocyanins is essential to understanding if they are active in the human body. The stability of anthocyanins was analyzed using an *in vitro* digestion system that simulated the physiochemical changes that occur in the stomach and upper gastrointestinal tract (McDougall et al. 2005). After 2 hours of simulated gastric conditions, red wine anthocyanins were found to be stable probably due to the well documented stability of the flavium cation form of anthocyanins in a low pH environment (Clifford 2000). Their stability under gastric conditions is important because the stomach may be a primary site for absorption (Passamonti et al. 2003; Talavera et al. 2003). Indeed, unmodified anthocyanins appeared rapidly in the plasma after oral administration due to apparent direct absorption in the stomach (Bub et al. 2001; Matsumoto et al. 2001). Following, pancreatic digestion was simulated by neutralizing the pH. After 2 hours, about half of the initial anthocyanin amount remained which was considered to be "colon available." In addition, vitisin derivatives, part of the pyranoanthocyanin family, were much more resistant to simulated pancreatic digestion (McDougall et al. 2005). This study shows that anthocyanins are potentially available for absorption by the intestines (Talavera et al. 2004) or subject to colon microflora metabolism.

Anthocyanin bioavailability is required for any effect in a specific tissue or organ. Therefore, anthocyanins must be efficiently absorbed from the gastrointestinal tract into the cardiovascular system for dispersion throughout the body. Studies have show that anthocyanins are rapidly absorbed from both stomach (Passamonti et al. 2003; Talavera et al. 2003) and small intestine (Talavera et al. 2004) and readily appear in rat and human plasma as unmodified glycosides (Miyazawa et al. 1999; Tsuda et al. 1999; Matsumoto et al. 2001). Oral consumption of anthocyanin-rich fruits, extracts or pure anthocyanins was linked to preventing or suppressing diseased states *in vivo* (Ramirez-Tortosa et al. 2001; Mazza et al. 2002) and increased antioxidant activity in serum (Serafini et al. 1998; Mazza et al. 2002; Bitsch et al. 2004). However, uptake of anthocyanins is relatively low based on serum concentrations (<1% of dose) (Lapidot et al. 1998; Bub et al. 2001; Frank et al. 2003) and correspondingly low urinary levels of intake or conjugated forms. Also, fecal recovery is very low. The apparent low bioavailability has made claims of their ability to exert beneficial effects throughout the body questionable.

As with other flavonoids, anthocyanins are subject to degradation by microorganisms. Cyanidin-3-*O*-glucoside (**ACN2**), peonidin-3-*O*-glucoside (**ACN2a**), cyanidin-3-*O*-rutinoside (**ACN2b**), cyanidin-3,5-*O*-diglucoside (**ACN2c**) and malvidin-3-*O*-glucoside (**ACN3a**) were incubated with pig gut microflora to determine the role of the methylation pattern of the aglycons or if the type of sugar moiety has any influence on hydrolysis (Keppler and Humpf 2005). All anthocyanins were degraded within 2 hours of incubation while the aglycons appeared at very low concentrations. Monoglucosides (**ACN2, ACN2b, ACN3a**) degraded rapidly while the diglucoside (**ACN2a**) underwent stepwise hydrolysis at a lower rate. Overall, the aglycon structure has no effect on the rate of hydrolysis while the presence of a rutinoside slowed degradation considerable.

Following the hydrolysis of the protective 3-*O*-glucosides, the aglycons are stable in acidic conditions but are unstable at neutral pH which is typical of physiological conditions. On the other hand, anthocyanins are rather stable at neutral pH and more so with increasing number of glycosylations (Fleschhut et al. 2006). Anthocyanidins can exist in four molecular forms depending on the pH. At neutral pH, the α-diketone form is predominant and easily decomposes to phenolic acids and aldehydes (Keppler and Humpf 2005). During the fermentation of anthocyanins with pig microflora, the aglycons were released but further degradation of the anthocyanidins followed. Cyanidin (**ACD2**) degraded into protocatechuic acid (**DEGh**), malvidin (**ACD3a**) into syringic acid (**DEGn**), and peonidin (**ACD2a**) into vanillic acid (**DEGo**) (Keppler and Humpf 2005). Syringic acid (**DEGn**) was further demethylated into gallic acid (**DEGp**) and vanillic acid (**DEGo**) into protocatechuic acid (**DEGh**). Since all aglycons have identical A-rings, phloroglucinaldehyde (**DEGq**) was formed in all incubations along with low concentrations of its oxidation product phloroglucinolcarboxylic acid (**DEGl**). After reaching a maximum, the low molecular weight degradation products of anthocyanins degraded slowly showing a higher chemical stability and lower degradation rate by intestinal microflora. These results show that the spontaneous or enzymatic degradation of anthocyanidins after bacterial deglycosylation is one limiting factor in the bioavailability of anthocyanins.

When using human gut microflora, cyanidin-3-*O*-gulcoside (**ACN2**) and cyanidin aglycone (**ACD2**) were identified as intermediates of the metabolism of cyanidin-3-*O*-rutinoside (**ACN2b**) (Aura et al. 2005). Furthermore, at early time points during the fermentation, formation of protocatechuic acid as the major metabolite for both cyanidin glucosides and three unidentified low molecular weight metabolites showed that human gut microflora metabolized anthocyanins. The degradation of complex anthocyanins was confirmed using human fecal microflora *in vitro* (Fleschhut et al. 2006). The human microflora was able to degrade anthocyanin mono-, di-, and even complex acylated glucosides. For example, Pelargonidin-3-*O*-sophoroside-5-*O*-glucoside acylated with ferulic acid (**ACN1a**) was degraded into 4-hydroxybenzoic acid (**DEGr**) as well as the hydroxycinnamic acids ferulic acid (**CA2a**), caffeic acid (**CA2**) and *p*-coumaric acid (**CA1**) (Fleschhut et al. 2006). These low molecular weight compounds are noteworthy because they have comparable antioxidant activity and free radical scavenging capacity compared to the parent anthocyanin (RiceEvans et al. 1996; Ghiselli et al. 1998; Natella et al. 1999; Kawabata et al. 2002; Sroka and Cisowski 2003). Additionally, protocatechuic acid (**DEGh**) exhibited carcinogenesis inhibition in a hamster model (Ohnishi et al. 1997). The biological effectives from the degradation products can therefore not be ruled out.

7.2.8 Pyranoanthocyanins

For many millennia, *Saccharomyces cerevisiae* and other yeast strains have been employed in the art of winemaking and selected for their ability to complete fermentation of media sugar-rich but poor in proteins. However, the selection of yeast is not solely based on the yeast's glycolytic ability to convert sugars into CO_2 and ethanol. During fermentation yeast also produces secondary metabolites (Boulton et al. 1996; Pretorius 2000) capable of interacting with the grape must in various ways. For example, the production of volatile vinylphenols have significant influence on wine aroma (Chatonnet et al. 1993) as well as higher alcohols and esters (Mauricio et al. 1997; Valero et al. 2002) and alcohol acetates (Stashenko et al. 1992). The smoothness of wine can be effected by yeast excreting polysaccharides (Dupin et al. 2000; Feuillat 2003) or producing large quantities of glycerol (Omori et al. 1996). Furthermore, secondary metabolites like pyruvic acid and acetaldehyde has been linked to the changes in coloration by reacting with anthocyanins and anthocyanin adducts during and after fermentation resulting in pyranoanthocyanins (Bakker and Timberlake 1997; Morata et al. 2003, 2006; Lee et al. 2004).

Pyranoanthocyanins are a relatively newly identified class of stable anthocyanin-derived compounds identified in red grape wines (Fulcrand et al. 1996b; Bakker and Timberlake 1997; Pozo-Bayon et al. 2004). The formation of pyranoanthocyanins is a result of cycloaddition between the anthocyanin C-4 and the hydroxyl group at the C-5 position with a ethylenic bond of another molecule (Monagas et al. 2005). The most interesting properties of these molecules are their resistance to bleaching by SO_2 and greater color stability over the low to neutral pH range compared to malvidin 3-O-glucoside (**ACN3a**) (Bakker and Timberlake 1997). Pyranoanthocyanins also believed to contribute significantly to the color change of red wines. Vistins, commonly found group of pyranoanthocyanins in wine with orange-red color, express more color at wine pHs, than, for example, malvidin-3-O-glucoside (**ACN3a**). The color expressed by vitisin A (**PAC3a**) was measured to be about 11 and 14 times greater than malvidin-3-O-glucoside (**ACN3a**) in a model wine at pH 3 and 2, respectively (Romero and Bakker 1999).

Pyranoanthocyanins from wine were found to be more stable than anthocyanins following simulated gastrointestinal digestion (McDougall et al. 2005). Assuming pyranoanthocyanins have similar pharmacological activity as their parent anthocyanins, their enhanced gastrointestinal stability could lead to enhanced bioavailability and bio-effectiveness *in vivo*. Pyranoanthocyanins are primarily formed from anthocyanins reacting with various phenols, pyruvic acid or acetaldehyde. The following section is broken into three parts reflecting the origin of the pyrane ring.

7.2.8.1 Vinylphenol-Derived Pyranoanthocyanins

Anthocyanins (**ACN**) are known to react with 4-vinylphenols to produce pyranoanthocyanins (**PAC**) (Fulcrand et al. 1996b; Hayasaka and Asenstorfer 2002) in a rapid and quantitative manner (SarniManchado et al. 1996; Schwarz et al. 2003) (Figs. 7.28–7.29). The production of 4-vinylphenols are attributed to a yeast cinnamate decarboxylase (CD) (Chatonnet et al. 1993) encoded by the phenylacrylic acid decarboxylase gene (PAD1) (Clausen et al. 1994). Pad1p was shown to decarboxylate cinnamic acid (**CA0**), *p*-coumaric acid (**CA1**) and ferulic acid (**CA2a**) into styrene (**VP0**), 4-vinylphenol (**VP1**) and 4-vinylguaiacol (**VP2a**), respectively (Clausen et al. 1994), and provides yeast a resistance to phenylacrylic acids (Larsson et al. 2001). The production of 4-vinylphenol (**VP1**) and 4-vinylguaiacol (**VP2a**) during wine fermentation is well documented (Etievant 1981; Baumes et al. 1986; Chatonnet et al. 1993; Dugelay et al. 1993; Lao et al. 1997) and these volatile compounds are attributed to undesirable phenolic off-flavors in wine (Etievant 1981; Marullo et al. 2006).

The reaction between vinylphenols and anthocyanins (**ACN**) adducts results in a cycloadditon of the vinyl C-C double bond to the C-4 carbon and C-5 hydroxyl group on the anthocyanin followed by an oxidation step to form a pyrane ring (ring D) (Fulcrand et al. 1996b) characteristic for pyranoanthocyanins. The pyrane ring provides resistance to SO_2 bleaching and hydration of the flavium ring structure

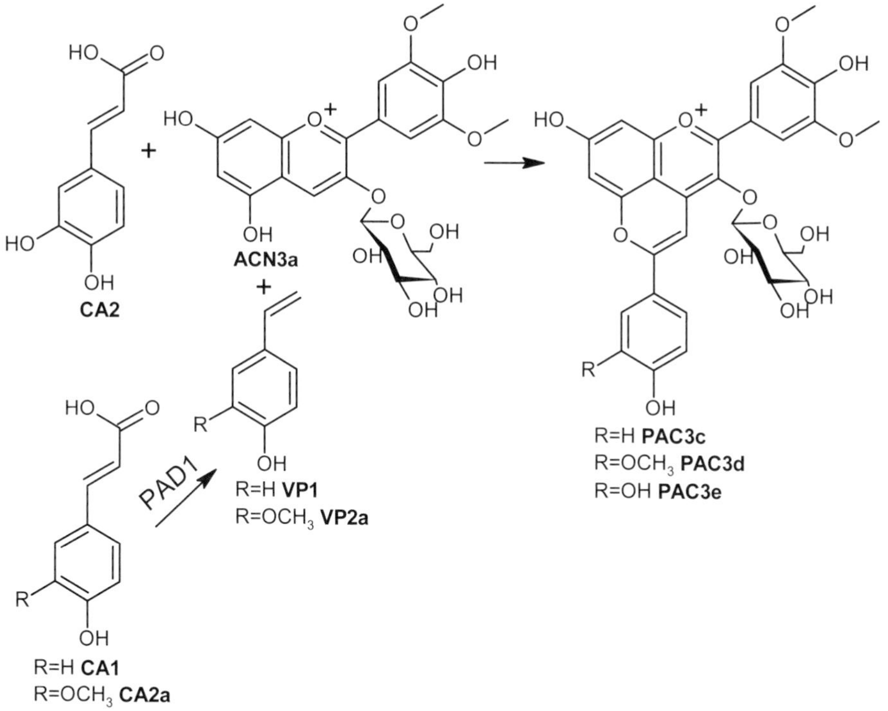

Fig. 7.28 Reaction between vinylphenols or cinnamic acids with malvidin-3-*O*-glucoside to form pyranoanthocyanin-vinylphenols

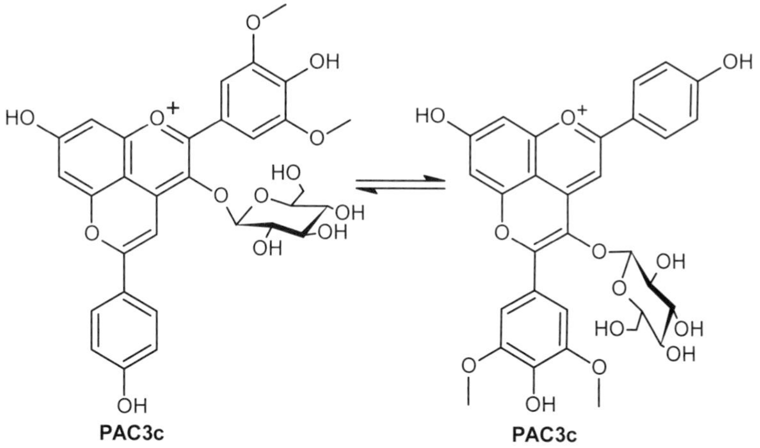

Fig. 7.29 Mesomeric forms of malvidin-3-*O*-glucoside 4-vinylphenol (**PAC3c**)

(Bakker and Timberlake 1997). The new orange-red pigments also have two mesomeric forms (**PAC3c**) with the malvidin-like form having a maximum color adsorption at 534 nm and a pelargonidin-like form with maximum adsorption at 506 nm (Fulcrand et al. 1996b). The reaction proceeds in wine at pH (pH 3–4) during aging (SarniManchado et al. 1996), and reduces the amount of volatile phenol content of red wines.

The first anthocyanin-vinylphenol was detected on the polymeric membranes used for cross-flow microfiltration of Carignane red wines (CameiradosSantos et al. 1996). It tentatively assigned as malvidin-3-*O*-glucoside 4-vinylphenol (**PAC3c**) and further characterized by UV-visible, mass and NMR spectroscopy (CameiradosSantos et al. 1996; Fulcrand et al. 1996b; Hayasaka and Asenstorfer 2002; Mateus et al. 2002a; Monagas et al. 2003; Pozo-Bayon et al. 2004). Malvidin-3-*O*-glucoside-vinylguaiacol (**PAC3d**) and anthocyanin-vinylphenol adducts derived from the condensation of 4-vinylphenols with malvidin-3-*O*-(6-*p*-coumaroyl)-glucoside, malvidin-3-*O*-(6-acetyl)-glucoside, malvidin-3-*O*-(6-caffeoyl)-glucoside, peonidin-3-*O*-glucoside, peonidin-3-*O*-(6-*p*-coumaroyl)-glucoside and petunidin-3-*O*-glucoside have also been reported in wine (Asenstorfer et al. 2001; Hayasaka and Asenstorfer 2002; Mateus et al. 2002a; Monagas et al. 2003; Wang et al. 2003; Alcalde-Eon et al. 2004; Pozo-Bayon et al. 2004).

Additional anthocyanin-vinylphenols have been identified including malvidin-3-*O*-glucoside and adducts with 4-vinylcatechol and 4-vinylsyringol in red wines (Hayasaka and Asenstorfer 2002; Monagas et al. 2003; Wang et al. 2003; Alcalde-Eon et al. 2004) and in rose sparkling wines (Pozo-Bayon et al. 2004). It was speculated that all anthocyanin-vinylphenols were derived from vinylphenols originating from cinnamic acids and anthocyanins found in wine follow the same mechanism (Hayasaka and Asenstorfer 2002; Hakansson et al. 2003). However, the Pad1p is incapable of converting caffeic acid (**CA2**) and sinapic acid (**CA2b**) into their corresponding vinylphenols (Chatonnet et al. 1993). Moreover, the decarboxylation of *p*-coumaric acid (**CA1**) is strongly inhibited by catechins and proanthocyanins (Chatonnet et al. 1993) found abundantly in red wines (Schwarz et al. 2003) which suggests that the mechanism involving 4-vinylphenols plays only a minor role in anthocyanin-vinylphenol formation. It was observed that the concentration of pinotin A (malvidin-3-*O*-glucoside-vinylcatechol) (**PAC3e**) was approximately ten times higher in aged wines (5–6 years old) than younger wines (< 1 year) (Schwarz et al. 2003). This suggests that the reaction between malvidin-3-*O*-glucoside (**ACN3a**) and 4-vinylcatechol (**VP2**) proceeds slowly over several years. However, according to Fulcrand et al. (1996b) and SarniManchado et al. (1996) the reaction between 4-vinylphenol (**VP1**) and malvidin 3-*O*-glucoside (**ACN3a**) proceeded to completion within hours. In agreement to this, synthesize 4-vinylcatechol (**VP2**) reacted almost quantitatively overnight with malvidin 3-*O*-glucoside (**ACN3a**) in a model wine solution according to Schwarz et al. (2003). Also, the detection of 4-vinylcatechol (**VP2**) in wine has not been reported (Schwarz and Winterhalter 2004). An alternative mechanism has been proposed involving free hydroxycinnamic acids and anthocyanins without enzymatic involvement and demonstrated that the reaction does indeed take place within several months (Schwarz et al. 2003; Schwarz and Winterhalter 2003). This mechanism involves the linkage of the nucleophilic C2 position on caffeic acid (**CA2**) with electrophilic C4 position on malvidin 3-*O*-glucoside (**ACN3a**) followed by cyclization involving the anthocyanin 5-OH and the caffeic acid C3 position to form the pyrane ring. Further oxidation and decarboxylation gives the final product (**PAC3e**).

Recently, a copigmentation study by Schwarz et al. (2005) explored the significance of the two different pathways involved in the formation of malvidin 3-*O*-glucoside-4-vinylphenol (**PAC3a**) and pinotin A (**PAC3e**). They examined the copigment effect of caffeic acid (**CA2**) and *p*-coumaric acid (**CA1**) when added to the grape must before alcoholic fermentation. The concentrations of the additives were determined following alcoholic fermentation by *S. cerevisiae* yeast followed by malolatic fermentation by *Oenococcus oeni* lactic acid bacteria and over several months of storage for two types of red wine. The concentrations of malvidin 3-*O*-glucoside-4-vinylphenol (**PAC3a**) and pinotin A (**PAC3e**) were determined after 21 months. Directly following alcoholic fermentation, the measured concentrations of hydroxycinnamic acids (**CA1** and **CA2**) were still significantly higher in the supplemented wines than their respective controls. However, the hydroxycinnamic acid concentration difference between wines with or without a pre-added hydroxycinnamic acid mostly vanished following malolatic fermentation due to the rapid hydrolysis of caftaric acid into caffeic acid (**CA2**) and coutaric acid into *p*-coumaric acid (**CA1**). Two different behaviors were observed regarding the formation of the two pyranoanthocyanins after 21 months. The final concentration of pinotin A (**PAC3e**) appeared to be correlated to the concentration of caffeic acid after malolatic fermentation suggesting that pinotin A (**PAC3e**) forms slowly during the months of storage. This is supported by a previous study (Schwarz et al. 2003). In contrast, the concentration of malvidin 3-*O*-glucoside-4-vinylphenol (**PAC3c**) was approximately ten times higher in wines when *p*-coumaric acid (**CA1**) was added prior to fermentations. This implies that the elevated concentrations of *p*-coumaric acid (**CA1**) is significant during yeast fermentation when presumably *p*-coumaric acid (**CA1**) is converted into 4-vinylphenol (**VP1**) enzymatically. Unfortunately the authors did not determine pyranoanthocyanins content during or directly following alcoholic yeast fermentation. In support of their claim, another study (Morata et al. 2006) observed malvidin 3-*O*-glucoside-4-vinylguaiacol (**PAC3b**) was only produced during *S. cerevisiae* fermentation but not during *Saccharomyces uvarum* fermentation. The phenylacrylic acid decarboxylase produced by *S. cerevisiae* was previously shown to convert ferulic acid (**CA2a**) into 4-vinylguaiacol (**VP2a**) (Chatonnet et al. 1993; Clausen et al. 1994).

7.2.8.2 Pyruvic Acid-Derived Pyranoanthocyanins

The reaction between malvidin-3-*O*-glucoside (**ACN3a**) with pyruvic acid (pyruvate) results in a common pyranoanthocyanin called vitisin A (**PAC3a**) (Fig. 7.30). The reaction is similar to the formation of anthocyanin-vinylphenol adducts. The pyrane ring of vitisin A (**PAC3a**) is the result of the cycloadditon of pyruvic acid with the C4 and the C5 hydroxyl group of the flavium ion followed by dehydration and rearomatization. The maximum light adsorption of vitisin A (**PACa**) is around 510 nm and a significant band around 370 nm compared to 534 nm for malvidin 3-*O*-glucoside (**ACN3a**) resulting in a color shift from a red to an orange hue (Bakker and Timberlake 1997; Fulcrand et al. 1998). Vitisin A (**PAC3a**) was first discovered,

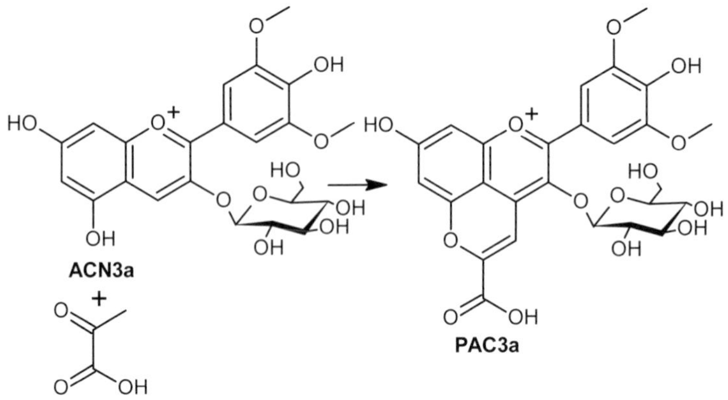

234   J.A. Chemler et al.

Fig. 7.30 Formation of pyranoanthocyanins from pyruvic acid

isolated and characterized in fortified red wine by Bakker and Timberlake (1997). Subsequently, Flucrand et al. (1998) also proposed a similar reaction mechanism but provided a slightly different chemical structure of vitisin A (**PAC3a**) based on NMR and MS/MS data. Further studies supported the structure proposed by Fulcrand et al. using ^{1}H and ^{13}C NMR (Mateus et al. 2001) and mass spectroscopy (Asenstorfer et al. 2001; Mateus et al. 2001; Hayasaka and Asenstorfer 2002). Malvidin-3-O-(6-p-coumaroyl)-glucoside pyruvate and malvidin-3-O-(6-acetyl)-glucoside pyruvate have also been identified in red and port wines (Revilla et al. 1999; Vivar-Quintana et al. 1999; Asenstorfer et al. 2001; Romero and Bakker 2001; Atanasova et al. 2002; Hayasaka and Asenstorfer 2002; Heier et al. 2002; Mateus et al. 2002a; Monagas et al. 2003; Morata et al. 2003; Pozo-Bayon et al. 2004). Other identified anthocyanin-pyruvate adducts include cyanidin, delphinidin, peonidin and petunidin forms in combination with their acylated glucosides such as p-coumaroyl or acetyl glucosides (Atanasova et al. 2002; Heier et al. 2002; Mateus et al. 2002a; Wang et al. 2003; Alcalde-Eon et al. 2004).

    S. cerevisiae produces pyruvic acid as an intermediate during the metabolism of glucose into ethanol and acetyl CoA. Pyruvic acid is excreted by yeast during fermentation and eventually reabsorbed and metabolized as the sugars are depleted (Whiting and Coggins 1960; Morata et al. 2003). The concentration of pyruvic acid was greatest when approximately half the sugars have been consumed (Whiting and Coggins 1960) and varied from 0 to 500 mg/L in wines (Radler 1992). The amount of vitisin A (**PACa**) formation follows the concentration of pyruvate during fermentation and most of the vitisin A (**PAC3a**) produced during fermentation occurred within the first few days (Morata et al. 2006). Indead, the maximum production of vitisin A (**PAC3a**) occurred when malvidin 3-O-glucoside (**ACN3a**) and pyruvate were at their highest concentrations which coincided with 20–85% glucose utilization (Asenstorfer et al. 2003). The ability to produce pyruvic acid varies depending on the strain of *S. cerevisiae* (Morata et al. 2003) and subject to improvement through metabolic engineering (Michnick et al. 1997). The type of yeast was also influential as *S. cerevisiae* produced over twice the amount of vitisin A (**PAC3a**) compared to *Saccharomyces uvarum* (Morata et al. 2006).

An analog with a phenol group has also been recently found having a hypsochromically shifted visible wavelength maximum of 538 nm (Mateus et al. 2006). The formation of the vinylpyranomalvidin 3-*O*-glucoside-phenol (**PAC3f**) is thought to involve malvidin 3-*O*-glucoside-pyruvate (**PAC3a**) reacting with 4-vinylphenol (**VP1**), a product of the enzymatic reaction of yeast cinnamate decarboxylase with *p*-coumaric acid (**CA1**) (Clausen et al. 1994). The mechanism follows that the resulting intermediate loses a formic acid and undergoes oxidation to yield the new anthocyanin-derived pigment (**PAC3f**) (Fig 7.31).

7.2.8.3 Acetaldehyde-Derived Pyranoanthocyanins

Another commonly found pyranoanthocyanin in wine is vitisin B (**PAC3b**) formed by the reaction between acetaldehyde and malvidin 3-*O*-glucoside (**ACN3a**) (Fig. 7.32). Acetaldehyde is a yeast glycolytic intermediate of the conversion of into ethanol (Liu and Pilone 2000). Alternatively, ethanol can be oxidized to form acetaldehyde in the presence of polyphenols during wine aging (Wildenradt and Singleton 1974).

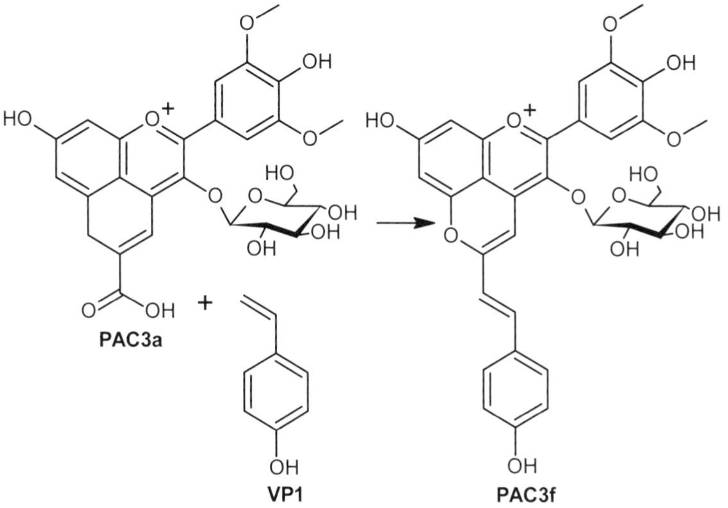

Fig. 7.31 Formation of vinylpyranomalvidin 3-*O*-glucoside-phenol (**PAC3f**)

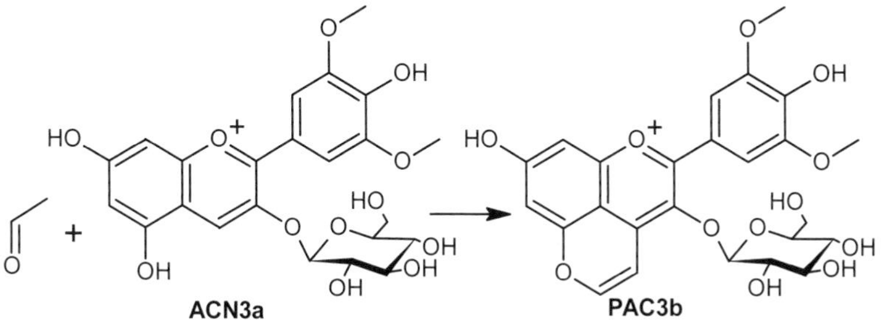

Fig. 7.32 Formation of pyranoanthocyanins from acetaldehyde

Acetaldehyde is a potent volatile compound found in many beverages and foods (Liu and Pilone 2000). Acetaldehyde plays an important role in the change of color of new wines. During yeast fermentation, vitisin B (**PACb**) forms initially at a slower rate than vitisin A (**PAC3a**) but forms in greater quantities towards the end of fermentation (Morata et al. 2006). Acetaldehyde is proportionally formed to the amount of sugars fermented and its accumulation was greatest at the end of fermentation. As with the relationship of pyruvate and vitisin A (**PAC3a**), the content of vitisin B (**PAC3b**) at the end of fermentation of grape must had a good correlation with acetaldehyde concentration (Morata et al. 2003).

Interactions between catechins and tannins with anthocyanins mediated by acetaldehyde have been documented (Fig. 7.33). Pyranoanthocyanin-vinylflavanols (**PAC3g,h**) were first reported in model wine solutions containing malvidin 3-O-glucoside (**ACN3a**), acetaldehyde) and (+)-catechin (**CAT2a**), (−)-epicatechin (**CAT2b**) or dicatechin (procyanidin B3) (**CAT2g**) (FranciaAricha et al. 1997) and later isolated and characterized in wines (Mateus et al. 2002b).

Pyruvate-derived pyranoanthocyanins are also capable of reacting with the vinyl group of an 8-vinyl-flavanol (**CAT2a,b**). The vinylflavanols (**CAT2h,i**) are speculated to be the result of the decomposition of ethyl-linked flavanol oligomers formed by an acetaldehyde-mediated condensation reaction of flavanols (Es-Safi et al. 1999), degradation of unstable ethanol adducts of flavanols resulting from the reaction between acetaldehyde and flavanols (Fulcrand et al. 1996a; Asenstorfer et al. 2001), or from the decomposition of anthocyanin-ethylflavanols (Mateus et al. 2003; Lee et al. 2004). This new group of blue anthocyanin-derived pigments, named

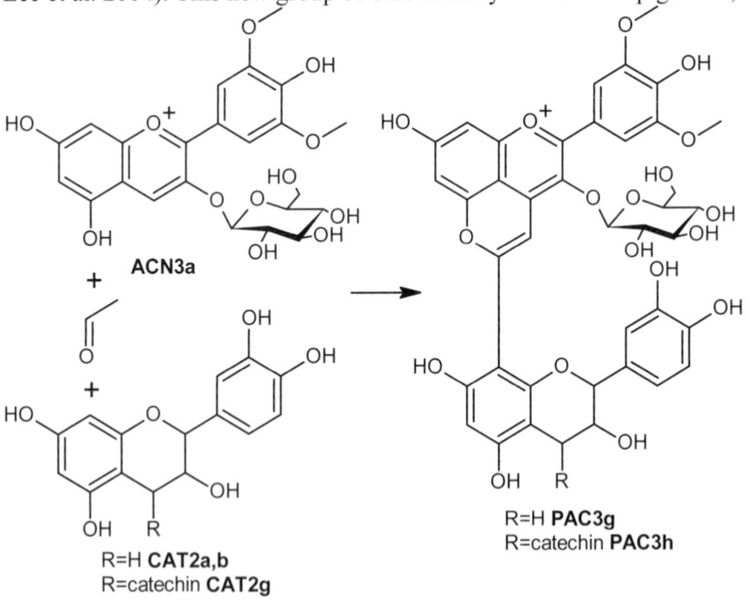

Fig. 7.33 Formation of pyranoanthocyanin-vinylflavanols

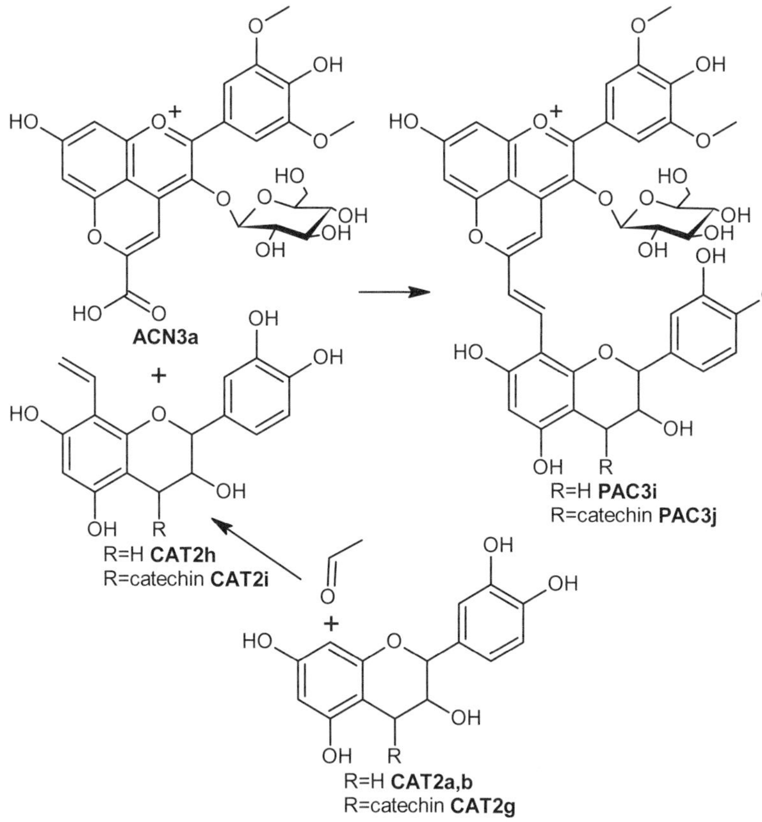

Fig. 7.34 Formation of portisins

portisins (**PAC3i,j**) for their occurrence in Port wine, were recently discovered (Mateus et al. 2003) (Fig. 7.34). These pigments were found to have a maximum UV-Vis adsorption about 575 nm at low pH, a bathochromatic shift from the parent anthocyanin to give a more bluish hue. These compounds are structurally comprised of a malvidin 3-*O*-glucoside-pyruvate and acetylated derivatives linked with a single catechin (**CAT2a**) (Mateus et al. 2004) or dimeric procyanidin (**CAT2g**) (Mateus et al. 2003) through a vinyl group. Similar pigments derived from different anthocyanins have been tentatively identified through LC-DAD-MS.

7.3 Conclusions

Flavonoids have long being established as one of the most numerous and promising class of phytochemicals. As such, significant advances have permitted the elucidation of the biochemical pathways that lead to the biosynthesis of various compounds with potential medicinal and nutraceutical applications.

In this chapter, we presented some recent metabolic engineering approaches that allow the biosynthesis of several flavonoid compounds at low cost and high yields in microorganisms. While key issues that mainly relate to production yields still need to be addressed, it is expected that such microbial fermentation technologies will soon provide a competitive alternative to current methods that rely on plant extraction or chemical synthesis.

At the same time, in this chapter we also presented an overview on flavonoid metabolism by microorganisms and their derivative molecules, especially from the gut flora. This is an area that is still unexplored, however we believe that it holds great promise not only because it plays a key role in the effect flavonoids have in human health but also for the development of new molecules as drug candidates in the future. In that respect, emerging technologies such as metagenomics can prove valuable for the elucidation of microbial metabolic pathways that will allow us to exploit to its full potential the chemical diversity of these polyphenolic compounds that nature has to offer.

References

Abdelrahim, I. and Y. J. Abulhajj (1990). Microbiological transformation of (+/–)-flavanone and (+/–)-isoflavanone. J. Nat. Prod. 53(3): 644–656.

Adlercreutz, H., K. Hockerstedt, C. Bannwart, S. Bloigu, E. Hamalainen, T. Fotsis and A. Ollus (1987). Effect of dietary-components, including lignans and phytoestrogens, on enterohepatic circulation and liver-metabolism of estrogens and on sex-hormone binding globulin (Shbg). J. Steroid Biochem. 27(4–6): 1135–1144.

Adlercreutz, H., P. I. Musey, T. Fotsis, C. Bannwart, K. Wahala, T. Makela, G. Brunow and T. Hase (1986). Identification of lignans and phytoestrogens in urine of chimpanzees. Clin. Chim. Acta 158(2): 147–154.

Akaza, H., N. Miyanaga, N. Takashima, S. Naito, Y. Hirao, T. Tsukamoto and M. Mori (2002). Is daidzein non-metabolizer a high risk for prostate cancer? A case-controlled study of serum soybean isoflavone concentration. Jpn. J. Clin. Oncol. 32(8): 296–300.

Alcalde-Eon, C., M. T. Escribano-Bailon, C. Santos-Buelga and J. C. Rivas-Gonzalo (2004). Separation of pyranoanthocyanins from red wine by column chromatography. Anal. Chim. Acta 513(1): 305–318.

Anyanwutaku, I. O., E. Zirbes and J. P. N. Rosazza (1992). Isoflavonoids from *Streptomycetes:* origins of genistein, 8-chlorogenistein, and 6,8-dichlorogenistein. J. Nat. Prod. 55(10): 1498–1504.

Arai, Y., M. Uehara, Y. Sato, M. Kimira, A. Eboshida, H. Adlercreutz and S. Watanabe (2000). Comparison of isoflavones among dietary intake, plasma concentration and urinary excretion for accurate estimation of phytoestrogen intake. J. Epidemiol. 10(2): 127–135.

Arora, A., M. G. Nair and G. M. Strasburg (1998). Antioxidant activities of isoflavones and their biological metabolites in a liposomal system. Arch. Biochem. Biophys. 356(2): 133–141.

Asenstorfer, R. E., Y. Hayasaka and G. P. Jones (2001). Isolation and structures of oligomeric wine pigments by bisulfite-mediated ion-exchange chromatography. J. Agr. Food Chem. 49(12): 5957–5963.

Asenstorfer, R. E., A. J. Markides, P. G. Iland and G. P. Jones (2003). Formation of vitisin A during red wine fermentation and maturation. Aust. J. Grape Wine Res. 9(1): 40–46.

Atanasova, V., H. Fulcrand, W. Cheynier and M. Moutounet (2002). Effect of oxygenation on polyphenol changes occurring in the course of wine-making. Anal. Chim. Acta 458(1): 15–27.

Atkinson, C., S. Berman, O. Humbert and J. W. Lampe (2004). *In vitro* incubation of human feces with daidzein and antibiotics suggests interindividual differences in the bacteria responsible for equol production. J. Nutr. 134(3): 596–599.

Aura, A. M., P. Martin-Lopez, K. A. O'Leary, G. Williamson, K. M. Oksman-Caldentey, K. Poutanen and C. Santos-Buelga (2005). *In vitro* metabolism of anthocyanins by human gut microflora. Eur. J. Nutr. 44(3): 133–142.

Aura, A. M., K. A. O'Leary, G. Williamson, M. Ojala, M. Bailey, R. Puupponen-Pimia, A. M. Nuutila, K. M. Oksman-Caldentey and K. Poutanen (2002). Quercetin derivatives are deconjugated and converted to hydroxyphenylacetic acids but not methylated by human fecal flora *in vitro*. J. Agr. Food Chem. 50(6): 1725–1730.

Aura, A. M., T. Seppänen-Laakso, I. Mattila, S. Guyot, C. M. G. C. Renard, J.-M. Souquet, V. Cheynier and K.-M. Oksman-Caldentey (2006). Microbial metabolism of (+)-catechin, (–)-epicatechin and proanthocyanidins by human fecal microbiota *in vitro*. Poly. Comm. 2006, 43–44.

Axelson, M., D. N. Kirk, R. D. Farrant, G. Cooley, A. M. Lawson and K. D. R. Setchell (1982). The identification of the weak estrogen equol [7-hydroxy-3-(4'-hydroxyphenyl)chroman] in human-urine. Biochem. J. 201(2): 353–357.

Bakker, J. and C. F. Timberlake (1997). Isolation, identification, and characterization of new color-stable anthocyanins occurring in some red wines. J. Agr. Food Chem. 45(1): 35–43.

Barnes, H. J., M. P. Arlotto and M. R. Waterman (1991). Expression and enzymatic activity of recombinant cytochrome P450 17 alpha-hydroxylase in *Escherichia coli*. Proc. Natl. Acad. Sci. USA 88(13): 5597–5601.

Barnes, S. (2003). Phyto-oestrogens and osteoporosis: what is a safe dose? Brit. J. Nutr. 89: S101-S108.

Baumes, R., R. Cordonnier, S. Nitz and F. Drawert (1986). Identification and determination of volatile constituents in wines from different vine cultivars. J. Sci. Food Agr. 37(9): 927–943.

Baur, J. A., K. J. Pearson, N. L. Price, H. A. Jamieson, C. Lerin, A. Kalra, V. V. Prabhu, J. S. Allard, G. Lopez-Lluch, K. Lewis, P. J. Pistell, S. Poosala, K. G. Becker, O. Boss, D. Gwinn, M. Wang, S. Ramaswamy, K. W. Fishbein, R. G. Spencer, E. G. Lakatta, D. Le Couteur, R. J. Shaw, P. Navas, P. Puigserver, D. K. Ingram, R. de Cabo and D. A. Sinclair (2006). Resveratrol improves health and survival of mice on a high-calorie diet. Nature 444(7117): 337–342.

Becker, J. V. W., G. O. Armstrong, M. J. Van der Merwe, M. G. Lambrechts, M. A. Vivier and I. S. Pretorius (2003). Metabolic engineering of *Saccharomyces cerevisiae* for the synthesis of the wine-related antioxidant resveratrol. FEMS Yeast Res. 4(1): 79–85.

Beld, M., C. Martin, H. Huits, A. R. Stuitje and A. G. Gerats (1989). Flavonoid synthesis in *Petunia hybrida*: partial characterization of dihydroflavonol-4-reductase genes. Plant Mol. Biol. 13(5): 491–502.

Birt, D. F., S. Hendrich and W. Wang (2001). Dietary agents in cancer prevention: flavonoids and isoflavonoids. Pharmacol Therapeut. 90(2–3): 157–177.

Bitsch, I., M. Janssen, M. Netzel, G. Strass and T. Frank (2004). Bioavailability of anthocyanidin-3-glycosides following consumption of elderberry extract and blackcurrant juice. Int. J Clin. Pharm. Th. 42(5): 293–300.

Blair, R. M., S. E. Appt, A. A. Franke and T. B. Clarkson (2003). Treatment with antibiotics reduces plasma equol concentration in cynomolgus monkeys (*Macaca fascicularis*). J. Nutr. 133(7): 2262–2267.

Block, G., B. Patterson and A. Subar (1992). Fruit, vegetables, and cancer prevention: a review of the epidemiological evidence. Nutr. Cancer 18(1): 1–29.

Bois, F., C. Beney, A. Boumendjel, A. M. Mariotte, G. Conseil and A. Di Pietro (1998). Halogenated chalcones with high-affinity binding to P-glycoprotein: Potential modulators of multidrug resistance. J. Med. Chem. 41(21): 4161–4164.

Bomser, J., D. L. Madhavi, K. Singletary and M. A. L. Smith (1996). *In vitro* anticancer activity of fruit extracts from *Vaccinium* species. Planta Med. 62(3): 212–216.

Boulton, R. B., V. L. Singleton, L. F. Bisson and R. E. Kunkee (1996). Yeast and biochemistry of ethanol fermentation. *Principles and Practices of Winemaking*. New York, Chapman and Hall: 127–137.

Bowey, E., H. Adlercreutz and I. Rowland (2003). Metabolism of isoflavones and lignans by the gut microflora: a study in germ-free and human flora associated rats. Food Chem. Toxicol. 41(5): 631–636.

Boyd, W. (2000). Natural colors as functional ingredients in healthy food. Cereal Foods World 45: 221–222.

Braune, A., M. Gutschow, W. Engst and M. Blaut (2001). Degradation of quercetin and luteolin by *Eubacterium ramulus*. Appl. Environ. Microb. 67(12): 5558–5567.

Brouillard, R. (1982). Chemical structure of anthocyanins. *Anthocyanins as food colors*. P. Markakis. New York, NY, Academic Press, Inc.

Brown, N. M. and K. D. R. Setchell (2001). Animal models impacted by phytoestrogens in commercial chow: implications for pathways influenced by hormones. Lab. Invest. 81(5): 735–747.

Bub, A., B. Watzl, D. Heeb, G. Rechkemmer and K. Briviba (2001). Malvidin-3-glucoside bioavailability in humans after ingestion of red wine, dealcoholized red wine and red grape juice. Eur. J. Nutr. 40(3): 113–120.

Burbulis, I. E. and B. Winkel-Shirley (1999). Interactions among enzymes of the Arabidopsis flavonoid biosynthetic pathway. Proc. Natl. Acad. Sci. USA 96(22): 12929–12934.

Caltagirone, S., C. Rossi, A. Poggi, F. O. Ranelletti, P. G. Natali, M. Brunetti, F. B. Aiello and M. Piantelli (2000). Flavonoids apigenin and quercetin inhibit melanoma growth and metastatic potential. Int. J. Cancer 87(4): 595–600.

CameiradosSantos, P. J., J. M. Brillouet, V. Cheynier and M. Moutounet (1996). Detection and partial characterisation of new anthocyanin-derived pigments in wine. J. Sci. Food Agr. 70(2): 204–208.

Cao, M. A., X. B. Sun, P. H. Zhao and C. S. Yuan (2006). Two new antibacterial flavanones from *Sophora flavescens*. Chinese Chem. Lett. 17(8): 1048–1050.

Cao, Y. H. and R. H. Cao (1999). Angiogenesis inhibited by drinking tea. Nature 398(6726): 381.

Cassidy, A., S. Bingham, J. Carlson and K. D. R. Setchell (1993). Biological effects of plant estrogens in premenopausal women. Faseb J. 7(4): A866.

Chang, Y. C. and M. G. Nair (1995). Metabolism of daidzein and genistein by intestinal bacteria. J. Nat. Prod. 58(12): 1892–1896.

Chatonnet, P., D. Dubourdieu, J. N. Boidron and V. Lavigne (1993). Synthesis of volatile phenols by *Saccharomyces cerevisiae* in wines. J. Sci. Food Agr. 62(2): 191–202.

Cho, J. (2006). Antioxidant and neuroprotective effects of hesperidin and its aglycone hesperetin. Arch. Pharm. Res. 29(8): 699–706.

Clausen, M., C. J. Lamb, R. Megnet and P. W. Doerner (1994). *Pad1* encodes phenylacrylic acid decarboxylase which confers resistance to cinnamic acid in *Saccharomyces cerevisiae*. Gene 142(1): 107–112.

Clifford, M. N. (2000). Anthocyanins – nature, occurrence and dietary burden. J. Sci. Food Agr. 80(7): 1063–1072.

Colantuoni, A., S. Bertuglia, M. J. Magistretti and L. Donato (1991). Effects of *Vaccinium myrtillus* anthocyanosides on arterial vasomotion. Drug Res. 41–2(9): 905–909.

Coldham, N. G., C. Darby, M. Hows, L. J. King, A. Q. Zhang and M. J. Sauer (2002). Comparative metabolism of genistin by human and rat gut microflora: detection and identification of the end-products of metabolism. Xenobiotica 32(1): 45–62.

Contreras, D. M., C. M. A. Ramírez, S. Guyot, I. Perraud-Gaime, S. Roussos and C. Augur (2006). Enzymatic degradation of catechin and procyanidin B2 by filamentous fungi. Poly. Comm. 2006, 77–78.

Cotelle, N., J. L. Bernier, J. P. Catteau, J. Pommery, J. C. Wallet and E. M. Gaydou (1996). Antioxidant properties of hydroxy-flavones. Free Radical Bio. Med. 20(1): 35–43.

Critchfield, J. W., J. E. Coligan, T. M. Folks and S. T. Butera (1997). Casein kinase II is a selective target of HIV-1 transcriptional inhibitors. Proc. Natl. Acad. Sci. USA 94(12): 6110–6115.

Czeczot, H., B. Tudek, J. Kusztelak, T. Szymczyk, B. Dobrowolska, G. Glinkowska, J. Malinowski and H. Strzelecka (1990). Isolation and studies of the mutagenic activity in the ames test of flavonoids naturally-occurring in medical herbs. Mutat. Res. 240(3): 209–216.

Das, S. and J. P. N. Rosazza (2006). Microbial and enzymatic transformations of flavonoids. J. Nat. Prod. 69(3): 499–508.

Day, A. J., F. J. Canada, J. C. Diaz, P. A. Kroon, R. Mclauchlan, C. B. Faulds, G. W. Plumb, M. R. A. Morgan and G. Williamson (2000). Dietary flavonoid and isoflavone glycosides are hydrolysed by the lactase site of lactase phlorizin hydrolase. FEBS Lett. 468(2–3): 166–170.

Decroos, K., S. Vanhemmens, S. Cattoir, N. Boon and W. Verstraete (2005). Isolation and characterisation of an equol-producing mixed microbial culture from a human faecal sample and its activity under gastrointestinal conditions. Arch. Microbiol. 183(1): 45–55.

Deweerd, K. A., A. Saxena, D. P. Nagle and J. M. Suflita (1988). Metabolism of the 18O-methoxy substituent of 3-methoxybenzoic acid and other unlabeled methoxybenzoic acids by anaerobic-bacteria. Appl. Environ. Microb. 54(5): 1237–1242.

Dhar, K. and J. P. N. Rosazza (2000). Purification and characterization of *Streptomyces griseus* catechol O-methyltransferase. Appl. Environ. Microb. 66(11): 4877–4882.

Dixon, R. A. and L. W. Sumner (2003). Legume natural products: understanding and manipulating complex pathways for human and animal health. Plant Physiol. 131(3): 878–885.

Dong, X. Y., E. L. Braun and E. Grotewold (2001). Functional conservation of plant secondary metabolic enzymes revealed by complementation of Arabidopsis flavonoid mutants with maize genes. Plant Physiol. 127(1): 46–57.

Dugelay, I., Z. Gunata, J. C. Sapis, R. Baumes and C. Bayonove (1993). Role of cinnamoyl esterase-activities from enzyme preparations on the formation of volatile phenols during winemaking. J. Agr. Food Chem. 41(11): 2092–2096.

Duncan, A. M., B. E. Merz-Demlow, X. Xu, W. R. Phipps and M. S. Kurzer (2000). Premenopausal equol excretors show plasma hormone profiles associated with lowered risk of breast cancer. Cancer Epidem. Biomar. 9(6): 581–586.

Dupin, I. V. S., B. M. McKinnon, C. Ryan, M. Boulay, A. J. Markides, G. P. Jones, P. J. Williams and E. J. Waters (2000). *Saccharomyces cerevisiae* mannoproteins that protect wine from protein haze: Their release during fermentation and lees contact and a proposal for their mechanism of action. J. Agr. Food Chem. 48(8): 3098–3105.

Es-Safi, N. E., H. Fulcrand, V. Cheynier and M. Moutounet (1999). Competition between (+)-catechin and (–)-epicatechin in acetaldehyde-induced polymerization of flavanols. J. Agr. Food Chem. 47(5): 2088–2095.

Etievant, P. X. (1981). Volatile phenol determination in wine. J. Agr. Food. Chem. 29(1): 65–67.

Ferrandiz, M. L. and M. J. Alcaraz (1991). Antiinflammatory Activity and Inhibition of Ara-chidonic-Acid Metabolism by Flavonoids. Agents Actions 32(3–4): 283–288.

Feuillat, M. (2003). Yeast macromolecules: origin, composition, and enological interest. Am. J. Enol. Viticult. 54(3): 211–213.

Fischer, D., K. Stich, L. Britsch and H. Grisebach (1988). Purification and characterization of (+)dihydroflavonol (3-hydroxyflavanone) 4-reductase from flowers of Dahlia variabilis. Arch. Biochem. Biophys. 264(1): 40–47.

Fischer, T. C., H. Halbwirth, B. Meisel, K. Stich and G. Forkmann (2003). Molecular cloning, substrate specificity of the functionally expressed dihydroflavonol 4-reductases from Malus domestica and Pyrus communis cultivars and the consequences for flavonoid me-tabolism. Arch. Biochem. Bioph. 412(2): 223–230.

Fleschhut, J., F. Kratzer, G. Rechkemmer and S. E. Kulling (2006). Stability and biotransfor-mation of various dietary anthocyanins in vitro. Eur. J. Nutr. 45(1): 7–18.

Forkmann, G. and W. Heller (1999). Comprehensive Natural Products Chemistry. Amster-dam, Elsevier.

Formica, J. V. and W. Regelson (1995). Review of the biology of quercetin and related biofla-vonoids. Food Chem. Toxicol. 33(12): 1061–1080.

FranciaAricha, E. M., M. T. Guerra, J. C. RivasGonzalo and C. SantosBuelga (1997). New anthocyanin pigments formed after condensation with flavanols. J. Agr. Food Chem. 45(6): 2262–2266.

Frank, T., M. Netzel, G. Strass, R. Bitsch and I. Bitsch (2003). Bioavailability of anthocya-nidin-3-glucosides following consumption of red wine and red grape juice. Can. J. Physiol. Pharm. 81(5): 423–435.

Frankenfeld, C. L., C. Atkinson, W. K. Thomas, E. L. Goode, A. Gonzalez, T. Jokela, K. Wahala, S. M. Schwartz, S. S. Li and J. W. Lampe (2004a). Familial correlations, segrega-tion analysis, and nongenetic correlates of soy isoflavone-metabolizing phenotypes. Exp. Biol. Med. 229(9): 902–913.

Frankenfeld, C. L., A. McTiernan, S. S. Tworoger, C. Atkinson, W. K. Thomas, F. Z. Stanc-zyk, S. M. Marcovina, D. S. Weigle, N. S. Weiss, V. L. Holt, S. M. Schwartz and J. W. Lampe (2004b). Serum steroid hormones, sex hormone-binding globulin concentrations, and urinary hydroxylated estrogen metabolites in post-menopausal women in relation to daidzein-metabolizing phenotypes. J. Steroid Biochem. 88(4–5): 399–408.

Fulcrand, H., C. Benabdeljalil, J. Rigaud, V. Cheynier and M. Moutounet (1998). A new class of wine pigments generated by reaction between pyruvic acid and grape anthocyanins. Phytochemistry 47(7): 1401–1407.

Fulcrand, H., T. Doco, N. E. EsSafi, V. Cheynier and M. Moutounet (1996a). Study of the acetaldehyde induced polymerisation of flavan-3-ols by liquid chromatography ion spray mass spectrometry. J. Chromatogr. A 752(1–2): 85–91.

Fulcrand, H., P. J. C. dosSantos, P. SarniManchado, V. Cheynier and J. FavreBonvin (1996b). Structure of new anthocyanin-derived wine pigments. J. Chem. Soc. Perk. Trans. 1(7): 735–739.

Genthner, B. R. S., C. L. Davis and M. P. Bryant (1981). Features of rumen and sewage-sludge strains of Eubacterium limosum, a methanol-utilizing and H_2-CO_2-utilizing species. Appl. Environ. Microb. 42(1): 12–19.

Gerhauser, C. (2005). Beer constituents as potential cancer chemopreventive agents. Eur. J. Cancer 41(13): 1941–1954.

Gerhauser, C., A. Alt, E. Heiss, A. Gamal-Eldeen, K. Klimo, J. Knauft, I. Neumann, H. R. Scherf, N. Frank, H. Bartsch and H. Becker (2002). Cancer chemopreventive activity of Xanthohumol, a natural product derived from hop. Mol. Cancer Ther. 1(11): 959–969.

Ghiselli, A., M. Nardini, A. Baldi and C. Scaccini (1998). Antioxidant activity of different phenolic fractions separated from an Italian red wine. J. Agr. Food Chem. 46(2): 361–367.

Gil, B., M. J. Sanz, M. C. Terencio, M. L. Ferrandiz, G. Bustos, M. Paya, R. Gunasegaran and M. J. Alcaraz (1994). Effects of flavonoids on Naja-Naja and human recombinant synovial phospholipases a(2) and inflammatory responses in mice. Life Sci. 54(20): Pl333-Pl338.

Goldin, B. R. (1990). Intestinal microflora – metabolism of drugs and carcinogens. Ann. Med. 22(1): 43–48.

Hagiwara, A., K. Miyashita, T. Nakanishi, M. Sano, S. Tamano, T. Kadota, T. Koda, M. Nakamura, K. Imaida, N. Ito and T. Shirai (2001). Pronounced inhibition by a natural anthocyanin, purple corn color, of 2-amino-1-methyl-6-phenylimidazo[4,5-b]pyridine (PhIP)associated colorectal carcinogenesis in male F344 rats pretreated with 1,2-dimethylhydrazine. Cancer Lett. 171(1): 17–25.

Hakansson, A. E., K. Pardon, Y. Hayasaka, M. de Sa and M. Herderich (2003). Structures and colour properties of new red wine pigments. Tetrahedron Lett. 44(26): 4887–4891.

Hall, R. D. and M. M. Yeoman (1986). Temporal and spatial heterogeneity in the accumulation of anthocyanins in cell-cultures of *Catharanthus roseus* (L) Don,G. J. Exp. Bot. 37(174): 48–60.

Hanagata, N., A. Ito, H. Uehara, F. Asari, T. Takeuchi and I. Karube (1993). Behavior of cell aggregate of *Carthamus tinctorius* L. cultured-cells and correlation with red pigment formation. J. Biotechnol. 30(3): 259–269.

Harborne, J. B. and C. A. Williams (2000). Advances in flavonoid research since 1992. Phytochemistry 55(6): 481–504.

Harris, G. K., A. Gupta, R. G. Nines, L. A. Kresty, S. G. Habib, W. L. Frankel, K. LaPerle, D. D. Gallaher, S. J. Schwartz and G. D. Stoner (2001). Effects of lyophilized black raspberries on azoxymethane-induced colon cancer and 8-hydroxy-2′-deoxyguanosine levels in the Fischer 344 rat. Nutr. Cancer 40(2): 125–133.

Haudenschild, C., M. Schalk, F. Karp and R. Croteau (2000). Functional expression of regiospecific cytochrome P450 limonene hydroxylases from mint (*Mentha* spp.) in *Escherichia coli* and *Saccharomyces cerevisiae*. Arch. Biochem. Biophys. 379(1): 127–136.

Havsteen, B. (1983). Flavonoids, a class of natural-products of high pharmacological potency. Biochem. Pharmacol. 32(7): 1141–1148.

Hayasaka, Y. and R. E. Asenstorfer (2002). Screening for potential pigments derived from anthocyanins in red wine using nanoelectrospray tandem mass spectrometry. J. Agr. Food Chem. 50(4): 756–761.

Heier, A., W. Blaas, A. Dross and R. Wittkowski (2002). Anthocyanin analysis by HPLC/ESI-MS. Am. J. Enol. Viticult. 53(1): 78–86.

Heinonen, S., K. Wahala and H. Adlercreutz (1999). Identification of isoflavone metabolites dihydrodaidzein, dihydrogenistein, 6′-OH-O-dma, and *cis*-4-OH-equol in human urine by gas chromatography-mass spectroscopy using authentic reference compounds. Anal. Biochem. 274(2): 211–219.

Heinonen, S. M., A. Hoikkala, K. Wahala and H. Adlercreutz (2003). Metabolism of the soy isoflavones daidzein, genistein and glycitein in human subjects. Identification of new metabolites having an intact isoflavonoid skeleton. J. Steroid Biochem. 87(4–5): 285–299.

Herath, W. H. M. W., D. Ferreira and I. A. Khan (2003). Microbial transformation of xanthohumol. Phytochemistry 62(5): 673–677.

Hollman, P. C. H. and I. C. W. Arts (2000). Flavonols, flavones and flavanols – nature, occurrence and dietary burden. J. Sci. Food Agr. 80(7): 1081–1093.

Hollman, P. C. H., J. M. P. vanTrijp, M. N. C. P. Buysman, M. S. VanderGaag, M. J. B. Mengelers, J. H. M. deVries and M. B. Katan (1997). Relative bioavailability of the antioxidant flavonoid quercetin from various foods in man. Febs Lett. 418(1–2): 152–156.

Hosny, M., K. Dhar and J. P. N. Rosazza (2001). Hydroxylations and methylations of quercetin, fisetin, and catechin by *Streptomyces griseus*. J. Nat. Prod. 64(4): 462–465.

Hosny, M. and J. P. N. Rosazza (1999). Novel isoflavone, cinnamic acid, and triterpenoid glycosides in soybean molasses. J. Nat. Prod. 62(6): 853–858.

Hotze, M., G. Schroder and J. Schroder (1995). Cinnamate 4-hydroxylase from *Catharanthus roseus*, and a strategy for the functional expression of plant cytochrome P-450 proteins as translational fusions with P-450 reductase in *Escherichia coli*. Febs Lett. 374(3): 345–350.

Howitz, K. T., K. J. Bitterman, H. Y. Cohen, D. W. Lamming, S. Lavu, J. G. Wood, R. E. Zipkin, P. Chung, A. Kisielewski, L. L. Zhang, B. Scherer and D. A. Sinclair (2003). Small molecule activators of sirtuins extend *Saccharomyces cerevisiae* lifespan. Nature 425(6954): 191–196.

Hur, H. G., R. D. Beger, T. M. Heinze, J. O. Lay, J. P. Freeman, J. Dore and F. Rafii (2002). Isolation of an anaerobic intestinal bacterium capable of cleaving the C-ring of the isoflavonoid daidzein. Arch. Microbiol. 178(1): 8–12.

Hur, H. G., J. O. Lay, R. D. Beger, J. P. Freeman and F. Rafii (2000). Isolation of human intestinal bacteria metabolizing the natural isoflavone glycosides daidzin and genistin. Arch. Microbiol. 174(6): 422–428.

Hur, H. G. and F. Rafii (2000). Biotransformation of the isoflavonoids biochanin A, formononetin, and glycitein by *Eubacterium limosum*. FEMS Microbiol. Lett. 192(1): 21–25.

Hutchins, A. M., J. L. Slavin and J. W. Lampe (1995). Urinary isoflavonoid phytoestrogen and lignan excretion after consumption of fermented and unfermented soy products. J. Am. Diet Assoc. 95(5): 545–551.

Hwang, E. I., M. Kaneko, Y. Ohnishi and S. Horinouchi (2003). Production of plant-specific flavanones by *Escherichia coli* containing an artificial gene cluster. Appl. Environ. Microb. 69(5): 2699–2706.

Ibrahim, A. R. and Y. J. Abulhajj (1989). Aromatic hydroxylation and sulfation of 5-hydroxyflavone by *Streptomyces fulvissimus*. Appl. Environ. Microb. 55(12): 3140–3142.

Ibrahim, A. R. S. (2000). Sulfation of naringenin by *Cunninghamella elegans*. Phytochemistry 53(2): 209–212.

Ibrahim, A. R. S. and Y. J. Abulhajj (1990). Microbiological transformation of flavone and isoflavone. Xenobiotica 20(4): 363–373.

Ibrahim, A. R. S., A. M. Galal, J. S. Mossa and F. S. El-Feraly (1997). Glucose-conjugation of the flavones of *Psiadia arabica* by *Cunninghamella elegans*. Phytochemistry 46(7): 1193–1195.

Ingram, D., K. Sanders, M. Kolybaba and D. Lopez (1997). Case-control study of phytooestrogens and breast cancer. Lancet 350(9083): 990–994.

Inoue, M., K. Tajima, K. Hirose, N. Hamajima, T. Takezaki, T. Kuroishi and S. Tominaga (1998). Tea and coffee consumption and the risk of digestive tract cancers: data from a comparative case-referent study in Japan. Cancer Causes Control 9(2): 209–216.

Jankun, J., S. H. Selman, R. Swiercz and E. SkrzypczakJankun (1997). Why drinking green tea could prevent cancer. Nature 387(6633): 561.

Jeon, K. S., G. E. Ji and I. K. Kwang (2002). Assay of beta-glucosidase activity of bifidobacteria and the hydrolysis of isoflavone glycosides by *Bifidobacterium* sp Int-57 in soymilk fermentation. J. Microbiol. Biotechn. 12(1): 8–13.

Jez, J. M., J. L. Ferrer, M. E. Bowman, R. A. Dixon and J. P. Noel (2000). Dissection of malonyl-coenzyme A decarboxylation from polyketide formation in the reaction mechanism of a plant polyketide synthase. Biochemistry-US 39(5): 890–902.

Joannou, G. E., G. E. Kelly, A. Y. Reeder, M. Waring and C. Nelson (1995). A urinary profile study of dietary phytoestrogens – the identification and mode of metabolism of new isoflavonoids. J. Steroid Biochem. Mol. Biol. 54(3–4): 167–184.

Johnson, E. T., S. Ryu, H. K. Yi, B. Shin, H. Cheong and G. Choi (2001). Alteration of a single amino acid changes the substrate specificity of dihydroflavonol 4-reductase. Plant J. 25(3): 325–333.

Jung, W., O. Yu, S. M. Lau, D. P. O'Keefe, J. Odell, G. Fader and B. McGonigle (2000). Identification and expression of isoflavone synthase, the key enzyme for biosynthesis of isoflavones in legumes. Nat. Biotechnol. 18(2): 208–212.

Juniewicz, P. E., S. P. Morell, A. Moser and L. L. Ewing (1988). Identification of phytoestrogens in the urine of male dogs. J. Steroid Biochem. Mol. Biol. 31(6): 987–994.

Justesen, U., E. Arrigoni, B. R. Larsen and R. Amado (2000). Degradation of flavonoid glycosides and aglycones during *in vitro* fermentation with human faecal flora. Food Sci. Technol. 33(6): 424–430.

Kahkonen, M. P. and M. Heinonen (2003). Antioxidant activity of anthocyanins and their aglycons. J. Agr. Food Chem. 51(3): 628–633.

Kamei, H., T. Kojima, M. Hasegawa, T. Koide, T. Umeda, T. Yukawa and K. Terabe (1995). Suppression of tumor-cell growth by anthocyanins *in vitro*. Cancer Invest. 13(6): 590–594.

Kang, S. Y., N. P. Seeram, M. G. Nair and L. D. Bourquin (2003). Tart cherry anthocyanins inhibit tumor development in Apc(Min) mice and reduce proliferation of human colon cancer cells. Cancer Lett. 194(1): 13–19.

Katsube, N., K. Iwashita, T. Tsushida, K. Yamaki and M. Kobori (2003). Induction of apoptosis in cancer cells by bilberry (*Vaccinium myrtillus*) and the anthocyanins. J. Agr. Food Chem. 51(1): 68–75.

Kawabata, J., Y. Okamoto, A. Kodama, T. Makimoto and T. Kasai (2002). Oxidative dimers produced from protocatechuic and gallic esters in the DPPH radical scavenging reaction. J. Agr. Food Chem. 50(19): 5468–5471.

Kelly, G. E., G. E. Joannou, A. Y. Reeder, C. Nelson and M. A. Waring (1995). The variable metabolic response to dietary isoflavones in humans. Proc. Soc. Exp. Biol. Med. 208(1): 40–43.

Keppler, K. and H. U. Humpf (2005). Metabolism of anthocyanins and their phenolic degradation products by the intestinal microflora. Bioorgan. Med. Chem. 13(17): 5195–5205.

Khan, M. S. Y. and S. M. Hasan (2003). Synthesis, antiinflammatory and antibacterial activity of some new flavonoidal derivatives. Indian J. Chem. B 42(8): 1970–1974.

Kim, D. H., E. A. Jung, I. S. Sohng, J. A. Han, T. H. Kim and M. J. Han (1998). Intestinal bacterial metabolism of flavonoids and its relation to some biological activities. Arch. Pharm. Res. 21(1): 17–23.

Kim, H. J. and I. S. Lee (2006). Microbial metabolism of the prenylated chalcone xanthohumol. J. Nat. Prod. 69(10): 1522–1524.

Klus, K. and W. Barz (1998). Formation of polyhydroxylated isoflavones from the isoflavones genistein and biochanin A by bacteria isolated from tempe. Phytochemistry 47(6): 1045–1048.

Kobayashi, Y., M. Akita, K. Sakamoto, H. F. Liu, T. Shigeoka, T. Koyano, M. Kawamura and T. Furuya (1993). Large-scale production of anthocyanin by *Aralia cordata* cell-suspension cultures. Appl. Microbiol. Biot. 40(2–3): 215–218.

Koide, T., U. Hashimoto, H. Kamei, T. Kojima, M. Hasegawa and K. Terabe (1997). Antitumor effect of anthocyanin fractions extracted from red soybeans and red beans *in vitro* and *in vivo*. Cancer Biother. Radio. 12(4): 277–280.

Koide, T., H. Kamei, Y. Hashimoto, T. Kojima and M. Hasegawa (1996). Antitumor effect of hydrolyzed anthocyanin from grape rinds and red rice. Cancer Biother. Radio. 11(4): 273–277.

Kostrzewa-Suslow, E., J. Dmochowska-Gladysz, A. Bialonska, Z. Ciunik and W. Rymowicz (2006). Microbial transformations of flavanone and 6-hydroxyflavanone by *Aspergillus niger* strains. J. Mol. Catal. B-Enzym. 39(1–4): 18–23.

Krishnamurty, H. G., K. J. Cheng, G. A. Jones, F. J. Simpson and J. E. Watkin (1970). Identification of products produced by the anaerobic degradation of rutin and related flavonoids by *Butyrivibrio* sp. C_3. Can. J. Microb. 16(8): 759–767.

Kuhnau, J. (1976). The flavonoids. A class of semi-essential food components. Their role in human nutrition. World Rev. Nutr. Diet 24: 117–191.

Kuiper, G. G. J. M., J. G. Lemmen, B. Carlsson, J. C. Corton, S. H. Safe, P. T. van der Saag, P. van der Burg and J. A. Gustafsson (1998). Interaction of estrogenic chemicals and phytoestrogens with estrogen receptor beta. Endocrinology 139(10): 4252–4263.

Lala, G., M. Malik, C. Zhao, J. He, Y. Kwon, M. M. Giusti and B. A. Magnuson (2006). Anthocyanin-rich extracts inhibit multiple biomarkers of colon cancer in rats. Nutr. Cancer 54(1): 84–93.

Lamartiniere, C. A., J. Wang, M. Smith-Johnson and I. E. Eltoum (2002). Daidzein: bioavailability, potential for reproductive toxicity, and breast cancer chemoprevention in female rats. Toxicol. Sci. 65(2): 228–238.

Lampe, J. W., S. C. Karr, A. M. Hutchins and J. L. Slavin (1998). Urinary equol excretion with a soy challenge: influence of habitual diet. Proc. Soc. Exp. Biol. Med. 217(3): 335–339.

Lamson, D. W. and M. S. Brignall (2000). Antioxidants and cancer, part 3: quercetin. Altern. Med. Rev. 5(3): 196–208.

Lao, C. L., E. Lopez-Tamames, R. M. Lamuela-Raventos, S. Buxaderas and M. D. C. De La Torre-Boronat (1997). Pectic enzyme treatment effects on quality of white grape musts and wines. J. Food Sci. 62(6): 1142–1144.

Lapidot, T., S. Harel, R. Granit and J. Kanner (1998). Bioavailability of red wine anthocyanins as detected in human urine. J. Agr. Food Chem. 46(10): 4297–4302.

Larsson, S., N. O. Nilvebrant and L. J. Jonsson (2001). Effect of overexpression of *Saccharomyces cerevisiae* Pad1p on the resistance to phenylacrylic acids and lignocellulose hydrolysates under aerobic and oxygen-limited conditions. Appl. Microbiol. Biot. 57(1–2): 167–174.

Lee, D. F., E. E. Swinny and G. P. Jones (2004). NMR identification of ethyl-linked anthocyanin-flavanol pigments formed in model wine ferments. Tetrahedron Lett. 45(8): 1 671–1674.

Leonard, E., J. Chemler, K. H. Lim and M. A. Koffas (2006a). Expression of a soluble flavone synthase allows the biosynthesis of phytoestrogen derivatives in *Escherichia coli*. Appl. Microbiol. Biotech. 70(1): 85–91.

Leonard, E., Y. Yan and M. A. Koffas (2006b). Functional expression of a P450 flavonoid hydroxylase for the biosynthesis of plant-specific hydroxylated flavonols in *Escherichia coli*. Meta. Eng. 8(2): 172–181.

Leonard, E., Y. J. Yan, K. H. Lim and M. A. G. Koffas (2005). Investigation of two distinct flavone synthases for plant-specific flavone biosynthesis in *Saccharomyces cerevisiae*. Appl. Environ. Microb. 71(12): 8241–8248.

Li, G., B. S. Min, C. J. Zheng, J. Lee, S. R. Oh, K. S. Ahn and H. K. Lee (2005). Neuroprotective and free radical scavenging activities of phenolic compounds from *Hovenia dulcis*. Arch. Pharm. Res. 28(7): 804–809.

Lietti, A., A. Cristoni and M. Picci (1976). Studies on *Vaccinium myrtillus* anthocyanosides .I. vasoprotective and antiinflammatory activity. Drug Res. 26(5): 829–832.

Liu, C. J., J. W. Blount, C. L. Steele and R. A. Dixon (2002). Bottlenecks for metabolic engineering of isoflavone glycoconjugates in Arabidopsis. Proc. Natl. Acad. Sci. USA 99(22): 14578–14583.

Liu, S. Q. and G. J. Pilone (2000). An overview of formation and roles of acetaldehyde in winemaking with emphasis on microbiological implications. Int. J. Food Sci. Tech. 35(1): 49–61.

Looman, A. C., J. Bodlaender, L. J. Comstock, D. Eaton, P. Jhurani, H. A. de Boer and P. H. van Knippenberg (1987). Influence of the codon following the AUG initiation codon on the expression of a modified *lacZ* gene in *Escherichia coli*. EMBO J. 6(8): 2489–2492.

Lozoya, E., H. Hoffmann, C. Douglas, W. Schulz, D. Scheel and K. Hahlbrock (1988). Primary structures and catalytic properties of isoenzymes encoded by the two 4-coumarate-CoA ligase genes in parsley. Eur. J. Biochem. 176(3): 661–667.

Lundh, T. (1995). Metabolism of estrogenic isoflavones in domestic-animals. Proc. Soc. Exp. Biol. Med. 208(1): 33–39.

Maatooq, G. T. and J. P. N. Rosazza (2005). Metabolism of daidzein by *Nocardia* species NRRL 5646 and *Mortierella isabellina* ATCC 38063. Phytochemistry 66(9): 1007–1011.

Macdonald, I. A., J. A. Mader and R. G. Bussard (1983). The role of rutin and quercitrin in stimulating flavonol glycosidase activity by cultured cell-free microbial preparations of human feces and saliva. Mutat. Res. 122(2): 95–102.

Magee, P. J. and I. R. Rowland (2004). Phyto-oestrogens, their mechanism of action: current evidence for a role in breast and prostate cancer. Brit. J. Nutr. 91(4): 513–531.

Martens, S. and G. Forkmann (1999). Cloning and expression of flavone synthase II from *Gerbera* hybrids. Plant J. 20(5): 611–618.

Martens, S., G. Forkmann, U. Matern and R. Lukacin (2001). Cloning of parsley flavone synthase I. Phytochemistry 58(1): 43–46.

Martens, S., T. Teeri and G. Forkmann (2002). Heterologous expression of dihydroflavonol 4-reductases from various plants. FEBS Lett. 531(3): 453–458.

Marullo, P., M. Bely, I. Masneuf-Pomarede, M. Pons, M. Aigle and D. Dubourdieu (2006). Breeding strategies for combining fermentative qualities and reducing off-flavor production in a wine yeast model. Fems Yeast Res. 6(2): 268–279.

Mateus, N., S. de Pascual-Teresa, J. C. Rivas-Gonzalo, C. Santos-Buelga and V. de Freitas (2002a). Structural diversity of anthocyanin-derived pigments in port wines. Food Chem. 76(3): 335–342.

Mateus, N., J. Oliveira, J. Pissarra, A. M. Gonzalez-Paramas, J. C. Rivas-Gonzalo, C. Santos-Buelga, A. M. S. Silva and V. de Freitas (2006). A new vinylpyranoanthocyanin pigment occurring in aged red wine. Food Chem. 97(4): 689–695.

Mateus, N., J. Oliveira, C. Santos-Buelga, A. M. S. Silva and V. de Freitas (2004). NMR structure characterization of a new vinylpyranoanthocyanin-catechin pigment (a portisin). Tetrahedron Lett. 45(17): 3455–3457.

Mateus, N., A. M. S. Silva, J. C. Rivas-Gonzalo, C. Santos-Buelga and V. De Freitas (2003). A new class of blue anthocyanin-derived pigments isolated from red wines. J. Agr. Food Chem. 51(7): 1919–1923.

Mateus, N., A. M. S. Silva, C. Santos-Buelga, J. C. Rivas-Gonzalo and V. de Freitas (2002b). Identification of anthocyanin-flavanol pigments in red wines by NMR and mass spectrometry. J .Agr. Food Chem. 50(7): 2110–2116.

Mateus, N., A. M. S. Silva, J. Vercauteren and V. de Freitas (2001). Occurrence of anthocyanin-derived pigments in red wines. J. Agr. Food Chem. 49(10): 4836–4840.

Matsumoto, H., H. Inaba, M. Kishi, S. Tominaga, M. Hirayama and T. Tsuda (2001). Orally administered delphinidin 3-rutinoside and cyanidin 3-rutinoside are directly absorbed in rats and humans and appear in the blood as the intact forms. J. Agr. Food Chem. 49(3): 1546–1551.

Mauricio, J. C., J. Moreno, L. Zea, J. M. Ortega and M. Medina (1997). The effects of grape must fermentation conditions on volatile alcohols and esters formed by *Saccharomyces cerevisiae*. J. Sci. Food Agr. 75(2): 155–160.

Mazza, G., C. D. Kay, T. Cottrell and B. J. Holub (2002). Absorption of anthocyanins from blueberries and serum antioxidant status in human subjects. J. Agr. Food Chem. 50(26): 7731–7737.

Mazza, G. and E. Miniati (1993). *Anthocyanins in Fruits, Vegetables, and Grains*. Boca Raton, FL, CRC Press.

McDougall, G. J., S. Fyffe, P. Dobson and D. Stewart (2005). Anthocyanins from red wine – their stability under simulated gastrointestinal digestion. Phytochemistry 66(21): 2540–2548.

Mclellan, M. R., R. W. Kime, C. Y. Lee and T. M. Long (1995). Effect of honey as an anti-browning agent in light raisin processing. J. Food Process. Pres. 19(1): 1–8.

Meselhy, M. R., N. Nakamura and M. Hattori (1997). Biotransformation of (–)-epicatecbin 3-*O*-gallate by human intestinal bacteria. Chem. Pharm. Bull. 45(5): 888–893.

Meyer, J. E., M. F. Pepin and M. A. L. Smith (2002). Anthocyanin production from *Vaccinium pahalae*: limitations of the physical micro environment. J. Biotechnol. 93(1): 45–57.

Michnick, S., J. L. Roustan, F. Remize, P. Barre and S. Dequin (1997). Modulation of glycerol and ethanol yields during alcoholic fermentation in *Saccharomyces cerevisiae* strains overexpressed or disrupted for GPD1 encoding glycerol 3-phosphate dehydrogenase. Yeast 13(9): 783–793.

Middleton, E. and C. Kandaswami (1992). Effects of flavonoids on immune and inflammatory cell functions. Biochem. Pharmacol. 43(6): 1167–1179.

Milligan, S., J. Kalita, V. Pocock, A. Heyerick, L. De Cooman, H. Rong and D. De Keukeleire (2002). Oestrogenic activity of the hop phyto-oestrogen, 8-prenylnaringenin. Reproduction 123(2): 235–242.

Milligan, S. R., J. C. Kalita, A. Heyerick, H. Rong, L. De Cooman and D. De Keukeleire (1999). Identification of a potent phytoestrogen in hops (*Humulus lupulus* L.) and beer. J. Clin. Endocr. Metab. 84(6): 2249–2252.

Miyake, Y., K. Yamamoto and T. Osawa (1997). Metabolism of antioxidant in lemon fruit (*Citrus limon* B-URM. f.) by human intestinal bacteria. J. Agr. Food Chem. 45(10): 3738–3742.

Miyazawa, T., K. Nakagawa, M. Kudo, K. Muraishi and K. Someya (1999). Direct intestinal absorption of red fruit anthocyanins, cyanidin-3-glucoside and cyanidin-3,5-diglucoside, into rats and humans. J. Agr. Food Chem. 47(3): 1083–1091.

Mol, J., E. Grotewold and R. Koes (1998). How genes paint flowers and seeds. Trends Plant Sci. 3(6): 212–217.

Molly, K., M. V. Woestyne and W. Verstraete (1993). Development of a 5-step multichamber reactor as a simulation of the human intestinal microbial ecosystem. Appl. Microbiol. Biotech. 39(2): 254–258.

Monagas, M., B. Bartolome and C. Gomez-Cordoves (2005). Updated knowledge about the presence of phenolic compounds in wine. Crit. Rev. Food Sci. 45(2): 85–118.

Monagas, M., V. Nunez, B. Bartolome and C. Gomez-Cordoves (2003). Anthocyanin-derived pigments in Graciano, Tempranillo, and Cabernet Sauvignon wines produced in Spain. Am. J. Enol. Viticult. 54(3): 163–169.

Monfort, S. L., M. A. Thompson, N. M. Czekala, L. H. Kasman, C. H. L. Shackleton and B. L. Lasley (1984). Identification of a non-steroidal estrogen, equol, in the urine of pregnant macaques – correlation with steroidal estrogen excretion. J. Steroid Biochem. 20(4): 869–876.

Moorthy, N. S. H. N., R. J. Singh, H. P. Singh and S. D. Gupta (2006). Synthesis, biological evaluation and *in silico* metabolic and toxicity prediction of some flavanone derivatives. Chem. Pharm. Bull. 54(10): 1384–1390.

Morata, A., A. C. Gomez-Cordoves, B. Colomo and J. A. Suarez (2003). Pyruvic acid and acetaldehyde production by different strains of *Saccharomyces cerevisiae*: relationship with vitisin A and B formation in red wines. J. Agr. Food Chem. 51(25): 7402–7409.

Morata, A., M. C. Gomez-Cordoves, F. Calderon and J. A. Suarez (2006). Effects of pH, temperature and SO$_2$ on the formation of pyranoanthocyanins during red wine fermentation with two species of *Saccharomyces*. Int. J. Food Microbiol. 106(2): 123–129.

Nakayama, T., T. Sato, Y. Fukui, K. Yonekura-Sakakibara, H. Hayashi, Y. Tanaka, T. Kusumi and T. Nishino (2001). Specificity analysis and mechanism of aurone synthesis catalyzed by aureusidin synthase, a polyphenol oxidase homolog responsible for flower coloration. FEBS Lett. 499(1–2): 107–111.

Natella, F., M. Nardini, M. Di Felice and C. Scaccini (1999). Benzoic and cinnamic acid derivatives as antioxidants: structure-activity relation. J. Agr. Food Chem. 47(4): 1453–1459.

Noda, Y., T. Kneyuki, K. Igarashi, A. Mori and L. Packer (2000). Antioxidant activity of nasunin, an anthocyanin in eggplant peels. Toxicology 148(2–3): 119–123.

Ohnishi, M., N. Yoshimi, T. Kawamori, N. Ino, Y. Hirose, T. Tanaka, J. Yamahara, H. Miyata and H. Mori (1997). Inhibitory effects of dietary protocatechuic acid and costunolide on 7,12-dimethylbenz[*a*]anthracene-induced hamster cheek pouch carcinogenesis. Jpn. J. Cancer Res. 88(2): 111–119.

Ohta, A., M. Uehara, K. Sakai, M. Takasaki, H. Adlercreutz, T. Morohashi and Y. Ishimi (2002). A combination of dietary fructooligosaccharides and isoflavone conjugates increases femoral bone mineral density and equol production in ovariectomized mice. J. Nutr. 132(7): 2048–2054.

Okuno, Y. and M. Miyazawa (2006). Microbial O-demethylation of sinesetin and antimutagenic activity of the metabolite. J. Chem. Technol. Biot. 81(1): 29–33.

Omori, T., K. Ogawa, Y. Umemoto, K. Yuki, Y. Kajihara, M. Shimoda and H. Wada (1996). Enhancement of glycerol production by brewing yeast (*Saccharomyces cerevisiae*) with heat shock treatment. J. Ferment. Bioeng. 82(2): 187–190.

Oszmianski, J. and C. Y. Lee (1990). Inhibition of polyphenol oxidase activity and browning by honey. J. Agr. Food Chem. 38(10): 1892–1895.

Pan, X. J., G. G. Niu and H. Z. Liu (2003). Microwave-assisted extraction of tea polyphenols and tea caffeine from green tea leaves. Chem. Eng. Process 42(2): 129–133.

Passamonti, S., U. Vrhovsek, A. Vanzo and F. Mattivi (2003). The stomach as a site for anthocyanins absorption from food. FEBS Lett. 544(1–3): 210–213.

Peterson, J. J., J. T. Dwyer, G. R. Beecher, S. A. Bhagwat, S. E. Gebhardt, D. B. Haytowitz and J. M. Holden (2006). Flavanones in oranges, tangerines (mandarins), tangors, and tangelos: a compilation and review of the data from the analytical literature. J. Food Compos. Anal. 19: S66–S73.

Possemiers, S., S. Bolca, C. Grootaert, T. Van de Wiele, W. Verstraete, D. De Keukeleire and A. Heyerick (2006). Metabolic activation of pro-estrogens from hops by intestinal microbiota results in increased exposure to the potent phytoestrogen 8-prenylnaringenin, *in vitro* and *in vivo*. Poly. Comm.: 51–52.

Possemiers, S., A. Heyerick, V. Robbens, D. De Keukeleire and W. Verstraete (2005). Activation of proestrogens from hops (*Humulus lupulus* L.) by intestinal microbiota; conversion of isoxanthohumol into 8-prenylnaringenin. J. Agr. Food. Chem. 53(16): 6281–6288.

Pozo-Bayon, M. A., M. Monagas, M. C. Polo and C. Gomez-Cordoves (2004). Occurrence of pyranoanthocyanins in sparkling wines manufactured with red grape varieties. J. Agr. Food Chem. 52(5): 1300–1306.

Pretorius, I. S. (2000). Tailoring wine yeast for the new millennium: novel approaches to the ancient art of winemaking. Yeast 16(8): 675–729.

Radler, F. (1992). Yeast. Metabolism of organic acids. In: G.H. Fleet (Ed.) *Wine Microbiology and Biotechnology*. Camberwell, Victoria, Australia, Hardwood Academic Publishers: 165–182.

Ralston, L., S. Subramanian, M. Matsuno and O. Yu (2005). Partial reconstruction of flavonoid and isoflavonoid biosynthesis in yeast using soybean type I and type II chalcone isomerases. Plant Physiol. 137(4): 1375–1388.

Ramirez-Tortosa, C., O. M. Andersen, P. T. Gardner, P. C. Morrice, S. G. Wood, S. J. Duthie, A. R. Collins and G. G. Duthie (2001). Anthocyanin-rich extract decreases indices of lipid peroxidation and DNA damage in vitamin E-depleted rats. Free Radical Bio. Med. 31(9): 1033–1037.

Rao, K. V. and N. T. Weisner (1981). Microbial transformation of quercetin by *Bacillus cereus*. Appl. Environ. Microb. 42(3): 450–452.

Renaud, S. and M. Delorgeril (1992). Wine, alcohol, platelets, and the French paradox for coronary heart-disease. Lancet 339(8808): 1523–1526.

Revilla, I., S. Perez-Magarino, M. L. Gonzalez-SanJose and S. Beltran (1999). Identification of anthocyanin derivatives in grape skin extracts and red wines by liquid chromatography with diode array and mass spectrometric detection. J. Chromatogr. A 847(1–2): 83–90.

Rice-Evans, C. A., N. J. Miller, P. G. Bolwell, P. M. Bramley and J. B. Pridham (1995). The relative antioxidant activities of plant-derived polyphenolic flavonoids. Free Radical Res. 22(4): 375–383.

RiceEvans, C. A., N. J. Miller and G. Paganga (1996). Structure-antioxidant activity relationships of flavonoids and phenolic acids. Free Radical Bio. Med. 20(7): 933–956.

Romero, C. and J. Bakker (1999). Interactions between grape anthocyanins and pyruvic acid, with effect of pH and acid concentration on anthocyanin composition and color in model solutions. J. Agr. Food Chem. 47(8): 3130–3139.

Romero, C. and J. Bakker (2001). Anthocyanin and colour evolution during maturation of four port wines: effect of pyruvic acid addition. J. Sci. Food Agr. 81(2): 252–260.

Rong, H., Y. Zhao, K. Lazou, D. De Keukeleire, S. R. Milligan and P. Sandra (2000). Quantitation of 8-prenylnaringenin, a novel phytoestrogen in hops (*Humulus lupulus* L.), hop products, and beers, by benchtop HPLC-MS using electrospray ionization. Chromatographia 51(9–10): 545–552.

Rowland, I., H. Wiseman, T. Sanders, H. Adlercreutz and E. Bowey (1999). Metabolism of oestrogens and phytoestrogens: role of the gut microflora. Biochem. Soc. Trans. 27(2): 304–308.

Rowland, I. R., H. Wiseman, T. A. Sanders, H. Adlercreutz and E. A. Bowey (2000). Interindividual variation in metabolism of soy isoflavones and lignans: influence of habitual diet on equol production by the gut microflora. Nutr. Cancer 36(1): 27–32.

Ruefer, C. E., C. Gerhauser, N. Frank, H. Becker and S. E. Kulling (2005). *In vitro* phase II metabolism of xanthohumol by human UDP-glucuronosyltransferases and sulfotransferases. Mol. Nutr. Food. Res. 49(9): 851–856.

Sakaguchi, M., K. Mihara and R. Sato (1984). Signal recognition particle is required for co-translational insertion of cytochrome P-450 into microsomal membranes. Proc. Natl. Acad. Sci. USA 81(11): 3361–3364.

Sanchez-Gonzalez, M. and J. P. N. Rosazza (2004). Microbial transformations of chalcones: hydroxylation, *O*-demethylation, and cyclization to flavanones. J. Nat. Prod. 67(4): 553–558.

SarniManchado, P., H. Fulcrand, J. M. Souquet, V. Cheynier and M. Moutounet (1996). Stability and color of unreported wine anthocyanin-derived pigments. J. Food Sci. 61(5): 938–941.

SatueGracia, M. T., M. Heinonen and E. N. Frankel (1997). Anthocyanins as antioxidants on human low-density lipoprotein and lecithin-liposome systems. J. Agr. Food Chem. 45(9): 3362–3367.

Scalbert, A. and G. Williamson (2000). Dietary intake and bioavailability of polyphenols. J. Nutr. 130(8): 2073s–2085s.

Schauder, B. and J. E. McCarthy (1989). The role of bases upstream of the Shine-Dalgarno region and in the coding sequence in the control of gene expression in *Escherichia coli*: translation and stability of mRNAs *in vivo*. Gene 78(1): 59–72.

Schneider, H., A. Schwiertz, M. D. Collins and M. Blaut (1999). Anaerobic transformation of quercetin-3-glucoside by bacteria from the human intestinal tract. Arch. Microbiol. 171(2): 81–91.

Schwarz, M., J. J. Picazo-Bacete, P. Winterhalter and I. Hermosin-Gutierrez (2005). Effect of copigments and grape cultivar on the color of red wines fermented after the addition of copigments. J. Agr. Food. Chem. 53(21): 8372–8381.

Schwarz, M., T. C. Wabnitz and P. Winterhalter (2003). Pathway leading to the formation of anthocyanin-vinylphenol adducts and related pigments in red wines. J. Agr. Food Chem. 51(12): 3682–3687.

Schwarz, M. and P. Winterhalter (2003). A novel synthetic route to substituted pyranoanthocyanins with unique colour properties. Tetrahedron Lett. 44(41): 7583–7587.

Schwarz, M. and P. Winterhalter (2004). Novel aged anthocyanins from Pinotage wines: Isolation, characterization, and pathway of formation. In: A.L. Waterhouse, J.A. Kennedy (Ed.): *Red Wine Color: Revealing the Mysteries*, ACS Symp. Ser. 886, American Chemical Society: Washington, DC, pp. 179–197.

Serafini, M., G. Maiani and A. Ferro-Luzzi (1998). Alcohol-free red wine enhances plasma antioxidant capacity in humans. J. Nutr. 128(6): 1003–1007.

Setchell, K. D. R., S. P. Borriello, P. Hulme, D. N. Kirk and M. Axelson (1984). Nonsteroidal estrogens of dietary origin – possible roles in hormone-dependent disease. Am. J. Clin. Nutr. 40(3): 569–578.

Setchell, K. D. R., C. Clerici, E. D. Lephart, S. J. Cole, C. Heenan, D. Castellani, B. E. Wolfe, L. Nechemias-Zimmer, N. M. Brown, T. D. Lund, R. J. Handa and J. E. Heubi (2005). S-Equol, a potent ligand for estrogen receptor beta, is the exclusive enantiomeric form of the soy isoflavone metabolite produced by human intestinal bacterial floral. Am. J. Clin. Nutr. 81(5): 1072–1079.

Setchell, K. D. R., L. ZimmerNechemias, J. N. Cai and J. E. Heubi (1997). Exposure of infants to phyto-oestrogens from soy-based infant formula. Lancet 350(9070): 23–27.

Setchell, K. D. R., L. Zimmer-Nechemias, J. N. Cai and J. E. Heubi (1998). Isoflavone content of infant formulas and the metabolic fate of these phytoestrogens in early life. Am. J. Clin. Nutr. 68(6): 1453s–1461s.

Setiawan, V. W., Z. F. Zhang, G. P. Yu, Q. Y. Lu, Y. L. Li, M. L. Lu, M. R. Wang, C. H. Guo, S. Z. Yu, R. C. Kurtz and C. C. Hsieh (2001). Protective effect of green tea on the risks of chronic gastritis and stomach cancer. Int. J. Cancer 92(4): 600–604.

Simons, A. L., M. Renouf, S. Hendrich and P. A. Murphy (2005a). Human gut microbial degradation of flavonoids: structure-function relationships. J. Agr. Food Chem. 53(10): 4258–4263.

Simons, A. L., M. Renouf, S. Hendrich and P. A. Murphy (2005b). Metabolism of glycitein (7,4'-dihydroxy-6-methoxy-isoflavone) by human gut microflora. J. Agr. Food. Chem. 53(22): 8519–8525.

Smith, M. A. L. and L. A. Spomer (1995). Vessels, gels, liquid media, and support systems. *Automation and Environmental Control in Plant Tissue Culture*. In: J. Aitken-Christie, T. Kozai and M. A. L. Smith (Eds). Dordrecht, The Netherlands, Kluwer Academic Publishers: 371–404.

Song, T. T., K. Barua, G. Buseman and P. A. Murphy (1998). Soy isoflavone analysis: quality control and a new internal standard. Am. J. Clin. Nutr. 68(6): 1474s–1479s.

Song, T. T., S. Hendrich and P. A. Murphy (1999). Estrogenic activity of glycitein, a soy isoflavone. J. Agr. Food Chem. 47(4): 1607–1610.

Sroka, Z. and W. Cisowski (2003). Hydrogen peroxide scavenging, antioxidant and anti-radical activity of some phenolic acids. Food Chem. Toxicol. 41(6): 753–758.

Stashenko, H., C. Macku and T. Shibamato (1992). Monitoring volatile chemicals formed from must during yeast fermentation. J. Agr. Food Chem. 40(11): 2257–2259.

Stevens, J. F., A. W. Taylor and M. L. Deinzer (1999). Quantitative analysis of xanthohumol and related prenylflavonoids in hops and beer by liquid chromatography tandem mass spectrometry. J. Chromatogr. A 832(1–2): 97–107.

Stich, E., K. Kloos, P. Cortona and S. Hake (1999). Color me natural. Nutr. World 2: 64–70.

Stintzing, F. C. and R. Carle (2004). Functional properties of anthocyanins and betalains in plants, food, and in human nutrition. Trends Food Sci. Technol. 15(1): 19–38.

Stormo, G. D., T. D. Schneider and L. M. Gold (1982). Characterization of translational initiation sites in E. coli. Nucleic Acids Res. 10(9): 2971–2996.

Takamatsu, S., A. M. Galal, S. A. Ross, D. Ferreira, M. A. ElSohly, A. R. S. Ibrahim and F. S. El-Feraly (2003). Antioxidant effect of flavonoids on DCF production in HL-60 cells. Phytother. Res. 17(8): 963–966.

Takamura, Y. and G. Nomura (1988). Changes in the intracellular concentration of acetyl-CoA and malonyl-CoA in relation to the carbon and energy metabolism of Escherichia coli K12. J. Gen. Microbiol. 134(8): 2249–2253.

Talavera, S., C. Felgines, O. Texier, C. Besson, J. L. Lamaison and C. Remesy (2003). Anthocyanins are efficiently absorbed from the stomach in anesthetized rats. J. Nutr. 133(12): 4178–4182.

Talavera, S., C. Felgines, O. Texier, C. Besson, C. Manach, J. L. Lamaison and C. Remesy (2004). Anthocyanins are efficiently absorbed from the small intestine in rats. J. Nutr. 134(9): 2275–2279.

Tsai, P. J., Y. Y. Hsieh and T. C. Huang (2004). Effect of sugar on anthocyanin degradation and water mobility in a roselle anthocyanin model system using O-17 NMR. J. Agr. Food Chem. 52(10): 3097–3099.

Tsangalis, D., J. F. Ashton, A. E. J. McGill and N. P. Shah (2002). Enzymic transformation of isoflavone phytoestrogens in soymilk by beta-glucosidase-producing bifidobacteria. J. Food Sci. 67(8): 3104–3113.

Tsangalis, D., J. F. Ashton, A. E. J. Mcgill and N. P. Shah (2003). Biotransformation of isoflavones by bifidobacteria in fermented soymilk supplemented with D-glucose and L-cysteine. J. Food Sci. 68(2): 623–631.

Tsuda, T., F. Horio and T. Osawa (1999). Absorption and metabolism of cyanidin 3-O-β-D-glucoside in rats. FEBS Lett. 449(2–3): 179–182.

Turner, N. J., B. M. Thomson and I. C. Shaw (2003). Bioactive isoflavones in functional foods: the importance of gut microflora on bioavailability. Nutr. Rev. 61(6): 204–213.

Udupa, S. R., A. Banerji and M. S. Chadha (1969). Microbiological transformations of flavonoids – II Transformations of (±)-flavanone Tetrahedron 25(22): 5415–5419.

Ueno, T., S. Uchiyama and N. Kikuchi (2002). The role of intestinal bacteria on biological effects of soy isoflavones in humans. J. Nutr. 132(3): 594s.

Valero, E., L. Moyano, M. C. Millan, M. Medina and J. M. Ortega (2002). Higher alcohols and esters production by Saccharomyces cerevisiae. Influence of the initial oxygenation of the grape must. Food Chem. 78(1): 57–61.

Vinson, J. A. and Y. A. Dabbagh (1998). Effect of green and black tea supplementation on lipids, lipid oxidation and fibrinogen in the hamster: mechanisms for the epidemiological benefits of tea drinking. FEBS Lett. 433(1–2): 44–46.

Viswanathan, M., S. K. Kim, A. Berdichevsky and L. Guarente (2005). A role for SIR-2.1 regulation of ER stress response genes in determining C. elegans life span. Dev. Cell 9(5): 605–615.

Vivar-Quintana, A. M., C. Santos-Buelga, E. Francia-Aricha and J. C. Rivas-Gonzalo (1999). Formation of anthocyanin-derived pigments in experimental red wines. Food Sci. Technol. Int. 5(4): 347–352.

Wang, H. B., E. J. Race and A. J. Shrikhande (2003). Anthocyanin transformation in Cabernet Sauvignon wine during aging. J. Agr. Food Chem. 51(27): 7989–7994.

Wang, H. J. and P. A. Murphy (1996). Mass balance study of isoflavones during soybean processing. J. Agr. Food Chem. 44(8): 2377–2383.

Wang, J. and G. Mazza (2002). Effects of anthocyanins and other phenolic compounds on the production of tumor necrosis factor alpha in LPS/IFN-gamma-activated RAW 264.7 macrophages. J. Agr. Food Chem. 50(15): 4183–4189.

Wang, L. Q., M. R. Meselhy, Y. Li, N. Nakamura, B. S. Min, G. W. Qin and M. Hattori (2001). The heterocyclic ring fission and dehydroxylation of catechins and related compounds by *Eubacterium* sp strain SDG-2, a human intestinal bacterium. Chem. Pharm. Bull. 49(12): 1640–1643.

Wang, R. F., W. W. Cao and C. E. Cerniglia (1996). PCR detection and quantitation of predominant anaerobic bacteria in human and animal fecal samples. Appl. Environ. Microb. 62(4): 1242–1247.

Wang, X. L., H. G. Hur, J. H. Lee, K. T. Kim and S. I. Kim (2005a). Enantioselective synthesis of S-equol from dihydrodaidzein by a newly isolated anaerobic human intestinal bacterium. Appl. Environ. Microb. 71(1): 214–219.

Wang, X. L., K. H. Shin, H. G. Hur and S. I. Kim (2005b). Enhanced biosynthesis of dihydrodaidzein and dihydrogenistein by a newly isolated bovine rumen anaerobic bacterium. J. Biotechnol. 115(3): 261–269.

Watanabe, S. and H. Adlercreutz (1998). Pharmacokinetics of soy phytoestrogens in humans. *Functional Foods for Disease Prevention*. T. Shibamato, J. Terao and T. Osawa. Washington, DC, American Chemical Society: 198–208.

Watts, K. T., P. C. Lee and C. Schmidt-Dannert (2004). Exploring recombinant flavonoid biosynthesis in metabolically engineered *Escherichia coli*. Chembiochem 5(4): 500–507.

Watts, K. T., P. C. Lee and C. Schmidt-Dannert (2006). Biosynthesis of plant-specific stilbene polyketides in metabolically engineered *Escherichia coli*. BMC Biotechnol. 6: 22.

Way, T. D., M. C. Kao and J. K. Lin (2004). Apigenin induces apoptosis through proteasomal degradation of HER2/neu in HER2/neu-overexpressing breast cancer cells via the phosphatidylinositol 3-kinase/Akt-dependent pathway. J. Biol. Chem. 279(6): 4479–4489.

Weisel, T., M. Baum, G. Eisenbrand, H. Dietrich, F. Will, J. P. Stockis, S. Kulling, C. Rufer, C. Johannes and C. Janzowski (2006). An anthocyanin/polyphenolic-rich fruit juice reduces oxidative DNA damage and increases glutathione level in healthy probands. Biotech. J. 1(4): 388–397.

Wellmann, E. (1975). UV dose-dependent induction of enzymes related to flavonoid biosynthesis in cell-suspension cultures of parsley. FEBS Lett. 51(1): 105–107.

Whiting, G. C. and P. A. Coggins (1960). Organic acid metabolism in cider and berry fermentations. III. Ket-acids in cider-apple juices and ciders. J. Sci. Food Agri. 11: 705–709.

Wildenradt, H. L. and V. L. Singleton (1974). Production of aldehydes as a result of oxidation of polyphenolic compounds and its relation to wine aging. Am. J. Enol. Viticult. 25(2): 119–126.

Williams, P. A., J. Cosme, V. Sridhar, E. F. Johnson and D. E. McRee (2000a). Mammalian microsomal cytochrome P450 monooxygenase: structural adaptations for membrane binding and functional diversity. Mol. Cell 5(1): 121–131.

Williams, P. A., J. Cosme, V. Sridhar, E. F. Johnson and D. E. McRee (2000b). Microsomal cytochrome P450 2C5: comparison to microbial P450s and unique features. J. Inorg. Biochem. 81(3): 183–190.

Williams, P. A., J. Cosme, D. M. Vinkovic, A. Ward, H. C. Angove, P. J. Day, C. Vonrhein, I. J. Tickle and H. Jhoti (2004). Crystal structures of human cytochrome P450 3A4 bound to metyrapone and progesterone. Science 305(5684): 683–686.

Williams, P. A., J. Cosme, A. Ward, H. C. Angove, D. Matak Vinkovic and H. Jhoti (2003). Crystal structure of human cytochrome P450 2C9 with bound warfarin. Nature 424(6947): 464–468.

Williamson, G., D. Barron, K. Shimoi and J. Terao (2005). *In vitro* biological properties of flavonoid conjugates found in vivo. Free Radical Res. 39(5): 457–469.

Willits, M. G., M. Giovanni, R. T. N. Prata, C. M. Kramer, V. De Luca, J. C. Steffens and G. Graser (2004). Bio-fermentation of modified flavonoids: an example of *in vivo* diversification of secondary metabolites. Phytochemistry 65(1): 31–41.

Winter, J., L. H. Moore, V. R. Dowell and V. D. Bokkenheuser (1989). C-ring cleavage of flavonoids by human intestinal bacteria. Appl. Environ. Microb. 55(5): 1203–1208.

Wood, J. G., B. Rogina, S. Lavu, K. Howitz, S. L. Helfand, M. Tatar and D. Sinclair (2004). Sirtuin activators mimic caloric restriction and delay ageing in metazoans. Nature 430(7000): 686–689.

Wrolstad, R. E. (2000). *Natural Food Colorants*. New York, Marcel Dekker, Inc.

Xie, D. Y., S. B. Sharma and R. A. Dixon (2004). Anthocyanidin reductases from *Medicago truncatula* and *Arabidopsis thaliana*. Arch. Biochem. Biophys. 422(1): 91–102.

Xu, X., K. S. Harris, H. J. Wang, P. A. Murphy and S. Hendrich (1995). Bioavailability of soybean isoflavones depends upon gut microflora in women. J. Nutr. 125(9): 2307–2315.

Xu, X., H. J. Wang, P. A. Murphy, L. Cook and S. Hendrich (1994). Daidzein is a more bioavailable soymilk isoflavone than is genistein in adult women. J. Nutr. 124(6): 825–832.

Yaipakdee, P. and L. W. Robertson (2001). Enzymatic halogenation of flavanones and flavones. Phytochemistry 57(3): 341–347.

Yamazaki, S., K. Sato, K. Suhara, M. Sakaguchi, K. Mihara and T. Omura (1993). Importance of the proline-rich region following signal-anchor sequence in the formation of correct conformation of microsomal cytochrome P–450s. J. Biochem. 114(5): 652–657.

Yan, Y., J. Chemler, L. Huang, S. Martens and M. A. Koffas (2005a). Metabolic engineering of anthocyanin biosynthesis in *Escherichia coli*. Appl. Environ. Microbiol. 71(7): 3617–3623.

Yan, Y., A. Kohli and M. A. Koffas (2005b). Biosynthesis of natural flavanones in *Saccharomyces cerevisiae*. Appl. Environ. Microbiol. 71(9): 5610–5613.

Yen, G. C. and H. Y. Chen (1995). Antioxidant activity of various tea extracts in relation to their antimutagenicity. J. Agr. Food Chem. 43(1): 27–32.

Yoshino, K., Y. Hara, M. Sano and I. Tomita (1994). Antioxidative effects of black tea theaflavins and thearubigin on lipid-peroxidation of rat-liver homogenates induced by tert-butyl hydroperoxide. Biol. Pharm. Bull. 17(1): 146–149.

Youdim, K. A., J. McDonald, W. Kalt and J. A. Joseph (2002). Potential role of dietary flavonoids in reducing microvascular endothelium vulnerability to oxidative and inflammatory insults. J. Nutr. Biochem. 13(5): 282–288.

Yu, O., J. Shi, A. O. Hession, C. A. Maxwell, B. McGonigle and J. T. Odell (2003). Metabolic engineering to increase isoflavone biosynthesis in soybean seed. Phytochemistry 63(7): 753–763.

Zhang, Y. J., S. K. Vareed and M. G. Nair (2005). Human tumor cell growth inhibition by nontoxic anthocyanidins, the pigments in fruits and vegetables. Life Sci. 76(13): 1465–1472.

Zhong, J. J., T. Seki, S. Kinoshita and T. Yoshida (1991). Effect of light irradiation on anthocyanin production by suspended culture of *Perilla frutescens*. Biotechnol. Bioeng. 38(6): 653–658.

8

Biosynthesis and Manipulation of Flavonoids in Forage Legumes

Susanne Rasmussen

AgResearch, Grasslands, Tennent Drive, Palmerston North, NZ,
Susanne.rasmussen@agresearch.co.nz

Abstract. Legumes are second only to the Gramineae in their importance to the food and pastoral industry, providing a large proportion of dietary proteins and oils to humans and animals. Their ability to associate with symbiotic rhizobia and the resulting fixation of atmospheric nitrogen makes legumes invaluable for the provision of nitrogen to agricultural systems. Two classes of flavonoids are of prominent importance in legumes: proanthocyanidins for their beneficial effects on animals and the environment; and isoflavonoids for their functions in plant protection and as symbiotic signaling molecules. This chapter presents an overview on the function, biosynthesis and structure of flavonols, anthocyanidins, proanthocyanidins and isoflavonoids in forage and other legumes. Strategies employing genetic modification or conventional breeding techniques to manipulate specific flavonoids in these plants are presented and discussed.

8.1 Introduction

Grain and forage legumes are grown on up to 15% of the Earth's arable surface and they account for 27% of the world's primary crop production, with grain legumes like bean, pea, chickpea and lentil providing 33% of dietary protein nitrogen to humans (Graham and Vance 2003). Soybean and peanut provide more than 35% of vegetable oil, and they are also rich sources of dietary protein for the chicken and pork industries. Major forage legumes used as animal feed are alfalfa (lucerne; *Medicago sativa*), clovers (*Trifolium* spp.), birdsfoot trefoil (*Lotus corniculatus*), and lupine (*Lupinus* spp.) which are either grown in mixed swards with forage grasses for grazing or as monocultures for hay and silage production.

The most important characteristic of legumes for agriculture is their ability to associate with rhizobial soil bacteria. These bacteria can infect legume roots and form nodular structures that are able to fix molecular nitrogen, the major component of the atmosphere (79%). Symbiotic N-fixation contributes about 20 Tg N year^{-1} (Tg = 10^{12} g) to crop production, compared to 80 Tg N year^{-1} by inorganic fertilizers (Smil 1999); it has been estimated that globally up to 90 Tg N$_2$ year^{-1} could be fixed by

K. Gould et al. (eds.), *Anthocyanins*, DOI: 10.1007/978-0-387-77335-3_8,
© Springer Science+Business Media, LLC 2009

legumes (Graham and Vance 2000). However, actual rates of symbiotic N-fixation are constrained by several extrinsic factors like drought (Serraj 2003), low phosphate availability (Edwards et al. 2006), acidic soils and associated toxic Al and Mn concentrations (Liao et al. 2006), as well as pests and diseases. Furthermore, when grown in mixed swards with grasses for grazing, the legume fraction typically declines over time as a result of complex dynamics between the two species (Schwinning and Parsons 1996).

The biosynthesis of flavonoids has been studied extensively in a wide range of plants (for reviews and further literature see: Winkel-Shirley 2001; Marles et al. 2003; Broun 2005; Dixon et al. 2005; Koes et al. 2005) and will be discussed in detail in other chapters of this book. Here, the focus will be on aspects of flavonoid biosynthesis and its manipulation relevant to legumes used as forage plants.

8.1.1 Proanthocyanidins

Considerable research efforts are currently directed towards the understanding and manipulation of proanthocyanidin (PA, syn. Condensed tannin, CT) biosynthesis in plants used for human consumption, because of their potential beneficial effects on human health including anticancer activity (see Dixon et al. 2005). But there is also growing interest in PAs as constituents of forage plants as they are associated with several beneficial effects on grazing animals related to their ability to complex proteins in the rumen resulting in an increase of essential amino acid uptake by ruminants and reduced occurrence of bloat in cattle (Aerts et al. 1999; McMahon et al. 2000). Anthelmintic effects of PAs on gastrointestinal parasites (Marley et al. 2006) and a reduction of methane production by ruminants (Tavendale et al. 2005) have also been reported.

However, legumes like *M. sativa* and clover accumulate only very low levels of foliar PAs, insufficient to prevent bloat, and the use of alternative species like *L. corniculatus* and *Onobrychis viciifolia* (sainfoin) as forage legumes has been suggested, although these plants do not persist well under grazing (Ramírez-Restrepo and Barry 2005). Major research projects are currently underway to create 'bloat-safe' clover and *M. sativa* plants by genetic modification (Jones et al. 2006; Xie et al. 2006).

8.1.2 Isoflavonoids

A second important group of flavonoids in forage legumes are isoflavonoids. They form a large diverse group of plant secondary metabolites involved in various processes ranging from plant disease resistance to symbiotic signaling. Isoflavonoids, especially genistein and daidzein from soybean, have recently become a major focus of research because of their perceived benefits to human health (e.g. prevention of hormone-dependent breast and prostrate cancer and osteoporosis). Their phytoestrogenic activity in mammals results from close structural resemblance to 17-β-estradiol and the ability to bind to estrogen receptors. Detailed information and further literature regarding these functions can be found in recent reviews (Piersen 2003; Cornwell et al. 2004; Dixon 2004; Beck et al. 2005). In grazing ruminants, however, this phytoestrogenic activity causes reduced fertility and other reproductive disorders in sheep and cattle feeding on pastures rich in red (*T. pratense*) or subterranean (*T. subterraneum*) clover species that accumulate high levels of isoflavonoids in leaves (Adams 1990; Adams 1995).

Isoflavonoids are usually associated with the subfamilies Papilionoideae and Leguminosae, but they have also been isolated from a range of other plant species (Reynaud et al. 2005). Isoflavonoid phytoalexins play important roles in plant disease resistance (Dakora and Phillips 1996), e.g. the pterocarpan (–) - medicarpin in *M. sativa* (Paiva et al. 1991; Baldridge et al. 1998) and the isoflavans sativan and vestitol in *L. corniculatus* (Bonde et al. 1973). Attempts to engineer specific isoflavonoid phytoalexins into legumes are currently employed to protect plants from diseases (Deavours and Dixon 2005).

8.1.3 Flavonoids as Signaling Molecules

Specific flavonoids are important regulators of symbiotic interactions between legume roots and nodule forming N-fixating rhizobia (Geurts and Bisseling 2002; Cooper 2004). Isoliquiritigenin and genistein in soybean root exudates have been identified as inducers of *nod* genes (Kape et al. 1992) and as activators of a protein secretion system (Süß et al. 2006) in *Bradyrhizobium japonicum*; it was also shown that silencing of isoflavone synthase significantly reduces nodulation (Subramanian et al. 2006). *M. sativa* and clover roots release a range of flavonoid related compounds which induce *nod* genes in *Rhizobium meliloti* (Maxwell and Phillips 1990); a recent study demonstrated that down-regulation of flavonoids by silencing chalcone synthase in *M. sativa* inhibits root nodule formation and prevents auxin transport regulation (Wasson et al. 2006). High nitrogen availability is known to inhibit the establishment of rhizobia-legume symbiosis, a possible mechanism might be reduced biosynthesis and exudation of flavonoids from legume roots under high nitrogen conditions (Coronado et al. 1995). Isoflavonoids have also been described to play important roles for the symbiotic association with mycorrhizal fungi in *M. sativa* (Volpin et al. 1995) and white clover (*Trifolium repens*; Nair et al. 1991).

8.2 Biosynthesis of Flavonoids in Legumes

Common precursors of the branched flavonoid pathway leading to flavonols, anthocyanins, proanthocyanidins and isoflavonoids are *p*-coumaroyl CoA and three molecules of malonyl CoA. These are condensed to naringenin chalcone by the polyketide synthase (type III PKS) chalcone synthase (CHS) and isomerised to the (2-*S*)-stereoisomer of naringenin by chalcone isomerase (CHI) (Fig. 8.1; Jez and Noel 2002; Austin and Noel 2003), although other CoA derivatives of e.g. cinnamic, caffeic, and ferulic acid can also be converted by CHS (Christensen et al. 1998; Jez et al. 2002). CHS has been cloned from several legumes including *Phaseolus vulgaris* (Ryder et al. 1987), *Glycine max* (soybean; Wingender et al. 1989), *Pisum sativum* (Ito et al. 1997), *M. sativa* (McKhann and Hirsch 1994) and *T. subterraneum* (Arioli et al. 1994). In all Leguminosae examined so far CHS is encoded by a small multi

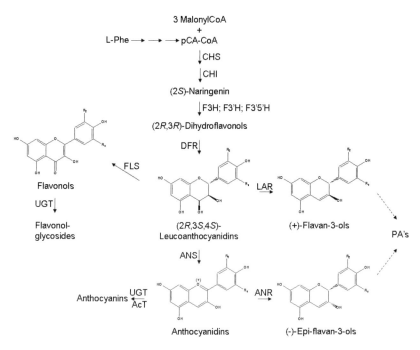

Fig. 8.1 Biosynthetic pathway to flavonols, anthocyanidins and proanthocyanidins

gene family of up to 8 members, but it is not known if the enzymes differ in substrate specificities. Interestingly, the subsequent enzyme CHI is encoded by two types of CHI genes in legumes (Shimada et al. 2003), the significance of this with respect to isoflavonoid biosynthesis will be discussed under 8.2.4.

(2S)-Naringenin is subsequently hydroxylated by flavanone 3β-hydroxylase (F3H), a 2-oxoglutarate-dependent dioxygenase, to form (2R, 3R)-dihydrokaempferol (DHK; Fig. 8.1, $R_1=R_2=H$). Genes encoding F3H have been cloned and characterized from *M. sativa* (Charrier et al. 1995) and *G. max* (Zabala and Vodkin 2005). DHK can be further hydroxylated by flavonoid 3′-hydroxylase (F3′H) and/ or flavonoid 3′,5′-hydroxylase (F3′5′H), both cytochrome P450-dependent monooxygenases, leading to dihydroquercetin (DHQ; $R_1=OH$, $R_2=H$) and dihydromyricetin (DHM; $R_1=R_2=OH$), respectively (Seitz et al. 2006). F3′H has been cloned from *G. max* (Toda et al. 2005), but no biochemical analysis of substrate specificities and enzyme activities has been performed. No F3′5′H has been cloned from legumes yet. Intermediates of the above steps usually do not accumulate, but are converted immediately to products as described below.

8.2.1 Biosynthesis of Flavonols in Legumes

At the level of dihydroflavonols the flavonoid pathway branches to the flavonols and anthocyanidins/proanthocyanidins, respectively. Flavonol synthase (FLS) catalyzes the conversion of dihydroflavonols to flavonols like kaempferol (K; Fig. 8.1, $R_1=R_2=H$), quercetin (Q; $R_1=OH$, $R_2=H$) and myricetin (M; $R_1=R_2=OH$). The enzyme belongs to the 2-oxoglutarate- and ferrous iron-dependent oxygenases oxidizing the C-ring of the flavonoid molecule (Turnbull et al. 2004). The resulting flavonols are immediately conjugated to a variety of sugars by UDPG-glycosyltransferases (UGT's), members of family 1 of the highly divergent, multigene family of glycosyltransferases (Paquette et al. 2003; Bowles et al. 2005).

UGTs often have low substrate specificity towards the aglycones (e.g. K, Q, M), but have strict substrate specificities for the sugar donors, such as UDP-glucose and UDP-galactose (Kubo et al. 2004). More than 300 glycosyltransferases are expressed in *M. truncatula* and several elicitor inducible UGTs have recently been cloned and functionally characterised, one of which seems to be involved *in vivo* in the glycosylation of triterpenes, but glycosylates isoflavonoids and the flavonol quercetin with higher efficiency *in vitro* (Achnine et al. 2005; Shao et al. 2005).

Major flavonols accumulating in *T. repens* leaves are Q- and K-3-*O*-galactosyl-(6″-*O*-xylosides) (Hofmann et al. 2000), which have been linked to abiotic stress tolerance including UV-B (Hofmann et al. 2001) and drought (Hofmann et al. 2003). Northern blot analysis of clover varieties with high levels of foliar flavonols and anthocyanins showed very high expression levels of CHS, F3H and F3′H, but not FLS compared to green leaved varieties with low flavonoid levels (Fig. 8.2A). Cold treatment resulted in the accumulation of foliar flavonoids and induced the expression of relevant biosynthetic genes as shown by microarray analysis (Fig. 8.2B; Rasmussen et al. 2006). This study also identified a potential candidate for a flavonoid-galactosyltranferase (Fig. 8.2C), which was also shown to be highly expressed in leaves of flavonoid accumulating clover genotypes (Fig. 8.2D).

8.2.2 Biosynthesis of Anthocyanins

Dihydroflavonol 4-reductase (DFR) catalyses the conversion of dihydroflavonols to leucoanthocyanidins, precursors for both anthocyanins and the monomeric units of proanthocyanidins. The reaction requires NADPH as co-factor and is stereospecific, resulting in (2*R*, 3*S*, 4*S*)-leucoanthocyanidins (Johnson et al. 2001; Martens et al. 2002). DFR has been cloned and characterised from a wide range of plants including legumes like *M. sativa* (Charrier et al. 1995), *M. truncatula* (Xie et al. 2004a), *L. corniculatus* (Paolocci et al. 2005) and *L. japonicus* (Shimada et al. 2005). The different members of the DFR gene family often differ in their substrate preferences and show specific expression patterns depending on tissue and developmental stages.

Anthocyanidin synthase (ANS), a 2-oxoglutarate-dependent oxygenase, has been postulated to catalyze the oxidation of the colourless leucoanthocyanidins to 2-flaven-3, 4-diols, which are spontaneously isomerized to 3-flaven-2, 3-diols (Nakajima et al. 2001). However, *in vitro* assays (Turnbull et al. 2000) and a mechanistic study of the oxidation reaction of ANS (Nakajima et al. 2006) showed that the oxidation

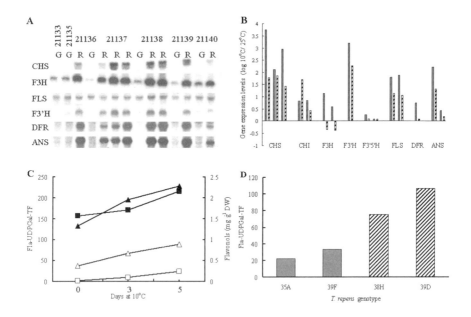

Fig. 8.2 Gene expression profiling in *T. repens* leaves. A - Northern blot analysis of flavonoid biosynthetic genes in leaves with low (G) and high (R) levels of flavonols and anthocyanins (F/A). B – Microarray analysis of cold treated (5 days, 10°C) leaves of genotypes with low (filled bars) and high (striped bars) F/A. C – Flavonol accumulation (■) and real time RT-PCR of a putative flavonoid-UDPgalactosyltransferase (▲) in cold-treated leaves with low (open symbols) and high (closed symbols) F/A. D – Real time RT-PCR of a putative flavonoid-UDPgalactosyltransferase in leaves with low (filled bars) and high (striped bars) F/A

of C-3 of leucocyanidin must be regarded as a 'side reaction' and that the preferential reaction of ANS is a C-4 oxidation leading to DHQ. Recently, it was shown that recombinant ANS from *Gerbera hybrida* incubated with (+)-catechin catalyzes the formation of mainly the 4, 4-dimer of oxidized (+)-catechin and only minor amounts of cyanidin and quercetin. This is the first report of an enzymatic dimerization of catechin monomers and might suggest a role for ANS beyond the formation of anthocyanins (Wellmann et al. 2006). Interestingly, if incubated in the presence of a UDP-glucose: flavonoid 3-*O*-glucosyltransferase from *Fragaria x ananassa*, mainly cyanidin 3-*O*-glucoside was formed with minor amounts of the dimerized catechin. *Gerbera* ANS did not convert (–)-epicatechin or any other catechin isomer. No legume recombinant ANS has been characterised so far and nothing is known about their preferential substrates and products.

Anthocyanidins are unstable and immediately conjugated to sugars and other molecules (Nakajima et al. 2001), followed by transport into the vacuole. No specific anthocyanidin UGTs or acyltransferases have been reported from legumes to date. Despite increasing knowledge about the important role of anthocyanins for plant protection against biotic and abiotic stresses (Gould 2004), very little is known about anthocyanins in legumes. Pelargonidin-, cyanidin- and delphinidin-3-O-galactosyl-(6″-xylosides) were identified by LC-ESI-MS in *T. repens* variants with red leaves and/ or flowers; cold treatment resulted in an increase of mainly C- and D-glycosides in the leaves, DFR and ANS were both highly expressed in red and cold-treated leaves (Fig. 8.2A,B; Rasmussen et al. 2006). Cold inducible anthocyanins ('blush') have been linked to the 'V'-locus of the clover genome and constitutive 'fleck' as well as whole leaf anthocyanins to the 'R'-locus (Quesenberry et al. 1991); mapping of these loci to identify underlying regulatory sequences is currently underway (A. Griffith, AgResearch, personal comm.).

8.2.3 Biosynthesis of Proanthocyanidins

Anthocyanidins are also precursors of 2, 3-*cis*-flavan-3-ols (e.g. (–)-epicatechin), which are important building blocks for PAs. These *cis*-flavan-3-ols are formed from anthocyanidins by anthocyanidin reductase (ANR), which has been cloned from *Arabidopsis thaliana* and *M. truncatula* (Xie et al. 2003, 2004b). In *A. thaliana* (–)-epicatechin is the exclusive PA monomer (Abrahams et al. 2002), but in many other species both (+)- and (–)-flavan-3-ols are polymerized to PAs.

Extracts of sainfoin and other legumes have been shown to catalyze the production of (+)-catechin from leucocyanidin in a NADPH-dependent reductase reaction (see e.g. Skadhauge et al. 1997), and several leucoanthocyanidin reductase encoding genes were recently cloned and characterized (Tanner et al. 2003; Pfeiffer et al. 2006). LAR from the PA-rich legume tree *Desmodium uncinatum* catalyzes the reduction of leucopelargonidin, leucocyanidin, and leucodelphinidin to afzelechin (Fig. 8.1; $R_1=R_2=H$), catechin ($R_1=OH$, $R_2=H$), and gallocatechin ($R_1=R_2=OH$), respectively. It is most closely related to the isoflavone reductase group of plant Reductase-Epimerase-Dehydrogenase (RED) proteins, but possesses an additional 65-amino acid C-terminal extension with unknown function (Tanner et al. 2003). No homologues of LAR have been found in *A. thaliana*, consistent with the exclusive presence of (–)-epicatechin derived CT building blocks. It is assumed that the monomers are transported into the vacuole and subsequently polymerized, but no enzymes or genes involved in this process have been identified so far. For an excellent discussion of possible stereospecific reaction mechanisms see Dixon et al. (2005).

Major forage legumes like clovers, *M. sativa* and lupine do not contain appreciable amounts of PAs in the leaves (Li et al. 1996), where they only accumulate in glandular trichomes (Fig. 8.3B). PAs isolated from *T. repens* flowers consist of mainly gallocatechin and epigallocatechin monomeric units (procyanidin (PC): prodelphinidin (PD) ratio = 1:99) (Foo et al. 2000), whereas floral PAs from *T. pratense* consist mainly of catechin and epicatechin monomeric units (PC: PD = 93:7) (Meagher et al. 2006). *L. pedunculatus* accumulates very high levels of foliar PAs (Fig. 8.3c) and, compared to *L. corniculatus,* has higher proportions of PD and a higher mean degree of polymerization (Sivakumaran et al. 2006).

Fig. 8.3 A- Leaves and flowers of common (*left*) and color variant (*right*) *T. repens*; B - DMACA staining of PA's in common (*left*) and color variant (*right*) *T. repens* flowers; C - DMACA staining of PA's in *T. repens* leaf trichomes; D - DMACA staining of PA's in *L. pedunculatus* (*top*) and *L. corniculatus* (*bottom*) leaves. See Plate 7 for colour version of these photographs

8.2.4 Biosynthesis of Isoflavonoids in Legumes

In addition to CHS forming 4,2',4',6'-tetrahydroxychalcone (chalcone) from *p*-coumaroyl-CoA, legumes possess an NADPH-dependent chalcone reductase (CHR; or better polyketide reductase, PKR) converting an intermediate of the multistep CHS reaction, probably a coumaroyl-trione, to 4,2',4'-trihydroxychalcone (6'-deoxychalcone (Fig. 8.4; Bomati et al. 2005). CHR has been cloned from several legumes including *G. max* (Welle et al. 1991), *M. sativa* (Ballance and Dixon 1995; Sallaud et al. 1995), and *Sesbania rostrata* (Goormachtig et al. 1999).

Deoxychalcones are further converted to 4',7-dihydroxyflavanone ((2*S*)-liquiritigenin) by a legume specific CHI (Dixon et al. 1988; Blyden et al. 1991; Wood and Davies 1994). Several CHIs are present in *L. japonicus* of which two encode the legume specific type II CHI and one the common type I CHI. Type I CHIs convert only 6'-hydroxychalcones to 5-hydroxyflavanone, but they do not use 6'-deoxychalcones as substrates (Shimada et al. 2003).

The 2-hydroxylation and aryl migration is catalysed by 2-hydroxyisoflavonoid synthase (IFS), a cytochrome P450 monooxygenase (Hashim et al. 1999). The 2-step reaction consists of hydroxylation of the flavanone molecule at C-2 and an

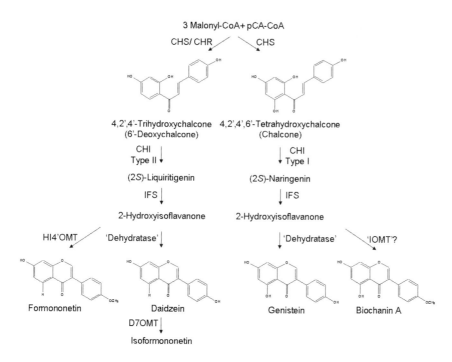

Fig. 8.4 Biosynthetic pathway to isoflavones

intramolecular 1,2-aryl migration from C-2 to C-3 to yield 2-hydroxyisoflavanone (Sawada et al. 2002). The enzyme from *G. max* expressed in insect cells converts liquiritigenin to daidzein and naringenin to genistein, but at a lower rate (Steele et al. 1999). Genes encoding IFS have been cloned from several legumes, including *Glycyrrhiza echinata* (licorice) (Akashi et al. 1999), *G. max*, mung bean, red clover (*Trifolium pratense*) and *M. sativa* (Jung et al. 2000). The immediate 2-hydroxyisoflavanone products are subsequently dehydrated, either spontaneously (Steele et al. 1999) or by a specific 2-hydroxyisoflavanone dehydratase (Akashi et al. 2005), introducing a double bond between C-2 and C-3.

Genistein and daidzein and their 7-*O*-glucosides and 7-*O*-glucosyl-6″-malonates are major isoflavones accumulating in *G. max*, and it has been proposed that the conjugates serve as pools for the release of aglycones, which may play important roles in defence reactions (Dakora and Phillips 1996). A β-glucosidase with high specific activity towards isoflavone conjugates was purified and characterized from white lupin (Piślewska et al. 2002) and *G. max* (Hsieh and Graham 2001), but genes encoding isoflavone specific β-glucosidases and malonylesterases have not been cloned so far.

8.2.5 Biosynthesis of *O*-methylated Isoflavonoids

Biochanin A and formononetin (7-hydroxy-4′-methoxyisoflavone) are 4′-*O*-methylated genistein and daidzein, respectively, with formononetin being the precursor for important phytoalexins such as medicarpin and vestitol. However, the biosynthesis of these 4′-*O*-methylated isoflavones remained elusive for some time. Elicitor induced accumulation of medicarpin in *M. sativa* cell cultures was accompanied by a strong induction of an isoflavone-*O*-methyltransferase (IOMT), but *in vitro* assays showed that it preferentially methylated daidzein at the 7-*O*-position of the A-ring, leading to isoformononetin (He et al. 1998). Subsequent studies showed that elicitation of *M. sativa* cell cultures overexpressing this IOMT led to a greater accumulation of formononetin and medicarpin compared to untransformed elicited cell cultures and that no isoformononetin was produced (He and Dixon 2000). Radiolabelling and isotope dilution experiments demonstrated that daidzein was not the substrate of IOMT *in vivo*. Based on these and confocal microscopy studies on IOMT-GFP fusion proteins, it was proposed that elicited IOMT colocalizes to the endoplasmic reticulum-associated IFS, where it methylates the labile 2,7,4′-trihydroxyisoflavone product of IFS (Liu and Dixon 2001). This hypothesis has been questioned, because two cDNAs for isoflavone-*O*-methyltransferases with differing substrate and regiospecificities were isolated from *G. echinata*, one encoding an *S*-adenosyl-L-methionine: daidzein 7-*O*-methyltransferase (D7OMT) producing isoformononetin and the other a SAM: 2,7,4′-trihydroxyisoflavanone 4′-*O*-methyltransferase (HI4′OMT) producing formononetin (Fig. 8.4; Akashi et al. 2003). It remains to be shown if this HI4′OMT is present in other legumes as well.

Formononetin and biochanin A and their 7-*O*-glucosides and 7-*O*-glucoside-6″-*O*-malonates are the major isoflavonoids accumulating in *T. pratense*, with minor amounts of genistein, daidzein and other isoflavone conjugates (Lin et al. 2000; Peng and Ye 2006). A UDP-glucose: formononetin 7-*O*-glucosyltransferase has been cloned from elicitor induced *G. echinata* suspension cells (Nagashima et al. 2004), the recombinant protein could also use daidzein as substrate, but not flavonols. Treatment of *T. pratense* roots with elicitors resulted in deconjugation and accumulation of free F, suggesting the induction of a *β*-glucosidase (Tebayashi et al. 2001), but no gene for an isoflavone specific glucosidase has been cloned from *T. pratense* so far. Formononetin in *T. pratense* leaves has been described as a feeding deterrent against adult clover root weevils (*Sitona lepidus*) (Gerard et al. 2005; Crush et al. 2006), but high formononetin levels in *T. repens* N-fixating nodules have also been reported to act as an attractant to *S. lepidus* neonatal larvae, which feed preferentially on active nodules. *S. lepidus* may have become tolerant to the toxic effects of formononetin, which now acts as a signaling molecule for nodule location, indicating coevolutionary processes between herbivores and their plant hosts (Johnson et al. 2005). Formononetin also plays an important role in root nodule organogenesis by accelerating auxin breakdown in *T. repens* nodules (Mathesius 2001).

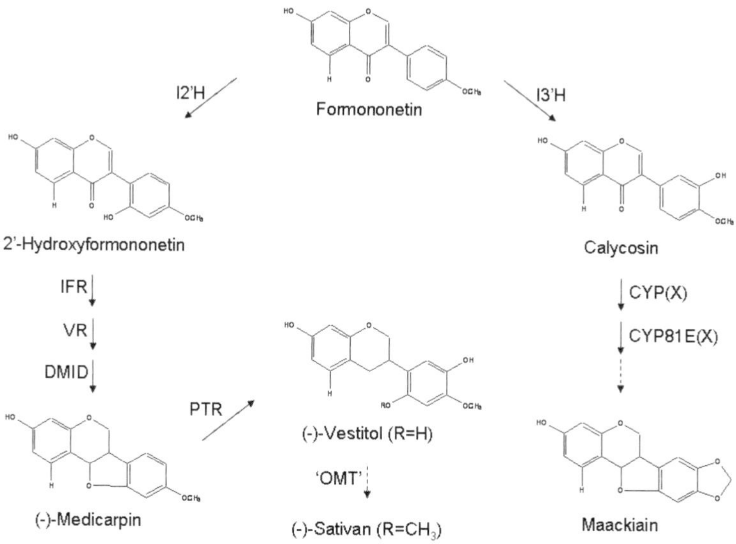

Fig. 8.5 Biosynthetic pathway to isoflavonoid phytoalexins

8.2.6 Biosynthesis of Phytoalexins and *Nod*-Inducers in Legumes

Major isoflavonoid phytoalexins in forage legumes are the pterocarpans medicarpin in *M. sativa* and maackiain in *T. pratense* and *T. subterraneum* (Fig. 8.5; Higgins 1972; Dewick and Ward 1978). Formononetin, the precursor of both pterocarpans, can be hydroxylated by an isoflavone 2′-hydroxylase (Akashi et al. 1998) to yield 2′-hydroxyformononetin, or by an isoflavone 3′-hydroxylase to yield calycosin; both enzymes belong to the cytochrome P450 81E subfamily (Liu et al. 2003). A third isoflavone hydroxylase (CYP81E(X)) isolated from *M. truncatula* belonging to the same subfamily catalyzes the 2′-hydroxylation of pseudobaptigenin, an intermediate in maackiain biosynthesis.

2′-Hydroxyformononetin is subsequently reduced by isoflavone reductase (IFR) to (3R)-vestitone. The enzyme is induced in *G. max* cell cultures (Fischer et al. 1990) and an elicitor inducible IFR has been cloned from *M. sativa* (Paiva et al. 1991). Vestitone is reduced by vestitone reductase (VR) to 7, 2′-dihydroxy-4′-methoxyisoflavonol, the direct precursor of (−)-medicarpin, an inducible VR has been cloned from *M. sativa* (Guo and Paiva 1995). The final step of (−)-medicarpin biosynthesis involves a 7, 2′-dihydroxy-4′-methoxyisoflavanol dehydratase (DMID), which seems to be associated with VR *in vivo* (Guo et al. 1994a, 1994b).

L. corniculatus and *L. japonicus* accumulate isoflavans (Fig. 8.5; vestitol and sativan) as phytoalexins (Bonde et al. 1973; Ingham 1977) and synthesis of isoflavans and pterocarpans appears to be closely related (Shimada et al. 2000). Recently pterocarpan reductases catalyzing the enantiospecific reductive cleavage of the dihydrofuran of (–)-medicarpin have been cloned and characterized *in vitro* from *L. japonicus* (Akashi et al. 2006).

4,4′-dihydroxy-2′-methoxychalcone in *M. sativa* root exudates has been described to act as a potent *nod* gene-inducer in *Sinorhizobium meliloti*. SAM: isoliquiritigenin 2′-O-methyltransferase (chalcone OMT) has been cloned and characterised from *M. sativa* (Maxwell et al. 1993). It is primarily expressed in root epidermal and cortical cells and to a lesser extent in root nodules. Interestingly, transcripts are strongly elicitor-induced in *M. sativa* cell cultures, but only low levels of methoxychalcone accumulate.

8.3 Manipulation of Flavonoids in Legumes

8.3.1 Manipulation of Proanthocyanidins in Legumes

Only few reports on the genetic modification of condensed tannins in forage legumes have been published. In general, attempts to alter PA levels in forage legumes either try to increase PAs to create 'bloat-safe' plants like *M. sativa* or clover or to decrease PAs to improve nutritional value and palatability in potential crop species like *Desmodium* and African browse species (Robbins and Morris 2000). It has been suggested that PA concentrations as low as 5 g/kg DM might be sufficient to prevent frothy bloat in cattle (Li et al. 1996). However, benefits relating to reduced N loss seem to require much higher concentrations of 20–40 g PA/kg DM (Aerts et al. 1999; Nguyen et al. 2005) and a relatively high degree of polymerization and prodelphinidin content (Porter and Woodruffe 1984; Horigome et al. 1988).

Transformation of *L. corniculatus* with parts of antisense *Antirrhinum majus* DFR resulted in reduced levels of PAs in 'hairy root' cultures (Carron et al. 1994), but this effect was genotype dependent and was not seen in lines derived from genotypes with high PA levels. PA monomeric composition also changed in some lines with reduced PA levels, resulting in an increase of PC: PD ratios. Regenerated plants (Robbins et al. 1998) from these transgenic hairy root cultures showed reduced levels of PAs in specific tissues, but again only in some lines. Two lines accumulated even higher levels of PAs in leaves and stems. Interestingly, the authors noted that crosses of some transgenic lines with untransformed *L. corniculatus* resulted in a strong underrepresentation of the transgene in the progeny. They also found that transgene-positive progeny did not show any significant reduction in mean tannin levels relative to control progeny.

Sense expression of *AmDFR* in *L. corniculatus* hairy root cultures resulted also in variable PA phenotypes with reduced, but also increased PA levels (Bavage et al. 1997). One line with increased PA levels accumulated increased levels of propelargonidin (PP) monomers, consistent with *A. majus* DFR substrate specificity for dihydrokaempferol. However, a subsequent study on plants regenerated from these hairy roots showed that only one line had an altered PC: PD ratio and none showed an increase in PP (Robbins et al. 2005). In this study the authors also tried to modify the hydroxylation pattern of PA monomeric units by introducing a F3′5′H gene from

Eustoma grandiflorum (Nielsen and Podivinsky 1997), which was expected to result in increased levels of prodelphinidin units. A few lines of root cultures had increased levels of PAs compared to control lines, but these lines had no altered monomeric composition.

Surprising results were achieved in *L. corniculatus* root cultures transformed with an antisense *Phaseolus vulgaris* stress-inducible CHS (Colliver et al. 1997). Glutathione-elicited root cultures of these transformants showed a significant increase in condensed tannin levels and a decrease in the phytoalexin vestitol. It was suggested that specific endogenous CHS genes involved in CT biosynthesis were up-regulated and CHS genes involved in isoflavonoid biosynthesis down-regulated by the heterologous antisense CHS construct.

Another strategy to alter PA/anthocyanin composition in *L. corniculatus* employed transformation with regulatory genes controlling the expression of anthocyanin/ PA pathway genes. *Sn*, encoding a maize (*Zea mays*) bHLH transcription factor (Tonelli et al. 1991), was introduced into *L. corniculatus* and plants were regenerated from hairy roots (Damiani et al. 1999; Robbins et al. 2003). Only a few transgenic lines derived from low CT genotypes had increased levels of PAs; the effects of *Sn* on anthocyanin accumulation were subtle and restricted to specific tissue regions. A reduction of PA's was observed in lines with multiple transgene integration events and probably due to suppression of endogenous homologues of *Sn*. A modification of PA composition was not reported. High light and the *Sn* transgene also acted synergistically in increasing PA levels in *L. corniculatus* leaves and both induced DFR and ANS transcript levels (Paolocci et al. 2005).

Transformation of *M. sativa* with the *Z. mays* bHLH transcription factor *Lc* (Ludwig et al. 1989) resulted in an increase of anthocyanin accumulation, but only under high light or at low temperatures (Ray et al. 2003). These conditions also enhanced CHS and F3H expression and LAR activity in *Lc* transformants compared to non-transformed plants and a slight increase in PAs in *M. sativa* leaves was reported, whereas the introduction of *Z. mays* transcription factors B-Peru (Radicella et al. 1991) and C1 (myb; Paz-Ares et al. 1990) did not result in any visible phenotype. Transformation of *T. repens* plants with several anthocyanin regulatory genes, both bHLH- and myb-transcription factors, resulted in various phenotypes with enhanced levels of anthocyanins in leaves, stolons and/or flowers, but effects on PAs were not reported (de Majnik et al. 2000).

A recent study (Xie et al. 2006), reporting the accumulation of PAs in normally PA-free tobacco (*Nicotiana tabacum*) tissues, involved the co-expression of PAP1 Myb transcription factor from *A. thaliana* (Borevitz et al. 2001) and ANR from *M. truncatula* (Xie et al. 2004). PAP1 transgene expression resulted in high levels of anthocyanin accumulation in most tissues, but no PA accumulation was observed. Crosses of *At*PAP1 with *Mt*ANR transformed plants accumulated less anthocyanins and showed increased levels of PA monomers (epicatechin and epigallocatechin) as well as a series of PA oligomers. *Mt*ANR was also transformed into *M. sativa*, resulting in reduced anthocyanins in the 'red spot' and the accumulation of low levels of PAs, presumably in cells of the 'red spot'. These results indicate that a high level of anthocyanidin biosynthesis is required for the accumulation of PAs using ANR. Interestingly the same group also reported on the metabolic engineering of PAs in *A.*

thaliana by co-expression of several transcription factors: TT2 – an R2R3 Myb transcription factor (Nesi et al. 2001), PAP1 and *Lc*. Those plants accumulated high levels of PAs in leaf tissue, but co-expression of all three genes resulted in early death of transgenic plants (Sharma and Dixon 2005).

The amounts of foliar PAs in both *M. sativa* and clover are very low, restricted to trichomes and show only very small natural variation, which makes it almost impossible to alter PA levels in these plants by conventional breeding methods. The creation of asymmetric somatic hybrid plants between *M. sativa* and *Onobrychis viciifolia* (high foliar PAs) was also not successful in introducing PAs into hybrid plants (Li et al. 1993). Breeding of *M. sativa* varieties with modified low initial rates of digestion resulted in the release of bloat-reducing cultivars, but this reduction in bloat incidences is unrelated to altered PA content (Coulman et al. 2000). However, breeding of *L. corniculatus* lines with increased PA levels of up to 14.5% DM and a modified morphology with increased persistence under grazing ('Creeping') was successful and new cultivars with these improved characteristics are currently developed (Sivakumaran et al. 2006).

8.3.2 Manipulation of Isoflavonoids in Legumes

Despite the ecological and agricultural importance of isoflavonoids, only few examples of genetic manipulation of isoflavones in legumes have been reported. *G. max* was transformed with IFS, but very few transgenic lines had significantly higher levels of seed isoflavones and some had reduced levels (Jung et al. 2003). A different strategy involved the activation of phenylpropanoid pathway genes by expression of *Z. mays* C1 (myb) and R (bHLH) transcription factors in *G. max* seeds (Yu et al. 2003). Reduced levels of genistein and increased levels of daidzein with an overall slight increase of isoflavone content were reported. A cosuppression of F3H to block the flavonol/anthocyanin branch in *G. max* seeds expressing *Z. mays* C1/R resulted in higher levels of isoflavones.

Constitutive expression of IFS from *M. truncatula* in *M. sativa* resulted in increased levels of genistein glucoside as well as the accumulation of biochanin A (4′-*O*-methylgenistein) and pratensein (3′-hydroxybiochanin A) glucosides (Deavours and Dixon 2005). *Mt*IFS was highly expressed in all organs, but genistein accumulated exclusively in the leaves. UV-B or infection with *Phoma medicaginis* induced the synthesis of additional isoflavonoids like formononetin and daidzein. Enhanced accumulation of medicarpin in *Mt*IFS expressing plants infected with *P. medicaginis* was also reported. Antisense suppression of *G. max* CHR resulted in reduced levels of daidzein- and glycetein-conjugates in *G. max* seeds, but results were variable and only slightly different compared to isoflavonoid variations in wild type plants (Maxwell et al. 2005).

Attempts to engineer isoflavonoids in non-leguminous plants try to confer the beneficial health effects of isoflavonoids into plants more common to the human diet in Western parts of the world. However, these attempts haven't been very successful so far in model plants like *A. thaliana*, *N. tabacum*, *Oryza sativa*, or *Zea mays*, usually only low levels of the glycosylated isoflavonoid genistein accumulated in transgenic plants transformed with IFS (Jung et al. 2000; Yu et al. 2000; Sreevidya et al.

2006). Overexpression of a *Pueraria montana* CHR resulted in a decrease of anthocyanins in *N. tabacum* flowers and the production of liquiritigenin, the precursor for daidzein (Joung et al. 2003).

Expression of *G. max* IFS in *A. thaliana tt6/tt3* double mutants (blocked in flavonol synthesis) resulted in considerably higher levels of glycosylated genistein, indicating strong competition between the endogenous flavonol synthesizing enzymes and introduced IFS for their common flavanone substrate (Liu et al. 2002). Recently, the construction of a bifunctional *G. max* IFS/CHI isomerase enzyme targeted to the endoplasmic reticulum membrane was reported to increase genistein and genistein glycoside production compared to *N. tabacum* plants transformed with a single *G. max* IFS construct (Tian and Dixon 2006). Co-expression of a *T. pratense* IFS and a rice P450 reductase in yeast resulted also in an increase of genistein production compared to yeast transformed with IFS alone (Dae et al. 2005).

Due to a high degree of natural variation of isoflavonoid levels in *T. pratense*, conventional breeding to alter isoflavonoid, especially formononetin, levels was very successful. Recently, cultivars with increased (Grasslands HF1; Rumball et al. 2005) and decreased formononetin (Grasslands G27; Rumball et al. 1997) were released. Grasslands G27 is supposed to increase ewe ovulation, conception and lambing; it also showed increased palatability. However, reductions in formononetin also lead to reduced disease tolerance of this cultivar.

T. repens shows only little natural variation in isoflavonoid content and attempts to increase isoflavones in *T. repens* were not successful so far. New routes for the introduction and modification of specific traits and metabolites usually not present in *T. repens* have been outlined in a recent paper (Williams et al. 2006). A phylogenetic analysis of approx. 200 *Trifolium* species using DNA sequencing revealed a 'white clover complex' comprising of species like *T. ambiguum, T. montanum, T. uniflorum, T. nigrescens* and others (Ellison et al. 2006). Based on this analysis a step-wise breeding strategy was developed to create interspecific hybrids and eight new fertile hybrid combinations were reported. A preliminary analysis of these plants indicates that relatively high levels of isoflavonoids accumulate in some of the hybrids.

Genetical metabolomics (metabolite profiling combined with quantitative trait locus analysis) is currently developed as a promising new tool useful for breeding plants with altered metabolic traits (Morreel et al. 2006). This approach might help to identify QTLs involved in flux control and to overcome current limitations in breeding.

8.4 Conclusions

Although some progress has been made in engineering the biosynthesis of specific flavonoids in plants, much more work needs to be done to create plants that accumulate these secondary metabolites in a consistent, reliable and specific manner. Major constraints faced by metabolic engineers are a lack of knowledge of all the genes and proteins involved in a particular biosynthetic pathway (e.g. final steps in PA biosynthesis, Dixon et al. 2005) and their co-localization (e.g. channeling of metabolites in metabolons; Winkel, 2004; Jørgensen et al. 2005; Kutchan 2005); lack of insights into cross-talks between metabolic networks and developmental regulation (e.g.

partial overlap of regulation of PA biosynthesis and seed and trichome development; Debeaujon et al. 2000; Winkel-Shirley 2001); and generally a lack of information about complex interaction dynamics between plants and their environment relating to secondary metabolite production (Hamilton et al. 2001). Currently, large research programmes employing 'omics' technologies (transcriptomics, proteomics and metabolomics) are dedicated to identify metabolic and regulatory networks to overcome some of these constraints (Fridman and Pichersky 2005).

However, considering that the production of plant secondary metabolites is associated with 'costs' to the plant (energy, C and N) and, in the absence of environmental threats (herbivores, pathogens, UV-B irradiation, drought), often results in reduced fitness (slower growth, reduced reproduction) compared to plants not synthesizing these metabolites (Mauricio et al. 1997; Karban et al. 1997; Karban et al. 1999), a careful analysis of costs/benefits to plants with increased production of secondary metabolites is necessary, if these plants are to be grown in an agricultural context. This need for multi-disciplinary research integrating 'omics' technologies, physiology, ecology and agronomy with a strong modeling component is increasingly recognized by the scientific community to overcome problems in manipulating complex traits (Wollenweber et al. 2005).

References

Abrahams, S., Tanner, G.J., Larkin, P.J. and Ashton, A.R. (2002) Identification and biochemical characterization of mutants in the proanthocyanidin pathway in *Arabidopsis*. Plant Physiol. 130, 561–576.

Achnine, L., Huhman, D.V., Farag, M.A., Sumner, L.W., Blount, J.W. and Dixon, R.A. (2005) Genomics-based selection and functional characterization of triterpene glycosyltransferases from the model legume *Medicago truncatula*. Plant J. 41, 875–887.

Adams, N.R. (1990) Permanent infertility in ewes exposed to plant oestrogens. Aust. Vet. J. 67, 197–201.

Adams, N.R. (1995) Detection of the effects of phytoestrogens on sheep and cattle. J. Animal. Sci. 73, 1509–1515.

Aerts, R.J., Barry, T.N. and McNabb, W.C. (1999) Polyphenols and agriculture: beneficial effects of proanthocyanidins in forages. Agricult. Ecosyst. Environm. 75, 1–12.

Akashi, T., Aoki, T. and Ayabe, S. (1998) CYP81E1, a cytochrome P450 cDNA of licorice (*Glycyrrhiza echinata* L.) encodes isoflavone 2′-hydroxylase. Biochem. Biophys. Res. Commun. 251, 67–70.

Akashi, T., Aoki, T. and Ayabe, S. (1999) Cloning and functional expression of a cytochrome P450 cDNA encoding 2-hydroxyisoflavanone synthase involved in biosynthesis of the isoflavonoid skeleton in licorice. Plant Physiol. 121, 821–828.

Akashi, T., Sawada, Y., Shimada, N., Sakurai, N., Aoki, T. and Ayabe, S. (2003) cDNA cloning and biochemical characterization of S-adenosyl-L-methionine: 2, 7, 4′-trihydroxyisoflavanone 4′-O-methyltransferase, a critical enzyme of the legume isoflavonoid phytoalexin pathway. Plant Cell Physiol. 44, 103–112.

Akashi, T., Aoki, T. and Ayabe, S. (2005) Molecular and biochemical characterization of 2-hydroxyisoflavanone dehydratase. Involvement of carboxylesterase-like proteins in leguminous isoflavone biosynthesis. Plant Physiol. 137, 882–891.

Akashi, T., Koshimizu, S., Aoki, T. and Ayabe, S. (2006) Identification of cDNAs encoding pterocarpan reductase involved in isoflavan phytoalexin biosynthesis in *Lotus japonicus* by EST mining. FEBS Lett. 580, 5666–5670.

Arioli, T., Howles, P.A., Weinman, J.J. and Rolfe, B.G. (1994) In *Trifolium subterraneum*, chalcone synthase is encoded by a multigene family. Gene. 138, 79–86.

Austin, M.B. and Noel, J.P. (2003) The chalcone synthase superfamily of type III polyketide synthases. Nat. Prod. Rep. 20, 79–110.

Baldridge, G.D., O'Neill, N.R., Samac, D.A. (1998) Alfalfa (*Medicago sativa L.*) resistance to the root lesion nematode, *Pratylenchus pendrans*: defense response gene mRNA and isoFlavonoid phytoaloxin levels in roots. Plant Mol. Biol. 38, 999–1010.

Ballance, G.M. and Dixon, R.A. (1995) *Medicago sativa* cDNAs encoding chalcone reductase. Plant Physiol. 107, 1027–1028.

Bavage, A., Davies, I.G., Robbins, M.P. and Morris, P. (1997) Expression of an *Antirrhinum* dihydroflavonol reductase gene results in changes in condensed tannin structure and accumulation in root cultures of *Lotus corniculatus* (bird's foot trefoil). Plant Mol. Biol. 35, 443–458.

Beck, V., Rohr, U. and Jungbauer, A. (2005) Phytoestrogens derived from red clover: an alternative to estrogen replacement therapy? J. Steroid Biochem. Mol. Biol. 94, 499–518.

Blyden, E.R., Doerner, P.W., Lamb, C.J. and Dixon, R.A. (1991) Sequence analysis of a chalcone isomerase cDNA of *Phaseolus vulgaris* L. Plant Mol. Biol. 16, 167–169.

Bomati, E.K., Austin, M.B., Bowman, M.E., Dixon, R.A. and Noel, J.P. (2005) Structural elucidation of chalcone reductase and implications for deoxychalcone biosynthesis. J. Biol. Chem. 280, 30496–30503.

Bonde, M.R., Millar, R.L. and Ingham, J.L. (1973) Induction and identification of sativan and vestitol as two phytoalexins from *Lotus corniculatus*. Phytochemistry 12, 2957–2959.

Borevitz, J., Xia, Y., Blount, J.W., Dixon, R.A. and Lamb, C. (2001) Activation tagging identifies a conserved MYB regulator of phenylpropanoid biosynthesis. Plant Cell 12, 2383–2393.

Bowles, D., Isayenkova, J., Lim, E.-K. and Poppenberger, B. (2005) Glycosyltransferases: managers of small molecules. Curr. Opin. Plant Biol. 8, 254–263.

Broun, P. (2005) Transcriptional control of flavonoid biosynthesis: a complex network of conserved regulators involved in multiple aspects of differentiation in Arabidopsis. Curr. Opin. Plant Biol. 8, 272–279.

Carron, T.R., Robbins, M.P. and Morris, P. (1994) Genetic modification of condensed tannin biosynthesis in *Lotus corniculatus*. 1. Heterologous antisense dihydroflavonol reductase down-regulates tannin accumulation in 'hairy root' cultures. Theor. Appl. Genet. 87, 1006–1015.

Charrier, B., Coronado, C, Kondorosi, A. and Ratet, P. (1995) Molecular characterization and expression of alfalfa (*Medicago sativa* L.) flavanone 3-hydroxylase and dihydroflavonol-4-reductase encoding genes. Plant Mol. Biol. 29, 773–786.

Christensen, A.B., Gregersen, P.L., Schröder, J. and Collinge, D.B. (1998) A chalcone synthase with an unusual substrate preference is expressed in barley leaves in response to UV light and pathogen attack. Plant Mol. Biol. 37, 849–857.

Colliver, S.P., Morris, P. and Robbins, M.P. (1997) Differential modification of flavonoid and isoflavonoid biosynthesis with an antisense chalcone synthase construct in transgenic *Lotus corniculatus*. Plant Mol. Biol. 35, 509–522.

Cooper, J.E. (2004) Multiple responses of rhizobia to flavonoids during legume root infection. Adv. Bot. Res. 41, 1–62.

Cornwell, T., Cohick, W. and Raskin, I. (2004) Dietary phytoestrogens and health. Phytochemistry 65, 995–1016.

Coronado, C., Zuanazzi, J.A.S., Sallaud, C., Quirion, J., Esnault, R., Husson, H., Kondorosi, A. and Ratet, P. (1995) Alfalfa root flavonoid production is nitrogen regulated. Plant Physiol. 108, 533–542.

Coulman, B., Goplen, B., Majak, W., McAllister, T., Cheng, K.-J., Berg, B., Hall, J., McCartney, D. and Acharya, S. (2000) A review of the development of a bloat-reduced alfalfa cultivar. Can. J. Plant Sci. 80, 487–491.

Crush, J.R., Gerard, P.J. and Rasmussen, S. (2006) Formononetin in clovers as a feeding deterrent against clover root weevil. . In: C.F. Mercer (Ed.) *Breeding for Success: Diversity in Action. Proceedings of the 13th Australasian Plant Breeding Conference*, Christchurch, New Zealand, pp. 1066–1072.

Dae, K., Bong, K., Hyo, L., Yoongho, L., Hor, H. and Joong-Hoon, A. (2005) Enhancement of isoflavone synthase activity by co-expression of P450 reductase from rice. Biotechnol. Lett. 27, 1291–1294.

Dakora, F.D. and Phillips, D.A. (1996) Diverse functions of isoflavonoids in legumes transcend anti-microbial definitions of phytoalexins. Physiol. Mol. Plant Pathol. 49, 1–20.

Damiani, F., Paolocci, F., Cluster, P.D., Arcioni, S., Tanner, G.J., Joseph, R.G., Li, Y.G., de Majnik, J. and Larkin, P.J. (1999) The maize transcription factor Sn alters proanthocyanidin synthesis in transgenic *Lotus corniculatus* plants. Aust. J. Plant Physiol. 26, 159–169.

Deavours, B.E. and Dixon, R.A. (2005) Metabolic engineering of isoflavonoid biosynthesis in alfalfa. Plant Physiol. 138, 2245–2259.

Debeaujon, I., Leon-Kloosterziel, K.M. and Koornneef, M. (2000) Influence of the testa on seed dormancy, germination and longevity in *Arabidopsis*. Plant Physiol. 122, 403–413.

Dewick, P.M. and Ward, D. (1978) Isoflavone precursors of the pterocarpans phytoalexin maackiain in *Trifolium pratense*. Phytochemistry 17, 1751–1754.

Dixon, R.A., Blyden, E.R., Robbins, M.P., Van Tunen, A.J. and Mol, J.N. (1988) Comparative biochemistry of chalcone isomerases. Phytochemistry 27, 2801–2808.

Dixon, R.A. (2004) Phytoestrogens. Annu. Rev. Plant Biol. 55, 225–261.

Dixon, R.A., Xie, D.-Y. and Sharma, S.B. (2005) Proanthocyanidins – a final frontier in flavonoid research? New Phytol. 165, 9–28.

Edwards, E.J., McCaffery, S. and Evans, J.R. (2006) Phosphorus availability and elevated CO_2 affect biological nitrogen fixation and nutrient fluxes in a clover-dominated sward. New Phytol. 169, 157–167.

Ellison, N.W., Liston, A., Steiner, J.J., Williams, W.M. and Taylor, N.L. (2006) Molecular phylogenetics of the clover genus (*Trifolium* – Leguminosae). Mol. Phylogenet. Evol. 39, 688–705.

Fischer, D., Ebenau-Jehle, C. and Grisebach, H. (1990) Phytoalexin synthesis in soybean: purification and characterization of NADPH: 2′-hydroxydaidzein oxidoreductase from elicitor-challenged soybean cell cultures. Arch. Biochem. Biophys. 276, 390–395.

Foo, L.Y., Lu, Y., Molan, A.L., Woodfield, D.R. and McNabb, W.C. (2000) The phenols and prodelphinidins of white clover flowers. Phytochemistry 54, 539–548.

Fridman, E. and Pichersky, E. (2005) Metabolomics, genomics, proteomics, and the identification of enzymes and their substrates and products. Curr. Opin. Plant Biol. 8, 242–248.

Gerard, P.J., Crush, J.R. and Hackell, D.I. (2005) Interaction between *Sitona lepidus* and red clover lines selected for formononetin content. Ann. Appl. Biol. 147, 173–181.

Geurts, R. and Bisseling, T. (2002) *Rhizobium* nod factor perception and signaling. Plant Cell 14, S239–S249.

Goormachtig, S., Lievens, S., Herman, S., van Montagu, M. and Holsters, M. (1999) Chalcone reductase-homologous transcripts accumulate during development of stem-borne nodules on the tropical legume *Sesbania rostrata*. Planta 209, 45–52.

Gould, K.S. (2004) Nature's Swiss army knife: the diverse protective roles of anthocyanins in leaves. J. Biomed. Biotechnol. 5, 314–320.

Graham, P.H. and Vance, C.P. (2000) Nitrogen fixation in perspective: an overview of research and extension needs. Field Crops Res. 65, 93–106.

Graham, P.H. and Vance C.P. (2003) Legumes: importance and constraints to greater use. Plant Physiol. 131, 872–877.

Guo, L., Dixon, R.A. and Paiva, N.L. (1994a) The 'pterocarpan synthase' of alfalfa: association and co-induction of vestitone reductase and 7,2′-dihydroxy-4′-methoxy-isoflavonol (DMI) dehydratase, the two final enzymes in medicarpin biosynthesis. FEBS Lett. 356, 221–225.

Guo, L., Dixon, R.A. and Paiva, N.L. (1994b) Conversion of vestitone to medicarpin in alfalfa (*Medicago sativa* L.) is catalyzed by two independent enzymes. J. Biol. Chem. 269, 22372–22378.

Guo, L. and Paiva, N.L. (1995) Molecular cloning and expression of alfalfa (*Medicago sativa* L.) vestitone reductase, the penultimate enzyme in medicarpin biosynthesis. Arch. Biochem. Biophys. 320, 353–360.

Hamilton, J.G., Zangerl, A.R., DeLucia, E.H. and Berenbaum, M.R. (2001) The carbon-nutrient balance hypothesis: its rise and fall. Ecol. Lett. 4, 86–95.

Hashim, M.F., Hakamatsuka, T., Ebizuka, Y. and Sankawa, U. (1999) Reaction mechanism of oxidative rearrangement of flavanone in isoflavone biosynthesis. FEBS Lett. 271, 219–222.

He, X., Reddy, J.T. and Dixon, R.A. (1998) Stress responses in alfalfa (*Medicago sativa* L.). XXII. cDNA cloning and characterization of an elicitor-inducible isoflavone 7-*O*-methyltransferase. Plant Mol. Biol. 36, 43–54.

He, X. and Dixon, R.A. (2000) Genetic manipulation of isoflavone 7-*O*-methyltransferase enhances biosynthesis of 4′-*O*-methylated isoflavonoid phytoalexins and disease resistance in alfalfa. Plant Cell 12, 1689–1702.

Higgins, V.J. (1972) Role of the phytoalexin medicarpin in three leaf spot diseases of alfalfa. Physiol. Plant Pathol. 2, 289–300.

Hofmann, R.W., Swinny, E.E., Bloor, S.J., Markham, K.R., Ryan, K.G., Campbell, B.D., Jordan, B.R. and Fountain, D.W. (2000) Responses of nine *Trifolium repens* L. populations to ultraviolet-B radiation: Differential flavonol glycoside accumulation and biomass production. Ann. Bot. 86, 527–537.

Hofmann, R.W., Campbell, B.D., Fountain, D.W., Jordan, B.R., Greer, D.H., Hunt, D.Y. and Hunt, C.L. (2001) Multivariate analysis of intraspecific responses to UV-B radiation in white clover (*Trifolium repens* L.). Plant, Cell Environ. 24, 917–927.

Hofmann, R.W., Campbell, B.D., Bloor, S.J., Swinny, E.E., Markham, K.R., Ryan, K.G. and Fountain, D.W. (2003) Responses to UV-B radiation in *Trifolium repens* L. – physiological links to plant productivity and water availability. Plant, Cell Environ. 26, 603–612.

Horigome, T., Kumar, R. and Okamoto, K. (1988) Effects of condensed tannins prepared from leaves of fodder plants on digestive enzymes *in vitro* and in the intestine of rats. Br. J. Nutr. 60, 275–285.

Hsieh, M.-C. and Graham, T.L. (2001) Partial purification and characterization of a soybean β-glucosidase with high specific activity towards isoflavone conjugates. Phytochemistry 58, 995–1005.

Ingham, J.L. (1977) Isoflavan phytoalexins from *Anthyllis*, *Lotus* and *Tetragonolobus*. Phytochemistry 16, 1279–1282.

Ito, M., Ichinose, Y., Kato, H., Shiraishi, T. and Yamada, T. (1997) Molecular evolution and functional relevance of the chalcone synthase genes of pea. Mol. Gen. Genet. 255, 28–37.

Jez, J.M., Bowman, M.E. and Noel, J.P. (2002) Expanding the biosynthetic repertoire of plant type III polyketide synthases by altering starter molecule specificity. Proc. Natl. Acad. Sci. U.S.A. 99, 5319–5324.

Jez, J.T. and Noel J.P. (2002) Reaction mechanism of chalcone isomerase – pH dependence, diffusion control, and product binding differences. J. Biol. Chem. 277, 1361–1369.

Johnson, E.T., Ryu, S., Yi, H., Shin, B., Cheong, H. and Choi, G. (2001) Alteration of a single amino acid changes the substrate specificity of dihydroflavonol 4-reductase. Plant J. 25, 325–333.

Johnson, S.N., Gregory, P.J., Greenham, J.R., Zhang, X. and Murray, P.J. (2005). Attractive properties of an isoflavonoid found in white clover root nodules on the clover root weevil. J. Chem. Ecol. 31, 2223–2229.

Jones, C.S., Williams, W.M., Hancock, K.R., Ellison, N.E., Scott, A.G., Collette, V.E., Jahufer, M.Z.Z., Richardson, K.A., Hay, M.J.M., Rasmussen, S., Jones, C.G. and Griffiths, A.G. (2006) Pastoral genomics – a foray into the clover genome. In: C.F. Mercer (Ed.) *Breeding for Success: Diversity in Action. Proceedings of the 13th Australasian Plant Breeding Conference*, Christchurch, New Zealand, XXX.

Jørgensen, K., Rasmussen, A.V., Morant, M., Nielsen, A.H., Bjarnholt, N., Zagrobelny, M., Bak, S. and Møller, B.L. (2005) Metabolon formation and metabolic channeling in the biosynthesis of plant natural products. Curr. Opin. Plant Biol. 8, 280–291.

Joung, J., Kasthuri, G.M., Park, J., Kang, W., Kim, H., Yoon, B., Joung, H. and Jeon, J. (2003) An Overexpression of chalcone reductase of *Pueraria montana* var. *lobata* alters biosynthesis of anthocyanin and 5′-deoxyflavonoids in transgenic tobacco. Biochem. Biophys. Res. Commun. 303, 326–331.

Jung, W., Yu, O., Lau, S.C., O'Keefe, D.P., Odell, J., Fader, G. and McGonigle, B. (2000) Identification and expression of isoflavone synthase, the key enzyme for biosynthesis of isoflavones in legumes. Nature Biotech. 18, 208–212.

Jung, W., Chung, I.-M. and Heo, H.-Y. (2003) Manipulating isoflavone levels in plants. Plant Biotechnol. J. 5, 149–155.

Kape, R., Parniske, M., Brandt, S. and Werner, D. (1992) Isoliquiritigenin, a strong nod gene- and glyceollin resistance-inducing flavonoid from soybean root exudate. Appl. Environm. Microbiol. 58, 1705–1710.

Karban, R., Agrawal, A.A. and Mangel, M. (1997) The benefits of induced defenses against herbivores. Ecol. 78, 1351–1355.

Karban, R., Agrawal, A.A., Thaler, J.S. and Adler, L.S. (1999) Induced plant responses and information content about risk of herbivory. Trends Ecol. Evol. 14, 443–447.

Koes, R., Verweij, W. and Quattrocchio, F. (2005) Flavonoids: a colorful model for the regulation and evolution of biochemical pathways. Trends in Plant Sci. 10, 236–242.

Kubo, A., Arai, Y., Nagashima, S. and Yoshikawa, T. (2004) Alteration of sugar donor specificities of plant glycosyltransferases by a single point mutation. Arch. Biochem. Biophys. 429, 198–203.

Kutchan, T.M. (2005) A role for intra- and intercellular translocation in natural product biosynthesis. Curr. Opin. Plant Biol. 8, 292–300.

Li, Y.G., Tanner, G. and Larkin, P. (1996) The DMACA-HCl. Protocol and the threshold proanthocyanidin content for bloat safety in forage legumes. J. Sci. Food Agric. 70, 89–101.

Li, Y.-G., Tanner, G.J., Delves, A.C. and Larkin, P.J. (1993) Asymmetric somatic hybrid plants between *Medicago sativa* L. (alfalfa, lucerne) and *Onobrychis viciifolia* Scop. (sanfoin). Theor. Appl. Genet. 87, 455–463.

Liao, H., Wan, H., Shaff, J., Wang, X., Yan, X. and Kochian, L.V. (2006) Phosphorus and aluminum interactions in soybean in relation to aluminum tolerance. Exudation of specific organic acids from different regions of the intact root system. Plant Physiol. 141, 674–684.

Lin, L.-Z., He, X.-G., Lindenmaier, M., Yang, J., Cleary, M., Qiu, S.-X. and Cordell, G.A. (2000) LC-ESI-MS study of the flavonoid glycoside malonates of red clover (*Trifolium pratense*). J. Agric. Food Chem. 48, 354–365.

Liu, C. and Dixon, R.A. (2001) Elicitor-induced association of isoflavone O-methyltransferase with endomembranes prevents the formation and 7-O-methylation of daidzein during isoflavonoid phytoalexin biosynthesis. Plant Cell 13, 2643–2658.

Liu, C, Blount, J.W., Steele, C.L. and Dixon, R.A. (2002) Bottlenecks for metabolic engineering of isoflavone glycoconjugates in Arabidopsis. Proc. Natl. Acad. Sci. U.S.A. 99, 14578–14583.

Liu, C., Huhman, D., Sumner, L.W. and Dixon, R.A. (2003) Regiospecific hydroxylation of isoflavones by cytochrome P450 81E enzymes from *Medicago truncatula*. Plant J. 36, 471–484.

Ludwig, S.R., Habera, L.F., Dellaporta, S.L. and Wessler, S.R. (1989) Lc, a member of the maize R gene family responsible for tissue-specific anthocyanin production, encodes a protein similar to transcriptional activators and contains the myc-homology region. Proc. Natl. Acad. Sci. U.S.A. 86, 7092–7096.

De Majnik, J., Weinman, J.J., Djordjevic, M.A., Rolfe, B.G., Tanner, G.J., Joseph, R.G. and Larkin, P.J. (2000) Anthocyanin regulatory gene expression in transgenic white clover can result in an altered pattern of pigmentation. Aust. J. Plant Physiol. 27, 659–667.

Marles, M.A.S., Ray, H., Ad Gruber, M.Y. (2003) New perspectives on proanthocyanidin biochemistry and molecular regulation. Phytochemistry 64, 367–383.

Marley, C.L., Cook, R., Barrett, J., Keatinge, R. and Lampkin, N.H. (2006) The effects of birdsfoot trefoil (*Lotus corniculatus*) and chicory (*Cichorium intybus*) when compared with perennial ryegrass (*Lolium perenne*) on ovine gastrointestinal parasite development, survival and migration. Vet. Parasitol. 138, 280–290.

Martens, S., Teeri, T. and Forkmann, G. (2002) Heterologous expression of dihydroflavonol 4-reductases from various plants. FEBS Lett. 531, 453–458.

Mathesius, U. (2001) Flavonoids induced in cells undergoing nodule organogenesis in white clover are regulators of auxin breakdown by peroxidase. J. Exp. Bot. 52, 419–426.

Mauricio, R., Rausher, M.D. and Burdick, D.S. (1997) Variation in the defense strategies of plants: are resistance and tolerance mutually exclusive? Ecol. 78, 1301–1311.

Maxwell, C.A. and Phillips, D.A. (1990) Concurrent synthesis and release of *nod*-gene-inducing flavonoids from alfalfa roots. Plant Physiol. 93, 1551–1558.

Maxwell, C.A., Harrison, M.J. and Dixon, R.A. (1993) Molecular characterization and expression of alfalfa isoliquiritigenin 2′-O-methyltransferase, an enzyme specifically involved in the biosynthesis of an inducer of *Rhizobium meliloti* nodulation genes. Plant J. 4, 971–981.

Maxwell, C.A., Restrepo-Hartwig, M.A., Hession, A.O. and McGonigle, B. (2005) Metabolic engineering of soybean for improved flavor and health benefits. In: J.T.Romeo (Ed.) *Recent Advances in Phytochemistry- Vol. 38. Secondary Metabolism in Model Systems*, Oxford: Elsevier Science Ltd. pp. 153–176.

McKhann, H.I. and Hirsch, A.M. (1994) Isolation of chalcone synthase and chalcone isomerase cDNAs from alfalfa (*Medicago sativa* L.): highest transcript levels occur in young roots and root tips. Plant Mol. Biol. 24, 767–777.

McMahon, L.R., McAllister, T.A., Berg, B.P., Majak, W., Acharya, S.N., Popp, J.D., Coulman, B.E., Wang, Y. and Cheng, K.-J. (2000) A review of the effects of forage condensed tannins on ruminal fermentation and bloat in grazing cattle. Can. J. Plant Sci. 80, 469–485.

Meagher, L.P., Widdup, K., Sivakumaran, S., Lucas, R. and Rumball, W. (2006) Floral *Trifolium* proanthocyanidins: Polyphenol formation and compositional diversity. J. Agric. Food Chem. 54, 5482–5488.

Morreel, K., Goeminne, G., Storme, V., Sterck, L., Ralph, J., Coppieters, W., Breyne, P., Steenackers, M., Georges, M., Messens, E. and Boerjan, W. (2006) Genetical metabolomics of flavonoid biosynthesis in *Populus*: a case study. Plant J. 47, 224–237.

Nagashima, S., Inagaki, R., Kubo, A., Hirotani, M. and Yoshikawa, T. (2004) cDNA cloning and expression of isoflavonoid-specific glucosyltransferase from *Glycyrrhiza echinata* cell-suspension cultures. Planta 218, 456–459.

Nair, M.G., Safir, G.R. and Siqueira, J.O. (1991) Isolation and identification of vesicular-arbuscular mycorrhiza-stimulatory compounds from clover (*Trifolium repens*) roots. Appl. Environ. Microbiol. 57, 434–439.

Nakajima, J., Tanaka, Y., Yamazaki, M. and Saito, K. (2001) Reaction mechanism from leucoanthocyanidin to anthocyanidin 3-glucoside, a key reaction for coloring in anthocyanin biosynthesis. J. Biol. Chem. 276, 25797–25803.

Nakajima, J., Sato, Y., Hoshino, T., Yamazaki, M. and Saito, K. (2006) Mechanistic study on the oxidation of anthocyanidin synthase by quantum mechanical calculation. J. Biol. Chem. 281, 21387–21398.

Nesi, N., Jond, C., Debeaujon, I., Caboche, M. and Lepiniec, L. (2001) The *Arabidopsis* TT2 gene encodes an R2R3 MYB domain protein that acts as a key determinant for proanthocyanidin accumulation in developing seed. Plant Cell 13, 2099–2114.

Nguyen, T.M., van Binh, D. and Ørskov, E.R. (2005) Effect of foliages containing condensed tannins on gastrointestinal parasites. Anim. Feed Sci. Techn. 121, 77–87.

Nielsen, K.M. and Podivinsky, E. (1997) cDNA cloning and endogenous expression of a flavonoid 3'5'-hydroxylase from petals of lisianthus (*Eustoma grandiflorum*). Plant Sci. 129, 167–174.

Paiva, N.L., Edwards, R., Sun, Y., Hrazdina, G. and Dixon, R.A. (1991) Stress responses in alfalfa (*Medicago sativa* L.) 11. Molecular cloning and expression of alfalfa isoflavone reductase, a key enzyme of isoflavonoid phytoalexin biosynthesis. Plant Mol. Biol. 17, 653–667.

Paolocci, P., Bovone, T., Tosti, N., Arcioni, S. and Damiani F. (2005) Light and an exogenous transcription factor qualitatively and quantitatively affect the biosynthetic pathway of condensed tannins in *Lotus corniculatus* leaves. J. Exp. Bot. 56, 1093–1103.

Paquette, S., Lindberg Møller, B. and Bak, S. (2003) On the origin of family 1 plant glycosyltransferases. Phytochemistry. 62, 399–413.

Paz-Ares, J., Ghosal, D. and Saedler, H. (1990) Molecular analysis of the C1-1 allele from *Zea mays*: a dominant mutant of the regulatory C1 locus. EMBO J. 9, 315–321.

Peng, Y.Y. and Ye, J.N. (2006) Determination of isoflavones in red clover by capillary electrophoresis with electrochemical detection. Fitoter. 77, 171–178.

Pfeiffer, J., Kühnel, C., Brandt, J., Duy, D., Punyasiri, P.A.N., Forkmann, G. and Fischer, T.C. (2006) Biosynthesis of flavan 3-ols by leucoanthocyanidin 4-reductases and anthocyanidin reductases in leaves of grape (*Vitis vinifera* L.), apple (*Malus x domestica* Borkh.) and other crops. Plant Physiol. Biochem. 44, 323–334.

Piersen, C.E. (2003) Phytoestrogens in botanical dietary supplements: implications for cancer. Integrat. Cancer Therap. 2, 120–138.

Piślewska, M., Bednarek, P., Stobiecki, M., Zieliński, M. and Wojtaszek, P. (2002) Cell wall-associated isoflavonoids and β-glucosidase activity in *Lupinus albus* plants responding to environmental stimuli. Plant, Cell Environ. 25, 29–40.

Porter, L.J. and Woodruffe, J. (1984) Haemanalysis: the relative astringency of proanthocyanidin polymers. Phytochemistry 23, 1255–1256.

Quesenberry, K.H., Smith, R.R., Taylor, N.L., Baltensperger, D.D. and Parrott, W.A. (1991) Genetic nomenclature in clovers and special-purpose legumes: I. Red and white clover. Crop Sci. 31, 861–867.

Radicella, J.P., Turks, D. and Chandler, V.L. (1991) Cloning and nucleotide sequence of a cDNA encoding B-Peru, a regulatory protein of the anthocyanin pathway in maize. Plant Mol. Biol. 17, 127–130.

Ramírez-Restrepo, C.A. and Barry, T.N. (2005) Alternative temperate forages containing secondary compounds for improving sustainable productivity in grazing ruminants. Anim. Feed Sci. Techn. 120, 179–201.

Rasmussen, S., Cao, M., Fraser, K., Koulman, A., Park-Ng, Z., Xue, H. and Lane, G. (2006) Cold stress in white clover – an integrated view of metabolome and transcriptome responses. In: C.F. Mercer (Ed.) *Breeding for Success: Diversity in Action. Proceedings of the 13th Australasian Plant Breeding Conference*, Christchurch, New Zealand, pp. 750–757.

Ray, H., Yu, M., Auser, P., Blahut-Beatty, L., McKersie, B., Bowley, S., Westcott, N., Coulman, B., Lloyd, A. and Gruber, M.Y. (2003) Expression of anthocyanins and proanthocyanidins after transformation of alfalfa with maize *Lc*. Plant Physiol. 132, 1448–1463.

Reynaud, J., Guilet, D., Terreux, R., Lussignol, M. and Walchshofer, N. (2005) Isoflavonoids in non-leguminous families: an update. Nat. Prod. Rep. 22, 504–515.

Robbins, M.P. and Morris, P. (2000) Metabolic engineering of condensed tannins and other phenolic pathways in forage and fodder crops. In: R. Verpoorte, A.W. Alfermann (Eds.) *Metabolic Engineering of Plant Secondary Metabolism*. Kluwer Academic Publishers, Dordrecht, The Netherlands, pp. 165–177.

Robbins, M.P., Bavage, A.D., Strudwicke, C. and Morris, P. (1998) Genetic manipulation of condensed tannins in higher plants. II. Analysis of birdsfoot trefoil plants harboring antisense dihydroflavonol reductase constructs. Plant Physiol. 116, 1133–1144.

Robbins, M.P., Paolocci, F., Hughes, J.W., Turchetti, V., Allison, G., Arcioni, S., Morris, P. and Damiani, F. (2003) Sn, a maize bHLH gene, modulates anthocyanin and condensed tannin pathways in *Lotus corniculatus*. J. Exp. Bot. 54, 239–248.

Robbins, M.P., Bavage, A.D., Allison, G., Davies, T., Hauck, B. and Morris, P. (2005) A comparison of two strategies to modify the hydroxylation of condensed tannin polymers in *Lotus corniculatus* L. Phytochemistry 66, 991–999.

Rumball, W., Keogh, R.G., Miller, J.E. and Claydon, R.B. (1997) 'Grasslands G27' red clover (*Trifolium pratense* L.). NZ J. Agricult. Res. 40, 369–372.

Rumball, W., Keogh, R.G. and Sparks, G.A. (2005) 'Grasslands HF1' red clover (*Trifolium pratense* L.) – a cultivar bred for isoflavone content. NZ J. Agricult. Res. 48, 345–347.

Ryder, T.B., Hedrick, S.A., Bell, J.N., Liang, X.W., Clouse, S.D. and Lamb C.J. (1987) Organization and differential activation of a gene family encoding the plant defense enzyme chalcone synthase in *Phaseolus vulgaris*. Mol. Gen. Genet. 210, 219–233.

Sallaud, C., El-Turk, J., Bigarré, L., Sevin, H., Welle, R. and Esnault, R. (1995) Nucleotide sequences of three chalcone reductase genes from alfalfa. Plant Physiol. 108, 869–870.

Sawada, Y., Kinoshita, K., Akashi, T., Aoki, T. and Ayabe, S. (2002) Key amino acid residues required for aryl migration catalysed by the cytochrome P450 2-hydroxyisoflavanone synthase. Plant J. 31, 555–564.

Schwinning, S. and Parsons, A.J. (1996) Analysis of the coexistence mechanisms for grasses and legumes in grazing systems. J. Ecol. 84, 799–813.

Seitz, C., Eder, C., Deiml, B., Kellner, S., Martens, S. and Forkmann, G. (2006) Cloning, functional identification and sequence analysis of flavonoid 3′-hydroxylase and flavonoid 3′, 5′-hydroxylase cDNAs reveals independent evolution of flavonoid 3′, 5′-hydroxylase in the Asteraceae family. Plant Mol. Biol. 61, 365–381.

Serraj, R. (2003) Effects of drought stress on legume symbiotic nitrogen fixation: physiological mechanisms. Indian J. Exp. Biol. 41, 1136–1141.

Shao, H., He, X., Achnine, L., Blount, J.W., Dixon, R.A. and Wang, X. (2005) Crystal structures of a multifunctional triterpene/flavonoid glycosyltransferases from *Medicago truncatula*. Plant Cell 17, 3141–3154.

Sharma, S.B. and Dixon, R.A. (2005) Metabolic engineering of proanthocyanidins by ectopic expression of transcription factors in *Arabidopsis thaliana*. Plant J. 44, 62–75.

Shimada, N., Akashi, T., Aoki, T. and Ayabe, S. (2000) Induction of isoflavonoid pathway in the model legume *Lotus japonicus*: molecular characterization of enzymes involved in phytoalexin biosynthesis. Plant Sci. 160, 37–47.

Shimada, N., Aoki, T., Sato, S., Nakamura, Y., Tabata, S. and Ayabe, S. (2003) A cluster of genes encodes the two types of chalcone isomerase involved in the biosynthesis of general flavonoids and legume-specific 5-deoxy(iso)flavonoids *in Lotus japonicus*. Plant Physiol. 131, 941–951.

Shimada, N., Sasaki, R., Sato, S., Kaneko, T., Tabata, S., Aoki, T. and Ayabe, S. (2005) A comprehensive analysis of six dihydroflavonol 4-reductases encoded by a gene cluster of the *Lotus japonicus* genome. J. Exp. Bot. 56, 2573–2585.

Sivakumaran, S., Rumball, W., Lane, G.A., Fraser, K., Foo, L.Y., Yu, M. and Meagher, L.P. (2006) Variation of proanthocyanidins in *Lotus* species. J. Chem. Ecol. 32, 1797–1816

Skadhauge, B., Gruber, M.Y., Thomsen, K.K. and von Wettstein, D. (1997) Leucocyanidin reductase activity of proanthocyanidins in developing legume tissues. Am. J. Bot. 84, 494–503.

Smil, V. (1999) Nitrogen in crop production. Global Biogeochem. Cycles 13, 647–667.

Sreevidya, V.S., Rao, C.S., Sullia, S.B., Ladha, J.K. and Reddy, P.M. (2006) Metabolic engineering of rice with soybean isoflavone synthase for promoting nodulation gene expression in rhizobia. J. Exp. Bot. 57, 1957–1969.

Steele, C.L., Gijzen, M., Qutob, D. and Dixon, R.A. (1999) Molecular characterization of the enzyme catalyzing the aryl migration reaction of isoflavonoid biosynthesis in soybean. Arch. Biochem. Biophys. 367, 146–150.

Subramanian, S., Stacey, G. and Yu, O. (2006) Endogenous isoflavones are essential for the establishment of symbiosis between soybean and *Bradyrhizobium japonicum*. Plant J. 48, 261–273.

Süß, C., Hempel, J., Zehner, S., Krause, A., Patschkowski, T. and Göttfert, M. (2006) Identification of genistein-inducible and type III-secreted proteins of *Bradyrhizobium japonicum*. J. Biotechnol. 126, 69–77.

Tanner, G.J., Francki, K.T., Abrahams, S., Watson, J.M., Larkins, P.J. and Ashton, A.R. (2003). Proanthocyanidin biosynthesis in plants. Purification of legume leucoanthocyanidin reductase and molecular cloning of its cDNA. J. Biol. Chem. 278, 31647–31656.

Tavendale, M.H., Meagher, L.P., Pacheco, D., Walker, N., Attwood, G.T. and Sivakumaran, S. (2005) Methane production from in vitro rumen incubations with *Lotus pedunculatus* and *Medicago sativa*, and effects of extractable condensed tannin fractions on methanogenesis. Anim. Feed Sci. Techn. 123–124, 403–419.

Tebayashi, S., Ishihara, A. and Iwamura, H. (2001) Elicitor-induced changes in isoflavonoid metabolism in red clover roots. J. Exp. Bot. 52, 681–689.

Tian, L. and Dixon, R.A. (2006) Engineering isoflavone metabolism with an artificial bifunctional enzyme. Planta 224, 496–507.

Toda, K., Akasaka, M., Dubouzet, E.G., Kawasaki, S. and Takahashi, R. (2005) Structure of flavonoid 3′-hydroxylase gene for pubescence color in soybean. Crop Sci. 45, 2212–2217.

Tonelli, C., Consonni, G., Dolfini, S.F., Dellaporta, S.L., Viotti, A. and Gavazzi, G. (1991) Genetic and molecular analysis of Sn, a light-inducible, tissue specific regulatory gene in maize. Mol. Gen. Genet. 225, 401–410.

Turnbull, J.J., Sobey, W.J., Aplin, R.T., Hassan, A., Firmin, J.L., Schofield, C.J. and Prescott, A.G. (2000) Are anthocyanidins the immediate products of anthocyanidin synthase? Chem. Commun. 2000, 2473–2474.

Turnbull, J.J., Nakajima, J., Welford, R.W.D., Yamazaki, M., Saito, K. and Schofield, C.J. (2004) Mechanistic studies on three 2-oxoglutarate-dependent oxygenases of flavonoid biosynthesis. J. Biol. Chem. 279, 1206–1216.

Volpin, H., Phillips, D.A., Okon, Y. and Kapulnik, Y. (1995) Suppression of an isoflavonoid phytoalexin defense response in mycorrhizal alfalfa roots. Plant Physiol. 108, 1449–1454.

Wasson, A.P., Pellerone, F.I. and Mathesius, U. (2006) Silencing the flavonoid pathway in *Medicago truncatula* inhibits root nodule formation and prevents auxin transport regulation by rhizobia. Plant Cell. 18, 1617–1629.

Welle, R., Schröder, G., Schlitz, E., Grisebach, H. and Schröder, J. (1991) Induced plant responses to pathogen attack. Analysis and heterologous expression of the key enzyme in the biosynthesis of phytoalexins in soybean (*Glycine max* L. Merr. Cv. Harosoy 63). Eur. J. Biochem. 196, 423–430.

Wellmann, F., Griesser, M., Schwab, W., Martens, S., Eisenreich, W., Matern, U. and Lukačin, R. (2006). Anthocyanidin synthase from *Gerbera hybrida* catalyzes the conversion of (+)-catechin to cyanidin and a novel procyanidin. FEBS Lett. 580, 1642–1648.

Williams, W.M., Verry, I.M. and Ellison, N.W. (2006) A phylogenetic approach to germplasm use in clover breeding. In: C.F. Mercer (Ed.) *Breeding for Success: Diversity in Action. Proceedings of the 13th Australasian Plant Breeding Conference*, Christchurch, New Zealand, pp. 966–971.

Wingender, R., Röhrig, H., Höricke, C., Wing, D. and Schell, J. (1989) Differential regulation of soybean chalcone synthase genes in plant defence, symbiosis and upon environmental stimuli. Mol. Gen. Genet. 218, 315–322.

Winkel, B.S.J. (2004) Metabolic channeling in plants. Annu. Rev. Plant Biol. 55, 85–107.

Winkel-Shirley, B. (2001) Flavonoid biosynthesis. A colorful model for genetics, biochemistry, cell biology, and biotechnology. Plant Physiol. 126, 485–493.

Wollenweber, B., Porter, J.R. and Lübberstedt, T. (2005) Need for multidisciplinary research towards a second green revolution. Curr. Opin. Plant Biol. 8, 337–341.

Wood, A.J. and Davies, E. (1994) A cDNA encoding chalcone isomerase from aged pea epicotyls. Plant Physiol. 104, 1465–1466.

Xie, D.Y., Sharma, S.B., Paiva, N.L., Ferreira, D. and Dixon, R.A. (2003) Role of anthocyanidin reductase, encoded by BANYULS in plant flavonoid biosynthesis. Science 299, 396–399.

Xie, D.Y., Jackson, L.A., Cooper, J.D., Ferreira, D. and Paiva, N.L. (2004a) Molecular and biochemical analysis of two cDNA clones encoding dihydroflavonol-4-reductase from *Medicago truncatula*. Plant Physiol. 134, 979–994.

Xie, D.Y., Sharma, S.B. and Dixon, R.A. (2004b) Anthocyanidin reductases from *Medicago truncatula* and *Arabidopsis thaliana*. Arch. Biochem. Biophys. 422, 91–102.

Xie, D.Y., Sharma, S.B., Wright, E., Wang, Z.Y. and Dixon, R.A. (2006) Metabolic engineering of proanthocyanidins through co-expression of anthocyanidin reductase and the PAP1 MYB transcription factor. Plant J. 45, 895–907.

Yu, O., Jung, W., Shi, J., Croes, R.A., Fader, G.M., McGonigle, B. and Odell, J.T. (2000) Production of the isoflavones genistein and daidzein in non-legume dicot and monocot tissues. Plant Physiol. 124, 781–793.

Yu, O., Shi, J., Hession, A.O., Maxwell, C.A., McGonigle, B. and Odell, J.T. (2003) Metabolic engineering to increase isoflavone biosynthesis in soybean seed. Phytochemistry 63, 753–763.

Zabala, G. and Vodkin, L.O. (2005). The wp mutation of *Glycine max* carries a gene-fragment-rich transposon of the CACTA superfamily. Plant Cell 17, 2619–2632.

9

Anthocyanins as Food Colorants

Nuno Mateus and Victor de Freitas

University of Porto, Department of Chemistry, CIQ, Rua do Campo Alegre, 687; 4169-007 Porto, Portugal. nbmateus@fc.up.pt, vfreitas@fc.up.pt

Abstract. The interest of the food industry in natural colorants replacing synthetic dyes has increased significantly over the last years, mainly due to safety issues. This chapter deals with the interest of using natural anthocyanins and their derivatives as food colorants. The importance of color for the acceptability of food products and the need for satisfying and attracting more consumers to a growing competitive market has resulted in the development of new pigments and new products. Several aspects are briefly discussed, such as anthocyanin natural sources, extraction and analytical methods, anthocyanin stability and color variations, and applications in food products. Also, the interest in anthocyanin derivatives is referred, reviewing the recent discoveries of new classes of anthocyanin-derived pigments presenting attracting color hues. Research in this field has been productive but industrial applications are yet to be made. Nevertheless, there are good perspectives for more applications of anthocyanins and derivatives in food products.

9.1 Introduction

Color appearance of food products is one of the major concerns of the food industry. In a certain way, color may act as a "fingerprint" of a food product, being related to its flavor and at the same time estimate its overall quality. Color is also the first attribute to be perceived in foods and beverages and is usually positively correlated with standards of quality by the consumer.

Bearing this, many efforts have been made by several researchers in association with industrial R&D groups aiming to find new ways of improving color stability (improving the "shelf life") and quality. Besides, the need of the food industry for introducing new products with appealing features has opened new research lines aiming to develop attractive food colorants with different and unusual hues and at the same time they should be regarded as natural and, whenever possible, health-promoting products.

The color of foodstuffs commonly matches with the flavor of food (lemon with yellow, strawberries with red, etc.), but the food industry does not rule out the possi-

K. Gould et al. (eds.), *Anthocyanins*, DOI: 10.1007/978-0-387-77335-3_9,

bility of introducing new products, in which color may contradict flavor, a concept that appears challenging but that mainly targets a younger population. Therefore, new compounds to be used as food colorants are being studied foreseeing their application in the food industry.

Food colorants may be organized in three different types, depending on their source (Henry 1996):

a) Natural colors, organic colorants derived from natural edible sources such as anthocyanins;

b) Nature-identical colors, manufactured by chemical synthesis so as to be identical to colorants found in nature (e.g. β-carotene and riboflavin);

c) Synthetic colors that do not occur in nature, produced by chemical synthesis (e.g. tartrazine and carmoisine).

9.2 Anthocyanins as a Food Colorant

The use of unnatural additives is becoming less popular among the consumers, especially due to psychological reasons as the consumer easily associates natural colorants to health benefits and synthetic colorants to toxic issues. Therefore, replacing synthetic dyes by natural colorants has become a major issue over the last years. Pigments from natural sources may display a wide range of colors and are usually safe. Among those pigments widespread in nature, anthocyanins assume a crucial role when dealing with natural colorants. A good review of anthocyanins as food colorants is the one made by Francis (1989).

According to the numbering system used by the *Codex Alimentarius Commission,* anthocyanins (any anthocyanin-derived colorant) are listed as a natural colorant by the European Union (EU) legislation as product E163. With respect to the US, the FDA (Food and Drug Administration) has a different list of "natural" colors that do not require certification (without any FD & C numbers), and anthocyanins can be obtained either from "grape color extract", "grape skin extract", "fruit juices or vegetable juices". Nevertheless, the use of anthocyanins in food products is restricted to some products varying among countries. Usually, the USA is the most restrictive country regarding the use of anthocyanin colorants.

Anthocyanins are interesting pigments regarding their chromatic features. The oldest anthocyanin extract used in the food industry is enocyanin (Dieci 1967) obtained from red grape pomace and marketed in Italy since 1879. Nowadays, grape extracts are often used as colorants for sugar confectionary, dairy products, ice creams, etc.

Besides their color features, anthocyanins have recently attracted even more interest due to their possible health attributes, such as a reduced risk of coronary diseases, reduced risk of stroke, anticarcinogen activity, anti-inflammatory effects and improved cognitive behaviour (Cao et al. 1997; Wang et al. 1997; Clifford 2000; Scalbert and Williamson 2000; Prior 2003).

On the other hand, chemical transformations of anthocyanins yielding new pigments with different and unique color hues appear to be challenging as well as profitable. Several research groups are driving their investigations towards the identification and characterization of pigments from different natural sources, especially red fruit extracts, aiming to use them as colorants.

Overall, anthocyanins are interesting natural compounds to be used as food colorants as they may display a wide range of attractive colors, they are water-soluble and have health-promoting effects.

9.3 Anthocyanins: Structure and Natural Sources

Anthocyanins are the largest group of water-soluble pigments widespread in the plant kingdom. They are responsible for the colors displayed by many flowers, fruits and leaves of angiosperms. These natural pigments are usually associated with red fruits but also occur in vegetables, roots, legumes and cereals (Mazza and Miniati 1993; Markakis 1982; Francis 1982, 1989).

Chemically, these flavonoids naturally occur as glycosides of flavylium or 2-phenylbenzopyrylium salts and are most commonly based on six anthocyanidins: pelargonidin, cyanidin, peonidin, delphinidin, petunidin and malvidin. The sugar moieties vary but are commonly, glucose, rhamnose, galactose or arabinose. The sugar moiety may be a mono or disaccharide unit, and it may be acylated with a phenolic or aliphatic acid. These compounds differ in the methoxyl and hydroxyl substitution pattern of ring B (Fig. 9.1).

While there are six common anthocyanidins, there are a total of 539 anthocyanins reported to be isolated from plants, 277 of which have been identified after 1992 (Andersen and Jordheim 2005). The more widespread anthocyanins in fruits are glycosilated in the 3-OH position (3-*O*-monoglycosides) and, in less extension, in both positions 3-OH and 5-OH (3,5-*O*-diglycosides). Their chromatic features are importantly affected by the substitution pattern of ring B. For example, the increase of the hydroxylation pattern results in a bathochromic shift from red to a violet colour (Pelargonidin → Cyanidin → Delphinidin). The nature of sugar (e.g. glucose, arabinose, rutinose, sambubiose), acylated or not, and its position in the aglycone skeleton are also important structural factors that influence the color hue of these pigments. All these structural factors vary between species.

Concerning the natural sources of these pigments, grapes are amongst the best sources of anthocyanins, in a long list that also includes red cabbage, blood orange and several red fruit berries (elderberries, blackberries, blueberries, chokeberries, raspberries, etc.) (Kühnau 1976; Wilska-Jeszka et al. 1992; Skrede et al. 1992; Torre and Barrit 1977; Gao and Mazza 1995; Bakker and Timberlake 1985; Bajaj et al. 1990) (Table 9.1).

Consumption data reported in the literature by Kühnau refer that in the USA the average daily intake of anthocyanins is around 215 mg during the summer and 180 mg during the winter. These values are thought to be higher for regular red wine consumers. Indeed, despite not being the best source of anthocyanins compared to fruit berries, anthocyanins from red wine may complex with other wine components, thereby their consumption levels should be significantly higher than initially expected.

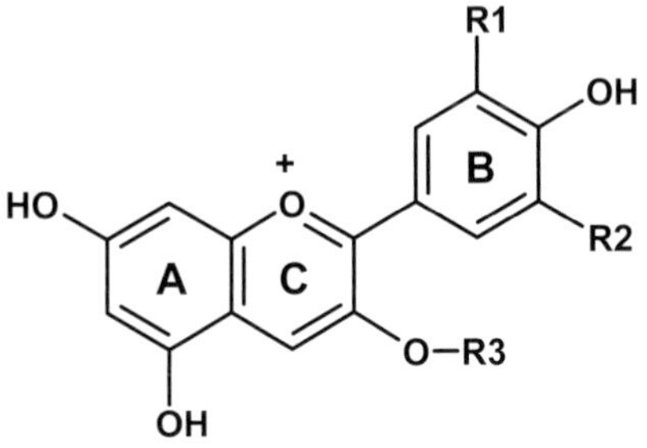

Anthocyanin	R1	R2	λ_{max} (nm)*	
			R3=H	R3=gluc
Delphinidin	OH	OH	546	541
Petunidin	OH	OCH$_3$	543	540
Malvidin	OCH$_3$	OCH$_3$	542	538
Cyanidin	OH	H	535	530
Peonidin	OCH$_3$	H	532	528
Pelargonidin	H	H	520	516

* In methanol with 0.01% HCl (Harborne 1984; Giusti et al. 1999)

Fig. 9.1 Structures of the major anthocyanin-3-*O*-glucoside presents in fruits and respective wavelength at the maximum of absorption in the visible region

Table 9.1 Average amount of anthocyanins in some foodstuffs (Clifford 2000)

Anthocyanin source	Amount (mg.litre^{-1} or mg.kg^{-1})
Blackberry	1150
Blueberry	825–4200
Boisenberry	1609
Cherry	20–4500
Chokeberry	5060–10000
Cranberry	600–2000
Cowberry	1000
Currant (black)	1300–4000
Elderberry	2000–10000
Red grapes	300–7500
Blood orange	2000
Plum	20–250
Sloe	1600
Strawberry	150–350
Raspberry (black)	1700–4277
Eggplant	7500
Onion	up to 250
Rhubarb	up to 2000
Red cabbage	250
Red wine	240–350
Port wine	140–1100

9.4 Extraction and Analysis of Anthocyanins

Anthocyanins have become one the most studied polyphenol group of compounds in R&D projects associated with the food industry. Consequently, methods for their extraction, purification and analysis have been refined. Since anthocyanins are not stable in neutral or alkaline solutions, acidic aqueous solvents have been used as extraction solvents in order to disrupt cell membranes and at the same time dissolve the water-soluble pigments. Usually, HCl (<1%) is chosen for acidulating the extraction solvent (Harborne 1984).

The most commonly used solvents are hydroalcoholic solutions containing ethanol or methanol, but it may also consist in *n*-butanol, cold acetone, propylene glycol, methanol/acetone/water mixtures or boiling water (Garcia-Viguera et al. 1998; Giusti et al. 1998). The addition of some organic acids has become useful for the extraction of complex polyacylated anthocyanins (Strack and Wray 1989). The addition of water will be depending on the nature of the sample as it may be useful to achieve complete anthocyanin extraction. For food applications, although having a lower extraction capacity and being difficult to eliminate afterwards, ethanol is usually preferred due to its low toxicity. After extraction, any concentration process should be performed at low temperatures (below 30°C) and preferentially *in vacuo* in order to minimize anthocyanin degradation.

Fractionation, separation and analysis of anthocyanins may be achieved through several techniques which include thin layer chromatography (TLC) (Strack and Wray 1989), several column chromatographies (CC) (Rivas-Gonzalo 2003), high performance liquid chromatography (HPLC) (Wulf and Nagel 1978; da Costa et al. 2000), high-speed counter current chromatography (HSCCC), high performance centrifugal partition chromatography (HPCPC) (Foucault and Chevolot 1998; Degenhardt and Winterhalter 2001; Schwarz et al. 2003), capillary electrophoresis (CE) and capillary zone electrophoresis (CZE) (Ichiyanagi et al. 2000). Different gels have been used in order to separate anthocyanins like silica gel, reversed-phase silica-gel, different kinds of polyamide, polymeric resins, etc.

The detection of anthocyanins can be achieved using UV-Vis spectroscopy (Wulf and Nagel 1978), spectrofluorometry, mass spectrometry (FAB-MS, HPLC-MS, ESI-MS) (Lopes da Silva et al. 2002; Mateus et al. 2002a; Favretto and Flamini 2000), NMR (Fossen and Andersen 1999) or infrared spectroscopy (Dambergs et al. 2006). The coupling of different analytical techniques allowed improving the separation as well as the sensitivity of detection of anthocyanins. HPLC coupled to nuclear magnetic resonance (NMR) is one of the recent examples of such achievement (Wolfender et al. 1998; Wolfender et al. 2003; Stintzing et al. 2004).

9.5 Anthocyanin Stability and Equilibrium Forms

The degradation of anthocyanins may occur during their extraction, food processing and storage. Therefore, a correct knowledge of the factors that govern anthocyanin stability seems to be decisive in terms of a putative application in food matrixes to serve as colorants. A delicate factor that is strictly related to the color displayed by anthocyanins is the fact that they co-exist in aqueous solution in equilibrium between five species depending on pH: flavylium cation, carbinol base, chalcone, quinonoidal base and anionic quinonoidal base (Brouillard and Delaporte 1977; Brouillard and Dubois 1977; Brouillard and Lang 1990; Brouillard 1982; Mistry et al. 1991) (Fig. 9.2).

The total anthocyanin red color is expressed only in strongly acidic pH. In solution at pH above 3.5, these natural compounds present little color expression. However, in natural media at physiological pH they may display more color intensity as a result of co-pigmentation phenomena (intra- and inter-molecular co-pigmentation)

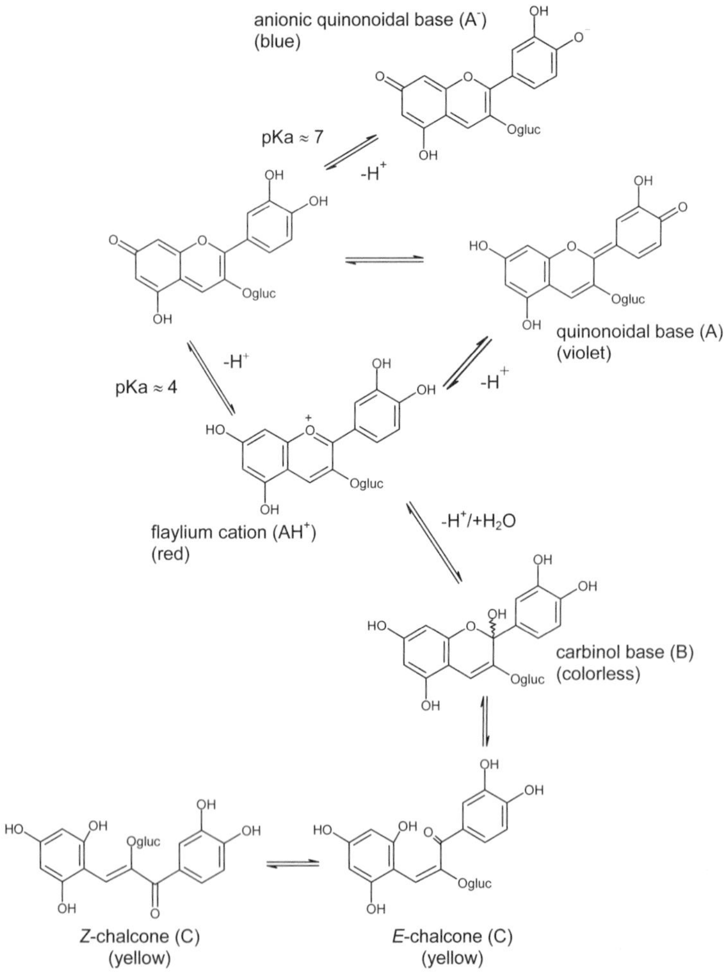

Fig. 9.2 Structural transformations of anthocyanins (cyanidin 3-O-glucoside) in aqueous solutions depending on pH (mod. after Brouillard and Lang 1990)

(Mazza and Brouillard 1990; Davies and Mazza 1993). This physical-chemical feature was initially described in the pioneer works of the Nobel Laureate Robert Robinson (Robinson and Robinson 1931) and firstly observed by another Nobel Laureate – Richard Willstatter – who attributed the wide range of natural colors associated with anthocyanins (namely from flowers) to pH variations in the plant cells (Willstatter and Zollinger 1916).

Features like acylation increase anthocyanin stability (Goto et al. 1978; Saito, et al. 1995; Haslam 1998; Figueiredo et al. 1999). Indeed, intramolecular co-pigmentation is characteristic of acylated anthocyanin structures and result from the stacking of the hydrophobic acyl moieties and the flavylium nucleus, thereby reducing anthocyanin hydrolysis (Dangles et al. 1993). The best example of such phenomenon is the heavenly-blue flower anthocyanin (Ipomoea tricolor) (Goto and Kondo 1991; Brouillard 1981, 1983). This event leads to an increase of the red color intensity and on the other hand a displacement of the equilibrium towards the quinonoidal base structures of violet-blue colors at neutral pH. The same events may occur through inter-molecular co-pigmentation involving colorless compounds such as metal ions, flavonols, etc. Concerning metals, some di- and trivalent cations may cause bathochromic shifts of the maximum absorption wavelengths causing a "blueing" effect of the color (e.g. magnesium, aluminium), through the formation of weak complexes with anthocyanins. This ability of anthocyanins to interact with colorless co-pigments is extremely relevant to the color stability in solution as it has been considered to be the first step in the formation of new anthocyanin-derived pigments (Brouillard and Dangles 1994). This will be focused further on when discussing the formation of anthocyanin-derived pigments in red wines.

Concerning temperature, anthocyanins become paler after heating as the equilibrium is displaced towards the colorless carbinol and chalcone forms. The degradation pathways are reported in the literature (Furtado et al. 1993). The stability of anthocyanins (or other derived pigments) regarding pH and temperature variations is thus one of the major obstacles for their application in food matrixes.

Moreover, the wide use of SO_2 as preservative in the food industry also leads to some color fluctuations as it reacts with anthocyanins to form a colorless addition adduct. This reaction is reversible and consequently SO_2 may be released as a gas, yielding the anthocyanin and restoring color. This reversibility is favored by heat. Resistance to SO_2 bleaching is thereby an important factor for a good food colorant. Bearing this, red grape extracts appear to be very interesting as they possess many oligomeric anthocyanins and anthocyanin-derived pigments which are more resistant to SO_2 bleaching. This higher resistance to SO_2 discoloration is due to the fact that the position for the nucleophilic attack of SO_2 in the anthocyanin structure (C-4 position of the pyranic ring C) is blocked (Berké et al. 1998).

Other factors that may affect color application forms like solubility, physical form, microbiological quality, light exposure, oxidation and the presence of metals and other substances may also limit the application of these compounds in foodstuffs (Markakis 1982). For instance, sugars are known to induce a color enhancement (through increasing anthocyanin intensity), especially in mildly acidic conditions (Lewis et al. 1995). But on the other hand, some studies have shown that ascorbic acid, glucose and fructose together with their degradation products may induce anthocyanin degradation catalyzed by oxygen, high temperature and metal ions (Es-Safi et al. 2002; Garcia-Viguera and Bridle 1999; Wrolstad et al.1990). The color desired may also be adjusted by adding anthocyanins from different red fruit sources.

In addition, enzymatic degradation may also influence anthocyanin stability in food matrixes (Sarni-Manchado et al. 1997).

9.6 Anthocyanin Extracts and Applications

Commercial applications of anthocyanins as food colorants include soft drinks, fruit preserves (jams, canned fruit), sugar confectionary (jellies), dairy products (essentially yogurts), dry mixes (acid dessert mixes and drink powders) and more rarely frozen products (ice cream) and few alcoholic drinks. Among these applications, soft drinks have been the main and ideal target for the use of anthocyanins as colorant. Bearing all the previously referred factors that govern and limit the application of anthocyanins in food matrixes, an acidic pH (below 3.5) appear to be necessary in order to obtain the desired red color. Application of anthocyanins in cloudy beverages (or other foodstuff) is not suitable as undesired side effects may occur ("blueing" effect and "thin layer" effect). Grape extracts are rich in anthocyanins complexed with other coumpounds and are thus more stable towards pH variations and in the presence of SO_2 than genuine monomeric anthocyanins, as already discussed. This is one of the reasons why grape extracts are widely used as anthocyanin sources of natural colorants. Besides, anthocyanins may be easily obtained in high quantities from grapes as these latter represent about a quarter of the annual fruit crop worldwide (Francis 1989).

Natural anthocyanins have powerful coloring properties as only small doses of anthocyanins are required to display the color desired in several food matrixes (e.g. 30–40 ppm for soft drinks and 20–60 ppm for fruit preserves). Generally, natural colorants have higher coloring capacities than synthetic ones, as seen from their relative absorptivities (Henry 1996). Anthocyanins are almost exclusively required to offer a red color to foodstuffs, but in some cases betalains (another group of natural compounds) are preferred. Over the past recent years, several approaches have been made aiming to apply different anthocyanin extracts to food matrixes. Besides grapes, other fruit sources like concentrated juice of blackcurrant, elderberry, cranberry and cherry have been assayed, and even plant extracts such as oxalis (*Oxalis triangularis*) (Pazmiño-Durán et al. 2001b).

Also, several vegetable extracts have been used to restore food color, including red cabbage (Dyrby et al. 2001; Shimizu et al. 1997), purple sweet potato (Odake et al. 1992; Terahara et al. 1999), red-fleshed potato (Rodriguez-Saonaet et al. 1998, 1999), radish (Rodriguez-Saona et al. 2001; Giusti and Wrolstad 1996a; Otsuki et al. 2002) and black carrot (Stintzing et al. 2002; Malien-Aubert et al. 2001). These vegetable extracts have been shown to be rich in acylated anthocyanins. As already referred from a molecular point of view in the previous section, acylation of anthocyanins improves the color stability during processing and storage (Giusti and Wrolstad 1996b; Bassa and Francis 1987; Teh and Francis 1988). Therefore, this has encouraged research on the application of these acylated anthocyanin-based food colorants. Some of these studies included the application of red cabbage extracts to soft drinks, which were shown to be extremely stable towards light and heat (Dyrby et al. 2001). In addition to temperature and light stability (Inami et al. 1996), pigment sources rich in acylated anthocyanins were equally found to be less susceptible to color changes through enzymatic activity (Rommel et al. 1992). The absence of non-anthocyanic phenolics in the extract to be tested as food colorant is also an important issue as these may be easily oxidized. This is one of the reasons why black carrot anthocyanin extracts constitute a good anthocyanin source from a technological point of view (Stintzing et al. 2002)

Less common fruits like the Brazilian Açai (*Euterpe oleracea*) were also tested as functional pigments for yogurts (Coisson et al. 2005). This fruit juice is already known for its high antioxidant and anti-radical properties (Bianchi et al. 2001), and their chromatic features (dark purple hue) make it appealing for food applications.

From an economic perspective, some of the best putative commercial sources of anthocyanins are those from which the pigment is a by-product of some industrial processes involving other value-added products, like grape skin extracts (Jackman and Smith 1996). For that reason, more attention has been paid to other potential anthocyanin-rich waste by-products, such as banana bracts (Pazmiño-Durán et al. 2001a) or purple corncobs (Jing and Giusti 2005).

Besides genuine anthocyanins as they are displayed by nature, other studies are currently focused on potential applications of anthocyanin derivatives, as discussed further on.

9.7 Anthocyanins and Derived Pigments: New Colors

As already referred, red wine may be a good source of anthocyanins. The levels of anthocyanins (and respective coumaroyl, caffeoyl and acetyl esters) found in young red wines together with the wines chemical complexity favors the occurrence of newly-formed anthocyanin-derived pigments. Indeed, anthocyanins are likely to react with other colorless polyphenols and non-polyphenolic compounds present in the wine such as ketonic compounds, metals, proteins, carbohydrates, etc Consequently, red wine is probably the foodstuff that presents the highest structural diversity of polyphenolic pigments in their genuine form and in other derivative structures. Structural changes of anthocyanins in red wines have been extensively studied over the last century aiming to comprehend red wine color evolution. Findings began with the pioneer works of Jurd and Somers (Jurd 1967; Jurd and Somers 1970; Somers 1966, 1971). During ageing, anthocyanins polymerize by reaction with other wine constituent, such as tannins, which is favored by the presence of acetaldehyde mainly arising form ethanol oxidation (Timberlake and Bridle 1976; Rivas-Gonzalo et al. 1995; Wildenradt and Singleton 1974).

The formation of this great diversity of anthocyanin-derived pigments in red wine arises essentially from the high reactivity of anthocyanins. These pigments are very reactive mainly due to the electronic deficiency of their flavylium nuclei. It is currently known that anthocyanins have some positions that are likely to react with different nucleophilic and electrophilic compounds (Fig. 9.3). Anthocyanins can

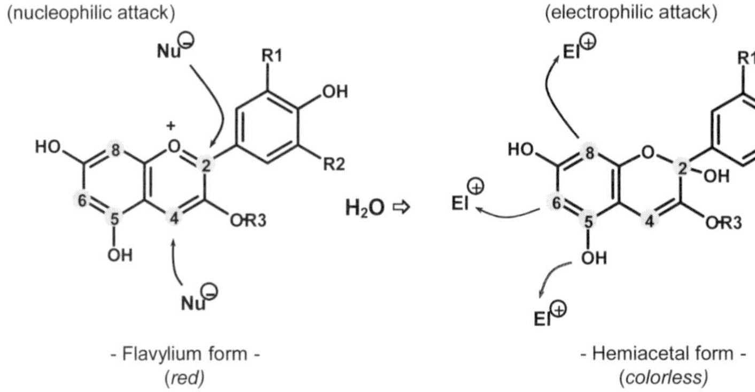

Fig. 9.3 Schematic representation of the main reactive position of anthocyanin structures (mod. after de Freitas and Mateus 2006)

undergo a nucleophilic attack at the positively charge carbons 2 and 4 of the pyranic ring C (hydration in carbon 2 gives rise to the colorless hemiacetal form). Despite its positive charge, anthocyanins were also shown to react with electrophilic compounds through its hydroxyl groups as well as carbon 6 and 8 of the phloroglucinol ring A, probably involving the uncharged hemiacetal form. The presence of a free 5-OH group is crucial for the reactivity of anthocyanins with other compounds that occur during anthocyanin-rich food processing and ageing, such as in red wine.

The growing analytical technology applied to food science (mass spectrometry and NMR techniques, coupled to HPLC) has allowed identifying several new classes of anthocyanin-derived pigments found in wines. These findings have helped to better understand some physical-chemical properties of anthocyanins and derived pigments. Over the last years, several anthocyanin-derived pigments occurring in red wines have been reported in the literature, especially those ones belonging to a group known as pyranoanthocyanins (Fig. 9.4). The major structural feature of these newly-formed pigments is the presence of an additional ring (ring D) in the chromophore, which allows a higher distribution of the positive charge (Cameira-dos-Santos et al.1996). Overall, the major new families of anthocyanin-derived pigments detected in red wines were described to result from:

a) Reaction between anthocyanins and small compounds (e.g. pyruvic and phenolic acids, acetaldehyde, *p*-vinylphenol) giving rise to new pyranoanthocyanin pigments (Fig. 9.4):
-Anthocyanin-pyruvic acid adducts (carboxypyranoanthocyanins) (Fulcrand et al. 1998);
-Pyranoanthocyanins (Vitisin B) (Bakker and Timberlake 1985);
-Pyranoanthocyanin-phenol pigments (Fulcrand et al. 1996; Schwarz et al. 2003);
b) Condensation between anthocyanins and flavanols mediated by aldehydes (e.g. acetaldehyde) (Fig. 9.5):
-Anthocyanin-ethyl-flavanol pigments (Timberlake and Bridle 1976);

-Pyranoanthocyanin-flavanol pigments (Francia-Aricha et al. 1997; Vivar-Quintana et al. 1999; Mateus et al. 2002, 2003a);
   c) Direct condensation between anthocyanins and flavanols (Fig. 9.6) (Remy et al. 2000).

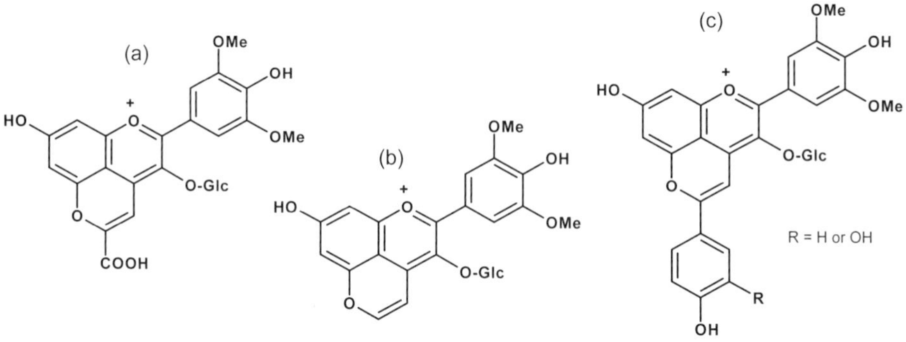

Fig. 9.4 Structure of (a) malvidin-3-glucoside-pyruvic acid adducts (Fulcrand et al. 1998), (b) vitisin B (Bakker and Timberlake, 1985) and (c) pyranomalvidin-3-glucoside-phenol pigments (Fulcrand et al. 1996; Schwarz et al. 2003)

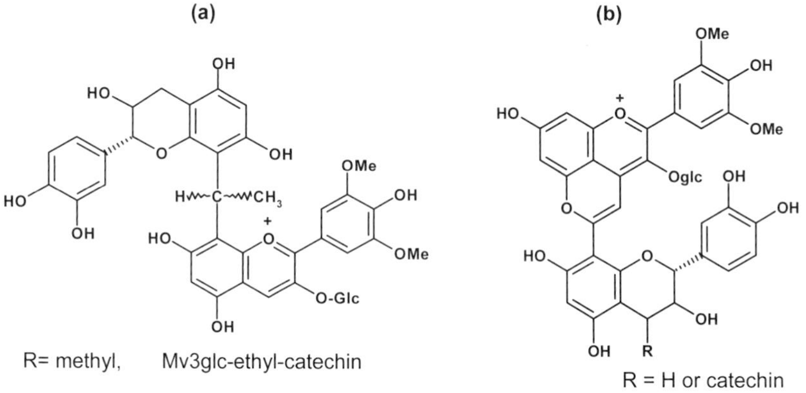

Fig. 9.5 Structure of (a) malvidin-3-glucoside-ethyl-flavanol pigments (Timberlake and Bridle, 1976) and (b) Pyranomalvidin-3-glucoside-flavanol pigments (Francia-Aricha et al. 1997; Vivar-Quintana et al. 1999; Mateus et al. 2002a and 2002b)

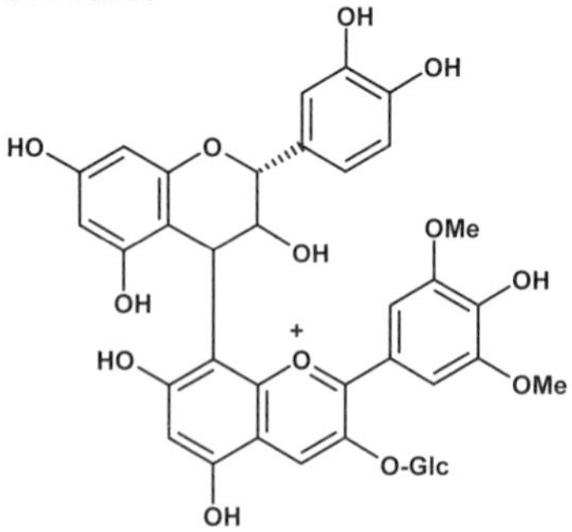

Fig. 9.6 Structure of a pigment resulting from direct condensation between anthocyanins and flavanols (Haslam 1980; Remy et al. 2000)

The color features of these pigments are attractive and nowadays their interest goes beyond oenology or wine chemistry. Most of these anthocyanin-derived pigments present more orange-like hues than their anthocyanin precursors. Besides their attractive colors, these anthocyanin-derived pigments seem to be more stable than their anthocyanin precursors regarding pH and SO_2 discoloration (de Freitas and Mateus 2006). Among these pigments, the anthocyanin-pyruvic acid adducts appear to be interesting for the food industry as they may be easily obtained through reaction of anthocyanin extracts with pyruvic acid. Their color features and their stability in aqueous environment foresee putative applications in some food products.

9.7.1 Portisins: New Blue Anthocyanin-Derived Food Colorants?

Recently, one new class of pyranoanthocyanins was found to naturally occur in red wine (Mateus et al. 2004b; Mateus et al. 2003b). These pigments present a complex structure in which a pyranoanthocyanin moiety is linked by a vinyl group to a catechin or a procyanidin dimer moiety. This new pigment class was named as portisins as these pigments were first and only reported in Port wines. This new class of pigments may be synthetised in the laboratory through reacting anthocyanin-pyruvic acid adducts and flavanols in the presence of acetaldehyde (Mateus et al. 2004b). Furthermore, other portisins resulting from reaction with other molecules like vinylphenol, phloroglucinol) have also been synthesized (Fig. 9.7).

Portisins bearing flavanols in their structure present unique bluish hues even under acidic conditions, which is very peculiar as anthocyanins are known to have blue colors only at high pH values due to their quinonoidal forms. The blue color is a very rare color to obtain as a result of a single pigment. In nature, the blue color could result from the complexation of anthocyanins with metals, or from co-pigmentation phenomena, as already reported in the literature (Saito et al. 1995).

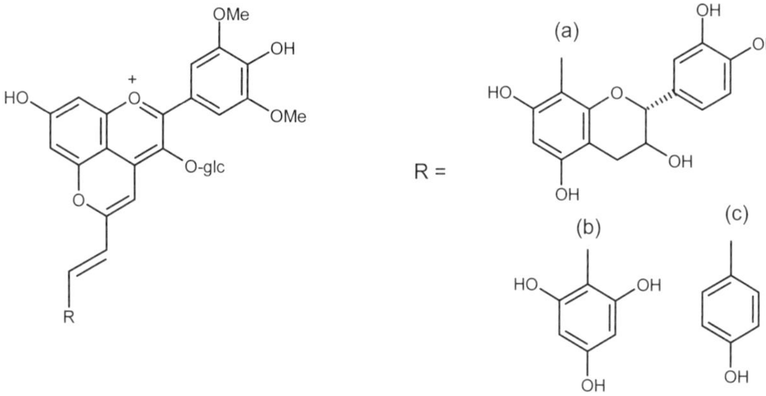

Fig. 9.7 Structure of different portisins (mv3glc=malvidin-3-O-glucoside): (a) vinylpyrano-mv3glc-p-phenol, (b) vinylpyrano-mv3glc-phloroglucinol, and (c) vinylpyrano-mv3glc-catechin (Mateus et al. 2004a; Mateus et al. 2006)

As already discussed, color displacements due to pH variations and SO_2 bleaching (being strictly related with the pigment stability) are important issues when developing new applications of food colorants. Bearing this, anthocyanin derivatives like anthocyanin-pyruvic acid adducts and portisins appear as interesting pigments to be tested in order to be applied as food colorants. Indeed, these two classes of anthocyanin-derived pigments were found to be much more stable towards pH variations and in the presence of SO_2 (up to 200 ppm), being these effects especially noticeable for portisins. The color displayed by portisins was proven to change only slightly with increasing pH up to neutral condition at which the anthocyanin solutions are almost colorless (Fig. 9.8) (Oliveira et al. 2006). This higher resistance to discoloration of portisins could be explained by a great protection of the chromophore moiety against the nucleophilic attack of water, which would give rise to the colorless hemiacetal form.

Concerning SO_2 bleaching, aqueous solution of the vinylpyrano-mv3glc-catechin showed a greater resistance to discoloration by the addition of increasing amounts of SO_2 (0–200 ppm) (Fig. 9.9). This outcome suggests a higher protection of the chromophore moiety against nucleophilic attack of SO_2, similarly to that described for water.

Furthermore, the structural features of portisins seem to be determinant for the color displayed by these anthocyanin-derived pigments. The increase in hydroxylation of the phenol ring (R group, Fig. 9.7) in the portisin structures contributes to a bathochromic shift of the wavelength of the maximum absorption (λ_{max}), changing the solution color from purple (vinylpyrano-mv3glc-p-phenol, λ_{max} = 553 nm) to a more bluish hue (vinylpyrano-mv3glc-phloroglucinol, λ_{max} =583 nm; vinylpyrano-mv3glc-catechin, λ_{max} =587 nm). Therefore, the development of the synthesis of this class of pigments may be directed towards different color hues. Increasing the hydroxylation pattern would have an even more "blueing" effect.

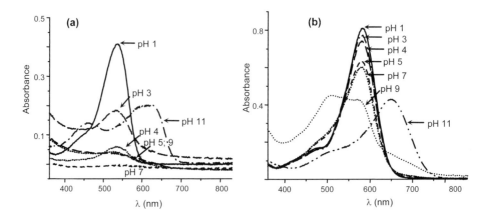

Fig. 9.8 Visible spectra of (a) mv3glc (33 μM) and (b) vinylpyrano-mv3glc-catechin (a portisin) (1.1μM) at different pH in a water/methanol (1:9, v/v). pH adjusted with HCl (de Freitas and Mateus 2006).

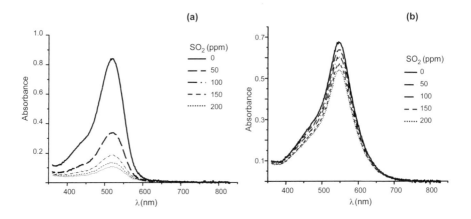

Fig. 9.9 Effect of the addition of increasing concentrations of SO₂ on the visible spectra of (a) mv3glc and (b) vinylpyrano-mv3glc-catechin (a portisin) at pH 3.0 in phosphate-citrate buffer. Pigment concentration of 0.08 mM (de Freitas and Mateus 2006).

In addition to these already appealing features, some portisins (namely vinylpyrano-mv3glc-phloroglucinol and vinylpyrano-mv3glc-catechin) were found to have a much higher molar extinction coefficient (ε) than genuine anthocyanins (approximately four to five times higher). For instance, the ε of vinylpyrano-mv3glc-catechin was estimated around 83000 $mol^{-1}dm^3cm^{-1}$, whilst the one of mv3glc determined in the same conditions is around 16000 $mol^{-1}dm^3cm^{-1}$. This color difference is particularly visible in an acidified alcoholic solution in which portisins display a much more intense blue color comparatively to the red color intensity of a malvidin-3-*O*-glucoside solution at the same concentration. This coloring capacity highlights the particular interest of using these pigments for the food industry, as only little amounts of portisins would be required to achieve the desired colors in the food matrixes.

The interest in this class of anthocyanin-derived pigments has thus gone beyond enology and they are now being assayed foreseeing a putative application as food colorants (Mateus et al. 2004a).

9.8 Future Perspectives for the Food Industry

The food industry has been using anthocyanins as food colorants through the addition of natural purified anthocyanin extracts to processed foodstuffs. The use of by-products of fruit and vegetable processing is challenging and requires cross-linking of different fields of research, which include food chemists, biochemists, nutritionists and toxicologists. Depending on the food matrix in which the anthocyanin extracts are intended to be used, other ingredients may be added in order to improve both solubility and color stability. With a correct formulation of different ingredients, as well as adequate processing and storage conditions of the food product, a wide range of stable and attractive color hues may be obtained for several food matrixes such as soft drinks, dairy and pastry products, chewing gums, dry powders, etc.

The growing competitive food market has forced food industries to search for new products able to attract more consumers, especially the younger generations. For these latter, appearance of the products is a crucial factor for their commercial success. Products such as "funny drinks" and dairy products with unusual and appealing colors like orange and blue appear to be interesting. The concept of a blue food or beverage seems odd and unappetizing, especially for older consumers, but cutting edge for younger consumers. Innovation driven by new attracting colors may be achieved by the use of anthocyanin-derived pigments displaying these unusual colors, like anthocyanin-pyruvic acid adducts and portisins. These compounds were found in food matrixes (red wine) and may be synthesized in the laboratory, thereby being classified as nature-identical colors. Their physical-chemical properties, especially their increased stability towards pH variations and SO_2 discoloration make them suitable targets to be developed as food colorants.

Besides the color appearance of new food products, efforts have been made in order to commercialize foods and beverages with health promoting attributes. These foodstuffs have been known as "functional foods". A definition has been attributed by the European consensus document (Bellisle et al. 1998) as follows: *A food can be regarded as "functional" if it is satisfactorily demonstrated to affect beneficially one or more target functions in the body, beyond adequate nutritional effects in a way that is relevant to either an improved state of health and well-being and/or reduction of risk of disease.* Indeed, health issues are always taken into account when developing new food products and the food industry have been trying replacing synthetic additives by natural or nature-like ones. This concern is the result of a commercial demand, namely by more apprehensive consumers regarding their diet. The typical consumer easily associates natural additives to health-promoting effects and synthetic additives to toxicological issues, especially allergies or digestive concerns. Anthocyanins are thus excellent candidates able to give the desired colors to some food products and at the same time please the consumer who sees these compounds as natural and with health benefits. The international scientific community can play an important role in the increase of acceptance of anthocyanins and especially anthocyanin-derived compounds by carrying out studies showing antioxidant and biological properties of these compounds. The more the public will be aware of health attributes of anthocyanin pigments, the more success anthocyanin containing-food products will have.

Replacing synthetic dyes by natural colors is a challenging task in which anthocyanin and derivatives may play an important role. Despite the number of factors that may limit the effective use of anthocyanin pigments in foodstuffs, the applications of these compounds is expected to grow significantly over the next years, especially if legislation restrictions concerning the use of several natural purified extracts were ended. More investigation studies will carry on and new insights yielded from different fields like chemistry or biotechnology, associated with industrial R&D projects developing industrial-scale extraction and purification processes, will probably increase the application of anthocyanins and anthocyanin-derived pigments in the food industry.

Acknowledgments

The authors would like to thank the Portuguese Foundation for Science and Technology (FCT), and the Portuguese Ministry of Agriculture (P.O. AGRO 386) for financial assistance.

References

Andersen, O.M. and Jordheim, M. (2005) The anthocyanins. In: O.M. Andersen and K.R. Markham (Eds.), *Flavonoids: Chemistry, Biochemistry and Applications*. CRC Press, pp. 471–552.

Bajaj, K. L., Kansal, B. D., Chadha, M. L. and Kaur, P. P. (1990) Chemical composition of some important varieties of egg plant (*Solanum melongena* L). Tropical Sci. 30, 255–261.

Bakker, J. and Timberlake, C. F. (1985) The distribution and content of anthocyanins in young Port wines as determined by high performance liquid chromatography. J. Sci. Food Agr. 36, 1325–1333.

Bassa, I. A. and Francis, F. J. (1987) Stability of anthocyanins from sweet potatoes in model beverages. J. Food Sci. 52, 1753–1754.

Bellisle, F., Blundell, J. E., Dye, L., Fantino, M., Fern, E., Fletcher, R. J., Lambert, J., Roberfoid, M., Specter, S., Westenhofer, J. and Westerterp-Plantenga, M. S. (1998) Functional food science and behaviour and psychological functions. Brit. J. Nutr. 80, S173-S193.

Berké, B., Chèze, C., Vercauteren J. and Deffieux G. (1998) Bisulfite addition to anthocyanins: Revisited structures of colourless adducts. Tetrahedron Lett. 39, 5771–5774.

Bianchi, L., Lazzé, C., Pizzala, R., Stivala, L. A., Savio, M. and Prosperi, E. (2001) Anthocyanins protect against oxidative damage in cell cultures. In: W. Pfannhauser, G. R. Fenwick and S. Khokhar (Eds.), *Biologically-Active Phytochemicals in Food*. The Royal Society of Chemistry, Cambridge, pp. 311–318.

Brouillard, R. (1982) Chemical structure of anthocyanins. In: P. Markakis (Ed.), *Anthocyanins as Food Colors*. Academic Press, New York, pp. 1–39.

Brouillard, R. (1983) The *in vivo* expression of anthocyanin colour in plants. Phytochemistry 22, 1311–1323.

Brouillard, R. and Dangles, O. (1994) Anthocyanin molecular interactions: the first step in the formation of new pigments during wine aging? Food Chem. 51, 365–371.

Brouillard, R. and Delaporte, B. (1977) Chemistry of anthocyanin pigments. 2. Kinetic and Thermodinamic study of proton transfer, hydration, and tautomeric reactions of malvidin 3-glucoside. J. Am. Chem. Soc. 99, 8461–8468.

Brouillard, R. and Dubois, J. E. (1977) Mechanism of the structural transformations of anthocyanins in acidic media. J. Am. Chem. Soc. 99, 1359–1364.

Brouillard, R. and Lang, J. (1990) The hemiacetal-*cis*-chalcone equilibrium of malvin, a natural anthocyanin. Canad. J. Chem. 68, 755–761.

Cameira-dos-Santos, P., Brillouet, J. M., Cheynier, V. and Moutounet, M. (1996) Detection and partial characterisation of new anthocyanin-derived pigments in wine. J. Sci. Food Agr. 70, 204–208.

Cao, G., Sofic, E. and Prior, R. L. (1997) Antioxidant and prooxidant behaviour of flavonoids: structure-activity relationships. Free Rad. Biol. Med. 22, 749–760.

Clifford, M. N. (2000) Anthocyanins – nature, occurrence and dietary burden. J. Sci. Food Agric. 80, 1063–1072.

Coisson, J. D., Travaglia, F., Piana, G., Capasso, M. and Arlorio, M. (2005) *Euterpe oleracea* juice as a functional pigment for yogurt. Food Res. Int. 38, 893–897.

da Costa, C. T., Horton, D. and Margolis, S. A. (2000) Analysis of anthocyanins in foods by liquid chromatography, liquid chromatography-mass spectrometry and capillary electrophoresis. J. Chromatogr. A. 881, 403–410.

Dambergs, R. G., Cozzolino, D., Cynkar, W. U., Janik, L. and Gishen, M. (2006) The determination of red grape quality parameters using the LOCAL algorithm. J. Near Infrared Spec. 14, 71–79.

Dangles, O., Saito, N. and Brouillard, R. (1993) Kinetic and thermodynamic control of flavylium hydration in the pelargonidin cinnamic acid complexation – origin of the extraordinary flower color diversity of Pharbitis- Nil. J. Am. Chem. Soc. 115, 3125–3132.

Davies, A. J. and Mazza, G. (1993) Copigmentation of simple and acylated anthocyanins with colourless phenolic compounds. J. Agr. Food Chem. 41, 716–720.

de Freitas, V. A. P. and Mateus, N. (2006) Chemical transformations of anthocyanins yielding a variety of colours (Review). Environm. Chem. Lett. 14, 175–183.

Degenhardt, A. and Winterhalter P. (2001) Isolation of natural pigments by high speed CCC. J. Liq. Chomatogr. 24, 1745–1764.

Dieci, E. (1967) Sull'enocianin tecnica. Riv. Vitic. Enol. 12, 567–573.

Dyrby, M., Westergaard, N. and Stapelfeldt, H. (2001) Light and heat sensitivity of red cabbage extract in soft drink model systems. Food Chem. 72, 431–437.

Es-Safi, N. E., Cheynier, V. and Moutounet, M. (2002) Interactions between cyanidin 3-O-glucosides and furfural derivatives and their impact on food color change. J. Agr. Food Chem. 50, 5586–5595.

Favretto, D. and Flamini, R. (2000) Application of electrospray ionization mass spectrometry to the study of grape anthocyanins. Am. J. Enol. Viticult. 51, 55–64.

Figueiredo, P., George, F., Tatsuzawa, F., Toki, K., Saito, N. and Brouillard, R. (1999) New features of intramolecular copigmentation by acetylated anthocyanins. Phytochemistry 1, 125–132.

Fossen, T. and Andersen, O. M. (1999) Delphinidin 3'-galloylgalactosides from blue flowers of Nymphaea caerulea. Phytochemistry 50, 1185–1188.

Foucault, A. P. and Chevolot, L. (1998) Counter-current chromatography: instrumentation, solvent selection and some recent applications to natural product purification. J. Chromatogr. A. 808, 3–22.

Francia-Aricha, E. M., Guerra, M. T., Rivas-Gonzalo, J. C. and Santos-Buelga, C. (1997) New anthocyanin pigments formed after condensation with flavanols. J. Agr. Food Chem. 45, 2262–2265.

Francis, F. J. (1982) Analysis of anthocyanins. In: P. Markakis (Ed.), *Anthocyanins as Food Colors*. Academic Press, New York, pp. 181–207.

Francis, F. J. (1989). Food colorants: anthocyanins. Crit. Rev. Food Sci. 28, 273–314.

Fulcrand, H., Benabdeljalil, C., Rigaud, J., Cheynier, V. and Moutounet, M. (1998) A new class of wine pigments generated by reaction between pyruvic acid and grape anthocyanins. Phytochemistry 47, 1401–1407.

Fulcrand, H., Cameira dos Santos, P. J., Sarni-Manchado, P., Cheynier, V. and Bonvin, J. F. (1996) Structure of new anthocyanin-derived wine pigments. J. Chem. Soc. Perkin Trans. 1, 735–739.

Furtado, P., Figueiredo, P., Chaves das Neves, H. and Pina, F. (1993) Photochemical and thermal degradation of anthocyanidins. J. Photochem. Photobiol. A: Chem. 75, 113–118.

Gao, L. and Mazza, G. (1995) Characterization, quantitation and distribution of anthocyanins and colourless phenolics in sweet cherries. J. Agr. Food Chem. 43, 343–346.

Garcia-Viguera, C. and Bridle, P. (1999) Influence of structure on colour stability of anthocyanins and flavylium salts with ascorbic acid. Food Chem. 64, 21–26.

Garcia-Viguera, C., Zafrilla, P. and Tomas-Barberan, F. A. (1998) The use of acetone as an extraction solvent for anthocyanins from strawberry fruit. Phytochem. Anal. 9, 274–277.

Giusti, M. M., Rodriguez-Saona, L. E., Baggett, J. R., Reed, G. L., Durst, R. W. and Wrolstad, R. E. (1998) Anthocyanin pigment composition of red radish cultivars as potential food colorants. J. Food Sci. 63, 219–224.

Giusti, M. M., Rodriguez-Saona, L. E. and Wrolstad, R. E. (1999) Molar absorptivity and color characteristics of acylated and non-acylated pelargonidin-based anthocyanins. J. Agr. Food Chem. 47, 4631–4637.

Giusti, M. M. and Wrolstad, R. E. (1996a) Characterization of red radish anthocyanins. J. Food Sci. 61, 322–326.

Giusti, M. M. and Wrolstad, R. E. (1996b) Radish anthocyanin extract as a natural red colorant for Maraschino cherries. J. Food Sci. 61, 688–694.

Goto, T. and Kondo, T. (1991) Structure and molecular stacking of anthocyanins – flower color variation. Angew. Chem. Int. Ed. Engl. 30, 17–33.

Goto, T., Takase, S. and Kondo, T. (1978) PMR spectra of natural acylated antocyanins. Determination of stereostructure of awobanin, shisonin and violanin. Tetrahedron Lett. 27, 2413–2416.

Harborne, J. B. (1984) *Phytochemical Methods. A Guide to Modern Techniques of Plant Analysis*, 2nd Ed. Chapman & Hall, London.

Haslam, E. (1980) In vino veritas: oligomeric procyanidins and the ageing of red wines. Phytochemistry 19, 2577–2582.

Haslam, E. (1988) *Practical Polyphenolics. From Structure to Molecular Recognition and Physiological Action*. Cambridge University Press, Cambridge, pp. 262–294.

Henry, B. S. (1996) Natural food colors. In: G. A. F. Hendry and J. D. Houghton (Eds.), *Natural Food Colorants*, Blackie Academic of Professional, pp. 40–79.

Hofmann, T. (1998) Studies on the influence of the solvent on the contribution of single Maillard reaction products to the total color of browned pentose/alanine solutions – a quantitative correlation using the color activity concept. J. Agr. Food Chem. 46, 3912–3917.

Ichiyanagi, T., Tateyama, C., Oikawa, K. and Konishi, T. (2000) Comparison of anthocyanin distribution in different blueberry sources by capillary zone electrophoresis. Biol. Pharm. Bull. 23, 492–497.

Inami, O., Tamura, I., Kikuzaki, H. and Nakatani, N (1996) Stability of anthocyanins of *Sambucus candensis* and *Sambucus nigra*. J. Agr. Food Chem. 44, 3090–3096.

Jackman, R. L. and Smith, J. L. (1996) Anthocyanins and betalains. In: G. A. F. Hendry and J. D. Houghton (Eds.), *Natural Food Colorants*, Blackie Academic of Professional, pp. 245–309.

Jing, P. and Giusti, M. M. (2005) Characterization of anthocyanin-rich waste from purple corncobs (*Zea mays L.*) and its application to color milk. J. Agr. Food Chem. 53, 8775–8781.

Jurd, L. (1967) Catechin flavylium salt condensation reactions. Tetrahedron 23, 1057–1064.

Jurd, L. and Somers, T. C. (1970) The formation of xanthylium salts from proanthocyanidins. Phytochemistry 9, 419–427.

Kühnau, J. (1976) The flavonoids. A class of semi-essential food components: their role in human nutrition. World Rev. Nutr. Diet. 24, 117–191.

Lewis, C. E., Walker, J. R. L. and Lancaster, J. E. (1995) Effect of polysacharides on the colour of anthocyanins. Food Chem. 54, 315–319.

Lopes-da-Silva, F., de Pascual-Teresa, S., Rivas-Gonzalo, J. and Santos-Buelga, C. (2002) Identification of anthocyanin pigments in strawberry (cv Camarosa) by LC using DAD and ESI-MS detection. Eur. Food Res. Technol. 214, 248–253.

Malien-Aubert, C., Dangles, O. and Amiot, M. J. (2001) Color stability of commercial anthocyanin-based extracts in relation to the phenolic composition. Protective effects by intra- and inter-molecular copigmentation. J. Agr. Food Chem. 49, 170–176.

Markakis, P. (1982) *Anthocyanins as Food Colors*. Academic Press, London.

Mateus, N., Carvalho, E., Carvalho, A., Melo, A., Santos-Buelga, C, Silva, A. M. S. and de Freitas, V. A. P. (2003a) Isolation and structural characterization of New Acylated Anthocyanin-Vinyl-Flavanol pigments occurring in aging red wines. J. Agr. Food Chem. 51, 277–282.

Mateus, N., Oliveira, J., González-Paramás, A. M., Santos-Buelga, C. and de Freitas, V. A. P. (2005) Screening of portisins (vinylpyranoanthocyanin pigments) in Port wines by LC/DAD-MS. Food Sci. Technol. Int. 11, 353–358.

Mateus N., Oliveira, J., Haettich-Motta, M. and de Freitas, V. A. P. (2004a) New family of bluish pyranoanthocyanins J. Biomed. Biotechnol. 299–305.

Mateus, N., Oliveira, J. Pissarra, J. González-Paramás, A. M., Rivas-Gonzalo, J. C., Santos-Buelga, C. and de Freitas, V. A. P. (2006) A new vinylpyranoanthocyanin pigment occurring in ageing red wine. Food Chem. 97, 689–695.

Mateus, N., Oliveira, J., Santos-Buelga, C., Silva, A. M. S. and de Freitas, V. A. P. (2004) NMR structure characterization of a new vinylpyranoanthocyanin-catechin pigment (a portisin). Tetrahedron Lett. 45, 3455–3457.

Mateus, N., Pascual-Teresa, S., Rivas-Gonzalo, J. C., Santos-Buelga, C. and de Freitas, V. A. P. (2002a) Structural diversity of anthocyanin-derived pigments detected in Port wines. Food Chem. 76, 335–342.

Mateus, N., Silva, A. M. S., Rivas-Gonzalo, J. C., Santos-Buelga, C. and de Freitas, V. A. P. (2003b) A new class of blue anthocyanins-derived pigments isolated from red wines. J. Agr. Food Chem. 51, 1919–1923.

Mateus, N., Silva, A. M. S., Rivas-Gonzalo, J. C., Santos-Buelga, C. and de Freitas, V. A. P. (2002b) Identification of anthocyanin-flavanol pigments in red wines by NMR and mass spectrometry. J. Agr. Food Chem. 50, 2110–2116.

Mazza, G. and Miniati, E. (1993) *Anthocyanins in Fruits, Vegetables and Grains*. CRC press, London.

Mazza, G. and Brouillard, R. (1990) The mechanism of copigmentation of anthocyanins in aqueous solutions. Phytochemistry 29, 1097–1102.

Mistry, T. V., Cai, Y., Lilley, T. H. and Haslam, E. (1991) Polyphenol interactions. Part 5 Anthocyanin co-pigmentation. J. Chem. Soc. Perkin Trans. 2, 1287–1296.

Odake, K., Terahara, N., Saito, N., Toki, K. and Honda, T. (1992) Chemical structures of two anthocyanins from purple sweet potato, *Ipomoe batatas*. Phytochemistry 31, 2127–2130.

Oliveira, J., de Freitas, V. A. P., Santos-Buelga, C., Silva, A. and Mateus, N. (2006) Chromatic and structural features of blue anthocyanin-derived pigments present in Port wine. Anal. Chim. Acta 563, 2–9.

Otsuki, T., Matsufuji, H., Takeda, M., Goda, Y. Acylated anthocyanins from red radish (*Raphanus sativus L.*). Phytochemistry 60(1), 79–87.

Pazmiño-Durán, E. A., Giusti, M. M., Wrolstad, R. E. and Gloria, M. B. A. (2001a) Anthocyanins from banana bracts (*Musa X paradisiaca*) as potential food colorants. Food Chem. 73, 327–332.

Pazmiño-Durán, E. A., Giusti, M. M., Wrolstad, R. E. and Gloria, M. B. A. (2001b) Anthocyanins from *Oxalis triangularis* as potential food colorants. Food Chem. 73, 211–216.

Prior, R. L. (2003) Absorption and metabolism of anthocyanins: potential health effects. In: Mark S. Meskin, Wayne R. Bidlack, Audra J. Davies, Douglas S. Lewis, and R. Keith Randolph (Eds.), *Phytochemicals: Mechanisms of Action*. Boca Raton, Fla., CRC press, London.

Remy, S., Fulcrand, H., Labarbe, B., Cheynier, V. and Moutounet, M. (2000) First confirmation in red wine of products resulting from direct anthocyanin-tannin reactions. J. Sci. Food Agric. 80, 745–751.

Rivas-Gonzalo, J. C. (2003) Analysis of anthocyanins. In: C. Santos-Buelga and G. Williamson (Eds.), *Methods in Polyphenol analysis*. The Royal Society of Chemistry, Cambridge, pp. 338–353.

Rivas-Gonzalo, J. C., Bravo-Haro, S. and Santos-Buelga, C. (1995) Detection of compounds formed through the reaction of malvidin-3-monoglucoside and catechin in the presence of acetaldehyde. J. Agr. Food Chem. 43, 1444–1449.

Robinson, G. M. and Robinson, R. A. (1931). A survey of anthocyanins. Biochem. J. 25, 1687–1705.

Rodriguez-Saona, L. E., Giusti M. M. and Wrolstad, R. E. (1998) Anthocyanin pigment composition of red-fleshed potatoes. J. Food Sci. 63, 458–465.

Rodriguez-Saona, L. E., Giusti M. M. and Wrolstad, R. E. (1999) Color and pigment stability of red radish and red-fleshed potato anthocyanins in juice model sytems. J. Food Sci. 64, 451–456.

Rodriguez-Saona, L. E., Giusti, M. M., Durst, R. W. and Wrolstad, R. E. (2001) Development and process optimization of red radish concentrate extract as potential natural red colorant. J. Food Process. Pres. 25, 165–182.

Rommel, A., Wrolstad, R. E. and Heatherbell, D. A. (1992) Blackberry juice and wine: processing and storage effects on anthocyanin composition, color and appearance. J. Food Sci. 57, 385–391.

Saito, N., Tatsuzawa, F., Yoda, K., Yokoi, M., Kasahara, K., Lida, S, Shigihara, A. and Honda, T. (1995) Acylated cyanidin glycosides in the violet-blue flowers of *Ipomoea purpurea*. Phytochemistry 40, 1283–1289.

Sarni-Manchado, P., Cheynier, V. and Moutounet, M. (1997) Reaction of enzymatically generated quinones with malvidin-3-glucoside. Phytochemistry 45, 1365–1369.

Scalbert, A. and Williamson, G. (2000) Dietary intake and bioavailability of polyphenols. J. Nutr. 130, 2073–2085.

Schwarz, M., Hillebrand, S., Habben, S., Degenhardt, A. and Winterhalter, P. (2003) Application of high-speed countercurrent chromatography to the large-scale isolation of anthocyanins. Biochem. Eng. J. 14, 179–189.

Schwarz, M., Wabnitz, T. C. and Winterhalter P. (2003) Pathway leading to the formation of anthocyanin-vinylphenol adducts and related pigments in red wines J. Agr. Food Chem. 51, 3682–3687.

Shimizu, T., Muroi, T., Ichi, T., Nakamura, M. and Yoshihira, K. (1997) Analysis of red cabbage colors in commercial food using high performance liquid chromatography with photodiode array detection-mass spectrometry. J. Food Hyg. Soc. Jap. 38, 34–38.

Skrede, G., Wrolstad, R. E., Lea, P. and Enersen, G. (1992) Color stability of strawberry and blackcurrant syrups. J. Food Sci. 57, 172–177.

Somers, T. C. (1971) The polymeric nature of wine pigments. Phytochemistry 10, 2175–2186.

Somers, T. C. (1966) Wine tannins isolation of condensed flavanoid pigments by gel-filtration. Nature 209, 368–370.

Stintzing, F. C., Conrad, J., Iris, K., Beifuss, U. and Carle R. (2004) Structural investigations on betacyanin pigments by LC NMR and 2D NMR spectroscopy. Phytochemistry 65, 415–422.

Stintzing, F. C., Stintzing, A. S., Carle, R., Frei, B. and Wrolstad, R. E. (2002) Color and antioxidant properties of cyaniding-based anthocyanin pigments. J. Agr. Food Chem. 50, 6172–6181.

Strack, D. and Wray V. (1989) Anthocyanins. In: P. M. Dey and J. B. Harborne (Eds.) *Methods in Plant Biochemistry-Vol 1*, Academic Press, New York, pp. 325–356.

Teh, L. S. and Francis, F. J. (1988) Stability of anthocyanins from *Zebrina pendula* and *Ipomoe tricolor* in a model beverage. J. Food Sci. 53, 1580–1581.

Terahara, N., Shimizu, T., Kato, Y., Nakamura, M., Maitani, T., Yamaguchi M.-A. and Goda, Y. (1999) Six diacylated anthocyanins from the storage root of purple sweet potato, *Ipomoe batatas*. Biosci. Biotechnol. Biochem. 63, 1420–1424.

Timberlake, C. F. and Bridle, P. (1976) Interactions between anthocyanins, phenolic compounds and acetaldehyde and their significance in red wines. Am. J. Enol. Vitic. 27, 97–105.

Torre, L. C. and Barrit, B. H. (1977) Quantitative evaluation of *Rubus* fruit anthocyanins pigments. J. Food Sci. 42, 488–490.

Vivar-Quintana, A. M., Santos-Buelga, C., Francia-Aricha, E. and Rivas-Gonzalo, J. C. (1999) Formation of anthocyanin-derived pigments in experimental red wines. Food Sci. Technol. Int. 5, 347–352.

Wang, H., Cao, G. and Prior, R. L. (1997) Oxygen radical absorbing capacity of anthocyanins. J. Agr. Food Chem. 45, 304–309.

Wildenradt, H. L. and Singleton, V. L. (1974) The production of aldehydes as a result of oxidation of polyphenolic compounds and its relation to wine aging. Am. J. Enol. Vitic. 26, 25–29.

Willstatter, R. and Zollinger, E. H. (1916) Uber die Farbstoffe der Weintraube und der Heidelbeere II. Justus Liebig´s Annalen. Chem. 412, 195–216.

Wilska-Jeszka, J., Bos, J. and Pawlak, M. (1992) Wild plant fruits as a source of catechins and proanthocyanidins. In: XVIe Journées Internationales Groupe Polyphenols and The Royal Society of Chemistry, Lisbon. DTA, Lisbon, pp. 246–250.

Wolfender, J.-L., Ndjoko, K. and Hostettmann, K. (2003) Application of LC-NMR in the structure elucidation of polyphenols. In: C. Santos-Buelga and G. Williamson (Eds.), *Methods in Polyphenol Analysis*, Royal Society of Chemistry, Cambridge, Chapter 6.

Wolfender, J.-L., Ndjoko, K. and Hostettmann, K. (1998) LC/NMR in natural products chemistry. Curr. Org. Chem. 2, 575.

Wrolstad, R. E., Skrede, G., Lea, P. and Enersen, G. (1990) Influence of sugar on anthocyanin pigment stability in frozen strawberries. J. Food Sci. 55, 1209–1212.

Wulf, L. W. and Nagel, C. W. (1978) High-pressure liquid chromatographic separation of anthocyanins of Vitis vinifera. Am. J. Enol. Vitic. 29, 42–49.

10

Interactions Between Flavonoids that Benefit Human Health

Mary Ann Lila

University of Illinois, Department of Natural Resources & Environmental Sciences,
imagemal@uiuc.edu

Abstract. Interactions between anthocyanin pigments and other flavonoids or other phyto-chemicals accumulating within a plant contribute significantly to the ability of natural plant extracts (ingested as food or pharmaceutical product) to protect human health or mitigate disease damage. Interactions are the rule, rather than the exception, for biologically-active plant-derived components. Given the complex, multi-faceted roles interacting phytochemicals play in the human body, the use of plant cell culture production and research models can contribute novel insights as to competing mechanisms of action, bioavailability, and distribution in situ.

10.1 Phytochemical Interactions

When science attempts to unravel the mechanisms by which natural products are able to inhibit disease symptoms or otherwise enhance human health, a daunting obstacle that complicates and obstructs research progress is invariably encountered: *the responsible natural chemicals seldom work independently*. Instead, a complex interplay between diverse chemical components modifies the pharmacological effects of functional foods, nature-derived supplements or pharmaceuticals, as they modulate biological processes in human metabolism.

The prevalence of these potentiating chemical interactions has been increasingly documented in recent years, especially for *plant*-derived extracts with bioactive properties. The bioactive constituents in edible plants are always ingested in the form of natural mixtures. The epidemiological evidence for health benefits of certain ethnic diets may actually be accounted for by specific interactions between phyto-chemicals in the most common traditional foods (Lila and Raskin 2005). In other cases, the harsh processing involved in formulation of some plant extracts into dietary supplements can disrupt the delicate associations between natural co-occurring phytochemicals, resulting in erratic or much-attenuated efficacy in the commercial product. Interactions make the whole realm of natural product science both fascinat-

K. Gould et al. (eds.), *Anthocyanins*, DOI: 10.1007/978-0-387-77335-3_10,

ing and bewildering at the same time, because it is no small feat of diagnosis to sort out the complexities of simultaneous contributions from multiple chemicals in a natural mixture. (Even single phytochemical compounds, once successfully isolated and purified from a mixed extract, can be sufficiently complex that they elude attempts at chemical synthesis, or structural characterization by existing analytical technology.)

The typical reductionist approaches used to gauge potency of drugs in the pharmaceutical industry, all based on single active ingredients, are not adequate to assess the pharmacological properties of natural products, given the multifaceted, multitiered interactions between components in a mixture and diverse mechanisms of action that can be simultaneously at work. Since natural interactions are difficult to pinpoint and define, commercial standardization and scientific authentication may elude some of the most powerful nature-derived mixtures.

Biologically-active natural chemicals, once ingested by a human or animal, may interact during metabolism to alter biological responses, with either positive or negative consequences. That is, these interactions can either (1) intensify, or (2) interfere with, the bioactive potency of chemicals in the extract. From an efficacy standpoint, some of the most interesting are positive interactions that intensify, or potentiate, a benefit for human health, beyond the benefits accrued by consuming a single natural chemical alone. The biological significance of natural compounds in situ can contrast sharply with that measured in bioassays of isolated purified components. Although some loss of activity may be attributable to decomposition of active compounds during isolation and purification, changes in activity may also be explained by subtraction, during the purification steps, of important components of interacting systems. As has been noted by toxicologists, biologists, and medical practitioners alike, potency is frequently attenuated when biologically active mixtures (extracts) are separated into purified components and administered separately (Lila and Raskin 2005; Liu 2003).

Interactions that potentiate the bioactivity in a mixture of natural chemicals can be classified as additive (the combined effects of the two chemicals acting together amount to the sum total of either one acting independently), or the interactions may result in a synergistic boost to biological activity (the potency is significantly greater than the sum of the biologically active phytochemicals if administered separately). The benefits associated with lycopene consumption for protection against human prostate carcinogenesis, for example, are amplified through both additive and synergistic interactions with various other tomato carotenoids, and with polyphenolic phytochemicals also abundant in the tomato fruit (Boileau et al. 2003; Campbell et al. 2006a, 2006b). Similarly, enhanced anticarcinogenic protection after co-administration of phytochemicals (including lycopene) in tomato fruits with phytochemicals present in broccoli indicate potentiation between these phytochemical mixtures (Canen-Adams et al. 2007). Synergy between ethanol and polyphenolic compounds from grapes was demonstrated in the inhibition of the inducible nitric oxide synthase pathway, suggesting potential health benefits for red wine in atherosclerosis and tumor initiation (Chan et al. 2000). Potentiating interactions between soy isoflavones genistein and daidzein synergistically inhibited neoplastic transformation (Franke et al. 1998). Nishida and Satoh (2004) demonstrated that extracts of

Ginkgo biloba increase peripheral circulation and slow age-related mental degradation as consequence of the interplay between inherent flavonoids (mostly proanthocyanidins and rutins) and terpene lactones (ginkgolides and bilobalide). These two classes of phytochemicals have distinct and complementary mechanisms of action that together result in potentiation: the flavonoids are antioxidants and enhance blood flow, whereas the terpene lactones are platelet activating factors. Mixtures of flavonoids and terpenes are also typical for many berry fruits and may account for their utility for improving circulation and slowing age-related cognitive losses (Joseph et al. 1999).

Dramatic evidence for the necessity of multicomponent mixtures in order to achieve bioactive potency has come from entomological studies. Extraction of plants with polar solvents will successfully isolate most compounds responsible for oviposition of insects, however, all biological activity disappears upon fractionation of these extracts (Honda 1990). As an example, methanolic extracts of white clover were able to induce oviposition by a pierid butterfly, and the active components were found to be water soluble, although fractionation of the extract led to loss of activity. Recombination of two separated fractions, one containing cyanogenic glucosides linamarin and lotaustralin and the other a group of ubiquitous carbohydrates, restored activity (Honda et al. 1997). In this case, the two groups of compounds had clear synergistic activity. In other examples, the active compounds included flavonoids and the glycoside of an amine, bufotenin (Nishida et al. 1990).

Tripterigium, a plant genus with powerful anti-inflammatory properties used an ingredient in Traditional Chinese Medicine, has been primarily considered on the basis of its main bioactive ingredient (the diterpene triepoxide triptolide). However, recent analysis has now revealed a series of diverse bioactive phytochemical classes that are collectively responsible for synergistically working in concert to give this herb efficacy against both autoimmune and inflammation-related conditions (Brinker et al. 2007). When synergies are evident, it is likely an indication that individual components in the biologically-active mixture are using different, complementary mechanisms of action to interact with human therapeutic targets (Lila and Raskin 2005). However, in general, the degree of potentiation must be experimentally quantified and statistically-demonstrated before assigning additive or synergistic labels to the observed phytochemical interaction (Mertens-Talcott et al. 2003).

Interactions are labeled as concomitant when one or more of the interacting phytochemicals does not appear to have any biological activity towards human health by itself, but the association of these phytochemicals together with other active compounds will improve the bioavailability, solubility or absorption of the bioactive constituents (Eder and Mehnert 1998; Gartner et al. 1997). A concomitant interaction may also improve the overall safety of an active phytochemical, when it is metabolized in the body; for example, whereas whole-soy protein containing diets have demonstrated significant chemoprotective benefits, researchers have shown that the isolated isoflavones genistin or genestein from soy can actually promote estrogen-dependent tumor proliferation, if not administered in concert with other soy phytochemicals in a natural mixture (Allred et al. 2001, 2004). Similarly, mixtures of mostly high molecular weight proanthocyanidins from grape provide selective cytotoxicity against liver cancer cell lines (no cytotoxicity to non-cancerous cells), but, if the proanthocyanidins are further subfractionated and purified, and lose natural associations with other natural co-occurring flavonoids, the selectivity is lost, the extracts become slightly cytotoxic even to normal cell lines (Jo et al. 2006b). Finally, phytochemical interactions can be considered negative (antagonistic or interfering) when certain constituents of a biological mixture inhibit the potency of the active compound(s). Adverse drug interactions generally fall into this latter category, and the interacting chemicals (natural or synthetic) are not usually produced by a single plant.

10.2 *Why* Are Multiple Bioactive Phytochemicals Usually Involved?

Plants have evolved the strategy of producing multiple phytochemicals, many with overlapping physiological roles but slightly different mechanisms of action, for good reason. As sessile organisms, plants must rely on chemical and physical adaptations through the production of secondary compounds for defense, protection, cell-to-cell signaling, and other adaptations. The secondary compounds (many of which also serve the health protective roles when ingested by animals) serve a myriad of supplemental roles in a plant's life cycle. The broad-spectrum array of chemicals that may be involved, and the interactions between these diverse chemicals, have co-evolved over an extraordinary time span along with the animals and insects and microbes that interface with stationary plants, resulting in many-layered, sometimes diabolical plant defensive strategies.

These chemicals typically occur in diverse families that interact synergistically or additively to broaden the range and the efficacy of survival tactics. By using this ingenious combined strategy, each plant is less susceptible to ultimate harm; if a pathogen evolves or mutates to overcome one defensive mechanism, there will be other, different chemical defenses still in place to prevent pathogen infestation. Within the plethora of co-occurring bioactive phytochemicals, each may employ a slightly different mechanism of action (modus operandi) for counteracting the pathogen, making it unlikely that any invading organism could overcome them all.

During a plant's life cycle, the mixed array of secondary phytochemicals plays a unique role in ecophysiology. Phytochemicals have *protective* roles when they act as shields between the delicate plant tissues and abiotic stress factors that threaten to damage them. Anthocyanin pigments are an excellent example of a stress-induced secondary compound which confers protection to the host plant. Vulnerability of the plant to UV light exposure is reduced by the pigments simultaneously scavenging free radicals and protecting chloroplasts from photodegradation (Gould 2004). Stress factors which trigger protective phytochemical production include cold temperature/frost, low (negative) water potential around the root system, UV-B irradiance, and mineral stress. Secondary products can assume an active *defensive* role when they act to disrupt feeding of an insect pest or prevent pathogen colonization. Many secondary phytochemicals produced by plants are classified as phytoalexins, which refers to non-specific toxins (bactericides or fungicides) produced by higher plants in response to pathogen attack (or similar stress). For example, lignans and isoflavones may accumulate in response to insect damage and will repel continued feeding on the foliage; these same secondary products can be fungistatic. Phytoalexins have been likened to antibodies produced in mammalian systems, because their production is directly triggered by elicitors. Similarly, the phytochemicals can inhibit herbivory from predators. Other examples of defensive secondary products are those produced in response to inter-plant competitive stress. Invasive plant species may succeed in nature in part due to their production and secretion of phytotoxic allelochemicals, which displace or discourage competing plant species.

Secondary phytochemicals also serve as attractants for the plant hosts, and the primary pigments including anthocyanins (in the flavonoid class) fall into this category. The phytochemicals guide potential pollinators to the flower, or entice animals to taste the fruit and therefore aid in dissemination of seed at a distant location. Here, the strategy of multiple interacting phytochemicals is again highly effective, because the pollinator or consumer of the fruit will be attracted in more than one way to visit the host plant, for example by the combination of color and scent. The secondary phytochemicals also may play a plethora of other roles for the plant host; they may mimic plant hormones and regulate growth and development, modulate gene expression, allow signal transduction with symbiotic organisms, and other functions. So, the phytochemicals contribute to a wealth of survival strategies and competitive advantages for the host plants.

In many cases, the plant won't produce the bioactive compounds unless the synthesis is triggered by some stress factor. The provocation can take the form of cold stimulus, excessive irradiance, drought, poor fertility, insect feeding, pathogen invasion, or other form of insult. As was noted above, in most of the cases, the accumulation of the secondary product will help the plant cope with the environmental insult, and continue to survive and thrive.

These same interactions that bolster a plant's resistance can similarly offer a many sided defense against human diseases by providing, typically, several different mechanisms of combating symptoms at once. In this era we have increasingly seen how rapidly cancerous cells or human pathogens can develop resistance to single ingredient pharmaceutical drugs. The plant's strategy, which relies on interaction of phytochemical complexes that target multiple molecules, not only circumvents resistance of plant pathogens, but of diseases or pathogens that invade human systems. The multifaceted strategies offered by the natural product are a means to broader spectrum protection.

10.3 Flavonoid Interactions that Potentiate Biological Activity

Interactions between flavonoids, including anthocyanins and other flavonoid classes, are increasingly cited as responsible for more intense potency of natural mixtures (as compared to purified compounds) in both in vitro and in vivo bioactivity trials. Seeram et al. (2004) determined that phytochemical constituents in the American cranberry, while individually efficacious against human carcinogenesis, provided maximum protection only when co-administered in natural mixtures. In this study, potential synergistic antiproliferative effects from mixtures of anthocyanins, proanthocyanidins, and flavonol glycosides were suggested. The research team also identified interferences that occurred between components in the extract including organic and phenolic acids and sugar. When semipurified polyphenolics were recombined together, a synergistic boost in activity occurred. In other research, combinations of two polyphenolic compounds from grapes (resveratrol and quercetin) demonstrated synergistic ability to induce apoptosis (activating caspase-3) in a human pancreatic carcinoma cell line (Mouria et al. 2002). Similarly, a mixed polyphenolic extract from red wine demonstrated stronger inhibition of DNA synthesis in oral squamous carcinoma cells than individual compounds, even when the concentrations of individually-administered quercetin or resveratrol were higher than those in the mixed extract (Elattar and Virji 1999).

Similarly, interactions between catechins in tea were responsible for protecting cells against damage induced by exposure to lead (Chen et al. 2002). When HepG2 cells were exposed to lead, the most protective influence was achieved by a synergistic interaction between epicatechin and epigallocatechin gallate (EGCG) components of green tea. EGCG administered at high doses in a purified form can damage cells as it will act as a reactive oxygen species (ROS) producer, yet at low concentrations or when mixed with other flavonoids, EGCG acts as an ROS scavenger. In many cases, the reactivity of a prominent flavonoid compound can be modulated by other flavonoids that co-occur in a natural protective mixture from a plant.

Mertens-Talcott et al. (2003) evaluated the interactions between two common grape polyphenolics, quercetin and ellagic acid, at low, physiologically-relevant concentrations, toward inhibition of a human leukemia cell line. Combination of these two compounds greatly reduced proliferation and viability, and induced apoptosis. The synergistic nature of this interaction was confirmed by an isobolographic analysis of the cell proliferation data. In a subsequent study, the research team determined that potentiation between these two polyphenols occurred for some mechanisms such as activation of MAP kinases, but, synergies were not apparent for other mechanisms of action (Mertens-Talcott et al. 2005).

The multiplicity of health benefits associated with consumption of anthocyanin-enriched pomegranate fruits and juices have been largely attributed to flavonoid content, however, the direct contribution of individual components or mixtures have not been established (Aviram et al. 2000, 2002; Hora et al. 2003). Recently, proliferation in DU 145 human prostate cancer cell lines was found to be suppressed by synergistic interactions between polyphenolic phytochemicals from anatomically discrete portions of the pomegranate fruit; fermented juice (quercetin, rutin, phenolic acids, ellagotannins, delphinidins), pericarp (polyphenols with estrogenic activity not found in juice like kaempferol and luteolin, and prodelphinidins), and seed oils (sterols, tocopherols, and steroids) (Lansky et al. 2005). These results suggested that the discrete compounds targeted different signaling pathways within the same cancer cells, and the effect was more than additive. Supra-additive, complementary and synergistic antiproliferative effects were demonstrated repeatedly, as well as demonstrations of synergism between components in suppressing invasiveness of PC-3 prostate cancer cells, or suppressing secretory phospholipase. The authors pointed out that the therapeutic advantage obtained by mixing compounds (rather than purifying and concentrating a dose of single actives) should provoke a radical overhauling of conventional strategies for drug development when mining the plant kingdom for active components.

Like the pomegranate research above, research on grape extracts has also demonstrated synergistic interactions between the fruit and the seed polyphenolics, for potentiating a protective effect against cardiovascular disease. Research involving both human and canine subjects demonstrated that maximal antiplatelet benefits were achieved only when extracts from seeds and fruit skins were combined (Shanmuganayagam et al. 2002). Synergistic inhibition of platelet function was also observed with flavonoid combinations of quercetin and catechin, which effectively inhibited intracellular production of hydrogen peroxide (Pignatelli et al. 2000).

Flavonoid interactions, as well as potential interactions between anthocyanins and lipophilic fruit constituents, appear to account for differences in the chemopreventive value of European and American elderberry fruits (Thole et al. 2006). European elderberry (*Sambucus nigra*) has been intensively bred for generations to favor high anthocyanin production, and the fruits have a recognized high antioxidant capacity. American elderberry (*Sambucus canadensis*) on the other hand is a wild species in North America. When compared side by side in a series of cancer-chemoprevention assays, aqueous acetone extracts from both species demonstrated significant ability to inhibit the initiation and promotion stages of carcinogenesis (strong induction of quinine reductase Phase II enzyme, and inhibition of cyclooxygenase-2, respectively). In addition, the American elderberry extracts were inhibitory towards the ornithine decarboxylase enzyme, indicative that this plant was potentially exhibiting an alternative mechanism to inhibit the promotion stage of carcinogenesis which was not shared by the European genotype. Structural chemical analysis showed that the American elderberry contained seven different species of anthocyanins, whereas the European genotype contained primarily four. The most active fractions from both genotypes were comprised not only of flavonoids but also of lipophilic compounds such as iridoid monoterpene glycosides, sesquiterpenes, and phytosterols; the different composition within mixtures from these extracts was linked to the differences in bioactive potential for chemoprevention.

In the same way, genotypic differences in the complement of interacting flavonoid components in blackberries from discrete geographical regions were determined to account for distinct differences in their antioxidant capacities, as gauged using two complementary antioxidant assays (Reyes-Carmona et al. 2005). A range of blackberry cultivars were monitored in this research, as in no case would the same cultivar be adapted to both the warm, dry climates in Mexico, and the shorter growing season and colder moister climates in the Pacific Northwest USA. Differences in the mixture of flavonoids in each extract were correlated with genotype.

When Black chokeberry fruits from the same species (*Aronia melanocarpa*) but from different climatic production regions (European cultivated plantations versus North American wild plantings) were directly compared, anthocyanins and other fruit flavonoids in bioactive mixtures were found to be markedly different between the two. Similar complements of anthocyanin pigments and proanthocyanidin oligomers were detected in both *Aronia* genotypes, however, HPLC-ESI-MS spectra indicated higher flavonoid concentration in the wild *Aronia*, and a predominance (up to 67%) of nonphenolic compounds in the berries from the cultivated genotype.

Consequently, the two genotypes also differed bioactive potential. Both genotypes exhibited inhibitory activity towards L1210 leukemia cell line, but only the wild genotype contained flavonoids capable of inhibiting topoisomerase II enzyme (a marker for the proliferation stage of carcinogenesis) (Sueiro et al. 2006).

One of the strongest advantages of flavonoid interactions is that multiple mechanisms to combat disease can be operative simultaneously in a single fruit extract, which broadens the protective effects. In support of this hypothesis, bioactivity-guided fractionation of wild blueberry extracts demonstrated that the broad chemoprotective benefits associated with these fruits are contributed by different flavonoid phytochemical components in different mixtures for each of the initiation, promotion, and proliferation stages of carcinogenesis (Kraft et al. 2005; Schmidt et al. 2004, 2006). Further analysis of food processing effects on wild blueberries demonstrated that certain physical processes that disrupt or inactivate some flavonoid constituents (in particular, heating of berry fruit components) can markedly attenuate the biological potency that is evident in fresh product, or fruit that is carefully frozen at –80°C before analysis (Schmidt et al. 2005).

While increased tomato and tomato product consumption has been significantly associated with reduced risk of prostate cancer in epidemiological studies (Miller et al. 2002), and the health benefits of tomatoes are usually assumed to be due to the carotenoid lycopene, recent research has clearly shown that lycopene alone does not account for tomato's protective action. A significant flavonoid content is also found in tomato fruits. The chemoprotective contributions of these compounds were tested by measuring the antiproliferative effects of flavonoids commonly present in tomato fruits (quercetin, kaempferol, and naringenin) in a mouse liver cancer cell line and a human prostate cancer cell line. Individual compounds, as well as combination treatments of flavonoids were evaluated to test for potential synergistic interactions in the inhibition of cancer cell growth. Dose-dependent inhibition of cancerous cell growth was demonstrated for individual flavonoid aglycones at 12.5–50 µM concentrations. A treatment combining all three flavonoid constituents (quercetin, kaempferol, and naringenin) suggested a potentiation effect that was not statistically significant, however, binary combinations demonstrated significant synergistic interactions in both of the experimental cell lines (Campbell et al. 2006a). The combination treatment of quercetin and kaempferol (the flavonols most concentrated in tomato skin) was particularly effective in cancer growth inhibition. Significant antiproliferative interactions between flavonols and flavanones were demonstrated. Flavonoid aglycones exhibited significant dose-dependent effects on cell proliferation, however, glycones rutin, quercetrin, and naringin did not provide similar inhibition.

Although flavonoids are typically found in glycosidic moieties in fruits and vegetables, these glycones are primarily deglycosylated in the small intestine, which is necessary for the absorption of many dietary flavonoids (Kroon et al. 2004). Matito et al. (2003) similarly demonstrated synergistic antiproliferative interactions for flavonoid constituents of grape including (+)-catechin, (–)-epicatechin, proanthocyanidin oligomers, and glycosylated flavonols, using Hepa-1c1c7 liver cancer cells.

Methanolic extraction and chromatographic fractionation of black Jamapa bean (*Phaseolus vulgaris*), which features anthocyanin pigments and proanthocyanidins in the seed coats, produced large quantities of natural mixtures of flavonoids suitable for testing bioactivity and phytochemical interactions (Aparico-Fernandez et al. 2005). When extracts were introduced to a carcinoma cell line (HeLa), higher molecular weight proanthocyanidin oligomers and polymers (rather than isolated anthocyanins or flavonol glycosides) were found to be primarily responsible for significant inhibition of cancer cell growth (Aparico-Fernandez et al. 2006). Extracts arrested the cell cycle of cancerous cells at the G1 phase, and resulted in an increase in apoptosis. Because the inhibitory potency (IC50 14.7) of the mixed methanolic extract was greater than for any single isolated constituents, it was hypothesized that the inhibitory activity is due to the complex of polyphenolics found naturally in these mixtures. Activity was partially attenuated when polyphenolic mixtures were further purified through subfractionation, potentially due to the loss of potentiating interactions (Aparico-Fernandez et al. 2006).

10.4 In Vitro Investigations of Flavonoid Interactions

Given the complex nature of flavonoids and their interactions within plants, as well as after ingestion into the body, it has been particularly advantageous to engineer streamlined systems for production and analysis of these phytochemicals. In vitro plant cell cultures provide a valuable vehicle for analysis in that flavonoids can be synthesized and accumulated in high concentrations within relatively simple, homogenous tissues, which greatly facilitates extraction, separation, and isolation of individual compounds. Interferences from enzymes, sugars, or adhering tissues that would complicate complete extraction from plant tissues can be avoided in cell culture production systems. One of the most advantageous attributes is that the flavonoids can be consistently biolabeled in the in vitro system, allowing their metabolism to be tracked after ingestion in animal models (Lila 2004).

Following the discovery that mixed polyphenolic fractions from grape (*Vitis* hybrid Bailey Alicant A) cell cultures provided highly potent catalytic inhibition of topoisomerase II (an enzyme causes DNA strands to relax prior to replication, and which is dramatically overexpressed in proliferating tumor cells; Jo et al. 2005), further investigations centered on elucidation of possible interactions between the phytochemicals in these fractions. When the cell culture crude extract was fractionated on a Toyopearl matrix, the 4th and 6th subfractions (TP4 and TP6) were potent inhibitors of the topoisomerase II enzyme, in a dose-dependent manner (Fig. 10.1). TP4 was predominately comprised of the phytochemicals epicatechin gallate, myricetin, procyanidin B_2, and resveratrol (Fig. 10.2). Each of these constituents were potent topoisomerase II catalytic inhibitors when bioassayed individually. However, in addition, strong potentitating interactions between procyanidin B_2 (at each of two test concentrations) and each of the other three polyphenolic constituents were demonstrated (Jo et al. 2006a). Combinations of myricetin and resveratrol also demonstrated potentiating interactions, and an isobolographic analysis confirmed that this was a synergistic interaction at a molar ratio of 1:70 (Fig. 10.3). All of the subfractions isolated from the grape cell cultures contained natural mixtures.

314   M.A. Lila

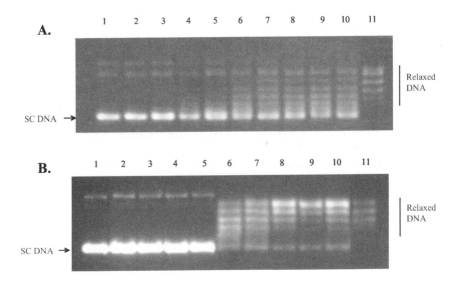

Fig. 10.1 A representative relaxation assay determined the inhibition of human DNA topoisomerase II enzyme catalytic activity by A). Toyopearl fraction 4 (TP4) and B). Toyopearl fraction 6 (TP6) from grape cell culture. Substrate supercoiled (SC) pRYG DNA (1 μL or 0.25 μg) was incubated with 4 units (2 μL) of human DNA topoisomerase II and steadily decreasing concentrations (from 5.0 to 0.05 μg/mL from left to right, lanes 2–10) of extracts from TP4 (A) or TP6 (B). Lane 1 was the SC pRYG DNA marker; lanes 2–10 show dose dependent inhibition of the enzyme by the polyphenolic fractions, lane 11 was the relaxed pRYG DNA marker. (Adapted from Jo et al. 2005, J. Ag. Food Chem. 53, 2489–2498.)

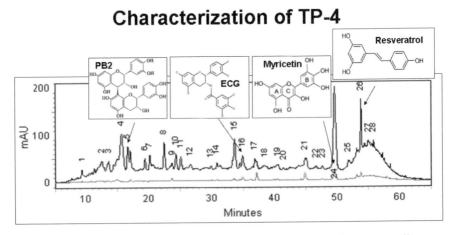

Fig. 10.2 Characterization of the primary components in fraction TP4, from grape cell suspension cultures. (From Jo, 2005, PhD thesis dissertation.)

Synergistic interaction of myricetin and resveratrol on catalytic inhibition of topo II

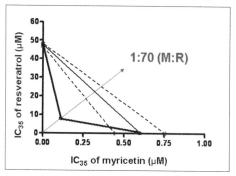

Fig. 10.3 IC_{35} isobologram for the synergistic interaction of myricetin and resveratrol (at a molar ratio of 1:70) with respect to its topoisomerase II catalytic inhibition. The line of additivity is shown with its 95% confidence limits. (Adapted from Jo et al., 2006a, J. Agric. Food Chem. 54, 2083–2087.)

Whereas the content of TP4 was dominated by the four components described above, analysis of the 2nd subfraction (TP2) revealed predominant anthocyanin content, whereas the 6th subfraction (TP6) was largely comprised of proanthocyanidins (Jo et al. 2005). In order to test for potentiation between polyphenolic classes, TP4 was combined with TP2 (1:2 ratio) or with TP6 (5:3 ratio) before administration in the topoisomerase II bioassay. (These ratios were based on the relative weight of each compound present naturally in the grape cell culture extract.) Calculated and experimental inhibition values (%) achieved by binary combinations were compared to investigate potentiating interactions. The natural mixtures were effective inhibitors of human DNA topoisomerase II in the catalytic inhibition assay, but combined fractions (TP2 + TP4, or TP4 + TP6) only showed additive inhibitory effects. Actually the anthocyanin-rich fraction (TP2) in combination with TP4 (composition defined previously) showed a clear tendency toward stronger potentiation (Fig. 10.4), but this interaction could not be statistically demonstrated (Jo 2005). Of course, potentiating interactions contributing to chemopreventive activity between different classes of polyphenols (as noted by Seeram et al. 2004) may be a result of mechanisms other than topoisomerase II catalytic inhibition.

As noted above, the phytochemical composition of another highly potent fraction from the grape cell cultures, TP6, consisted primarily of proanthocyanidins (dimers through hexamers), procyanidin digallates, anthocyanins, and procyanidins linked to cyanidin-3-O-glycosides (dimers to pentamers). This fraction not only exhibited strong inhibitory activity in the topoisomerase chemopreventive assay, but it also proved to be highly potent in a HepG2 liver cancer antiproliferation test (Jo et al. 2006b). Interestingly, the TP6 mixture was selectively cytotoxic to cancerous human

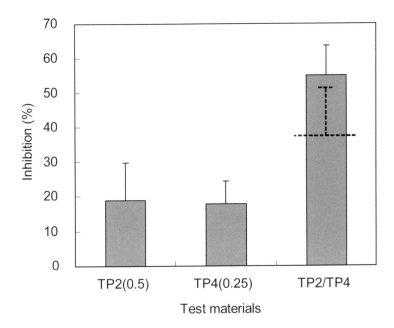

Fig. 10.4 Potential interaction between the components from an anthocyanin-rich fraction from grape cell culture (TP2), and the catechin, procyanidin dimer and flavanone-containing fraction (TP4) was tested at a 2:1 ratio based on the relative weight of each fraction present naturally in the grape cell culture extract. The dotted line indicates the level of inhibition of topoisomerase II enzyme expected for a purely additive interaction. (Adapted from Jo, 2005, PhD dissertation.)

liver cells (at a concentration of 20 μg/mL), but caused no cytotoxicity to non-cancerous cells even at very high concentrations (100 μg/mL). Previous work has similarly showed that grape seed extracts rich in procyanidins also showed selective cytotoxicity towards various human cancer cell lines including breast, lung, leukemia and prostate cancer cells, but did not inhibit normal human gastric mucosal and murine macrophage cells (Ye et al. 1999).

Using an open column method on Sephadex, it was possible to further subfractionate TP6 into 6 additional subfractions. Of the resulting subfractions, only one subfraction (TP6-5) showed anticancer activity higher than the parent fraction, however, TP6-5 did not exhibit the selectivity shown in the parent fraction, and did exhibit mild cytotoxicity even to non-cancerous cells (Table 10.1). Bioassays on all the subfractions indicated that the cytotoxicity against the HepG2 cells appeared to be proportional to the degree of polymerization of the proanthocyanidins. Also, the efficacy of these grape flavonoid fractions against the liver cancer cells was not at all linked to the ability of the same fractions to inhibit topoisomerase II enzyme; the combination of different polyphenols in the grape fractions clearly contributed greater than one anticancer mechanism of action simultaneously.

Table 10.1 Maximal toxicity (%) and ED_{50} of the mother fraction (TP6) and its potent chemo-preventive subfractions (TP6-4 to TP6-6), resveratrol and etoposide after 48 h of exposure to two different cells. Data represent means ± SEM (n = 6) from two independent studies. (Adapted from Jo et al. 2006b; Food Chem. Tox. 44, 1758–1767)

Cell line	HepG2			PK15		
	Max.	ED_{50}		Max.	ED_{50}	
Treatment	Toxicity (%)	µg/mL	µM[1]	Toxicity (%)	µg/mL	µM
TP-6	67.2 ± 1.8 [a]	49.6 ± 1.2 [b,c]	50.5 ± 1.2 [b,c]	ND[2]	> 100	> 101
TP-6-4	27.6 ± 5.1 [b]	70.2 ± 13.3 [a]	118.0 ± 22.4 [a]	ND	> 100	> 184
TP-6-5	71.8 ± 1.3 [a]	18.0 ± 0.8 [d]	14.1 ± 0.6 [c]	33.8 ± 2.6 [b]	36.9 ± 6.8 [a]	28.9 ± 5.3 [b]
TP-6-6	64.3 ± 4.7 [a]	53.3 ± 2.3 [a,b]	67.0 ± 4.9 [b]	ND	> 100	> 104
Resveratrol	68.6 ± 4.9 [a]	31.5 ± 4.0 [c,d]	138.0 ± 17.5 [a]	68.7 ± 8.4 [a]	46.1 ± 9.4 [a]	202.0 ± 36.8 [a]
Etoposide	33.6 ± 6.1 [b]	35.4 ± 9.7 [b,c,d]	60.1 ± 16.5 [b]	ND	> 100	> 169.9

[1] Calculated from estimated average molecular weight (MW) of polyphenols in grape cell culture fractions [TP-6 (MW, 982.6), TP-6-4 (MW, 543.0), TP-6-5 (MW, 1276.8), and TP-6-6 (MW, 959.4)], and MW of test compounds [Resveratrol (MW, 228.2) and etoposide (MW, 588.6)] based on LC-ESI/MS output.
[2] Not determined, since ED_{50} >100 ug/mL . Values with different letters (a, b) in a column are statistically different ($p < 0.05$) by ANOVA.

An additional advantage of plant cell culture systems for production and elucidation of biologically-active flavonoids is that through elicitation, the levels and the profiles of natural mixtures, including small molecules to high molecular weight oligomers and polymers can be controlled. Elicitation is a means to deliberately provoke the plant cell culture to trigger accumulation of secondary products of interest. For anthocyanins and other flavonoids accumulated in cell cultures, elicitors have included the signaling compound methyl jasmonate alone or in combination with ibuprofen or other elicitors, UV irradiance, osmotic or temperature stress, or other introduced stress factor (Fang et al. 1999; Meyer et al. 2002)

Even with the relative simplicity of the plant cell culture system for synthesis of biologically-active flavonoid compounds, the mechanisms of action, bioavailability, and interactions between compounds can be difficult to conclusively assign. One of the major challenges to measurement of flavonoid biological activity is the problem of selectively monitoring ingested phyochemicals from a food source in a complete, complex diet and discerning newly absorbed compounds of interest from the background levels already present in the body. By biolabeling flavonoid compounds as they are synthesized in vitro, a means to conclusively determine the mechanisms of biological action, absorption, clearance, and biodistribution, can be gauged. Both isotopically-labeled ^{13}C precursors to flavonoids, or ^{14}C radiolabeled precursors or sucrose (carbon source in the tissue culture medium) have successfully been introduced to flavonoid-accumulating plant cells in order to facilitate recovery of labeled mixtures that can be tracked in animal metabolism (Grusak et al. 2004; Krisa et al. 1999; Lila et al. 2005; Vitrac et al. 2002; Yousef et al. 2004). Cell cultures present an inherently efficient vehicle for introduction of labeled precursor, because the label can be introduced at a point when cells are not rapidly dividing, but instead, are synthesizing and accumulating metabolites. Vitrac et al. (2002) and Krisa et al. (1999) introduced a biolabeled immediate precursor to anthocyanin pigments to grape cell cultures and recovered anthocyanins with sufficient levels of label to allow tracking after intubation in an animal model.

Recognizing that metabolizing, living cells respire, Grusak et al. (2004) custom-built a transparent plexiglass chamber to facilitate safe, efficient radiolabeling of grape (*Vitis*) and berry (*Vaccinium*) flavonoids while minimizing any danger of contamination of the experimental airspace with radiolabeled, respired $^{14}CO_2$. Label was introduced into the system using uniformly labeled sucrose in the cell culture medium. Using this system, natural interacting mixtures of flavonoid phytochemicals were produced with a high level of enrichment, more than sufficient to allow the metabolic fate of the flavonoids to be tracked post-ingestion (Yousef et al. 2004; Lila et al. 2005). In these experiments, the most abundant concentration of ^{14}C label was localized in the anthocyanins and higher molecular weight proanthocyanidin oligomers. Labeled cell culture extracts rich in particular flavonoid components (proanthocyanidin oligomers, anthocyanins, or phenolic acids) were separated out of the crude extract via vacuum chromatography and were available to administer in animal models.

In an initial study, a jugular catheter was implanted into a male Sprague Dawley rat, which was gavaged with a subfraction from *Vaccinium* cell culture which contained primarily proanthocyanidin oligomers (hexamers in addition to some lower MW oligomers) mixed with some anthocyanins including cyanidin-3-*O*-glucoside, petunidin-3-*O*-glucoside, peonidin-3-*O*-glucoside, and cyanidin-3-*O*-*p* coumarylglucoside (Yousef et al. 2004). With a total dosage of 16.7 µCi, time-course samples of blood were taken, revealing that the labeled extract was absorbed very quickly (Fig. 10.5). Whereas previous labeling studies have only been able to examine very high administered doses of single labeled flavonoids compounds, the advantage of plant cell culture labeling is that physiologically-relevant (normal dietary) levels of flavonoids could be easily followed in serum using simple scintillation counting (Lila et al. 2005; Mullen et al., 2002). Another investigation using a tissue culture production system for labeling flavonoids was able to subsequently purify labeled resveratrol from grape cell cultures, and follow the distribution of labeled resveratrol in the tissues of a mouse (Vitrac et al. 2003).

Very recently, through the use of biolabeled flavonoids produced in plant cell cultures in animal intubation studies, the distribution of flavonoids and metabolites in brain tissues has been investigated (Janle et al. 2007). Brains and peripheral tissues were harvested from male Sprague Dawley rats and sectioned in a brain matrix. Distribution was uniform, and label detected in the brain tissues conclusively demonstrated that the flavonoids from grape cells, or their metabolites, were able to cross the blood-brain barrier. Current research in this arena is working to determine which sections of the brain accumulate specific flavonoid constituents, which may have implications for the role of polyphenolics in alleviation of Alzheimers and related disease symptoms.

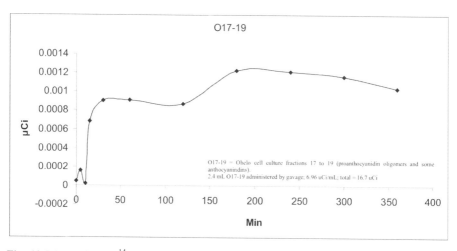

Fig. 10.5 Detection of ^{14}C label (µCi) in a rat's serum over time, following intubation with radiolabeled flavonoid fractions from *Vaccinium pahalae* (ohelo berry) cell cultures. (Adapted from Lila et al., 2005, J. Nutr. 135,1231–1235.)

10.5 Conclusion/Opportunities for Future Research

The recognized potentiating interactions between components within a natural phytochemical complex give strong support to the advocates of 'eating the whole (functional) food' rather than relying on extracts or formulations from foods which are sold in supplement form.   In the latter case, losses of interacting phytochemicals during the formulation phase of product development can result in greatly attenuated efficacy of the extract. It is well documented in the literature that flavonoids possess a wide variety of biological properties (as antioxidants, enzyme inhibitors, induction of detoxification enzymes, enhanced membrane stability, induction of apoptosis, arrest of cell cycle, etc.), which may account for the ability for a mixture of interacting flavonoids to provide enhanced synergistic therapeutic or protective action, through multiple pathways of intervention simultaneously. Various mechanisms of action of flavonoids on cancer cell growth or other therapeutic target may be complementary and overlapping, and a combination of these actions may be responsible for the overall efficacy of ingested natural phytochemicals. While an overwhelming body of new evidence has demonstrated potentiating interactions between flavonoids, and between flavonoids and other phytochemicals, progress on identifying the responsible mechanisms of action for the biological activity has lagged behind. In particular, additional in vivo bioassays are needed to investigate bioactivity in animal models. It will be critical to identify the flavonoids in natural mixtures that interact together to fortify biological activity for human heath. If programmed therapies involving key functional foods are to be considered for human health maintenance regimes, it is important to appreciate flavonoid interactions and how effective dosages can be prescribed. For these studies, the ability to radiolabel anthocyanin mix-

320   M.A. Lila

tures and other flavonoids is particularly advantageous, as it will allow investigation of kinetics, absorption, bioavailability, retention, and clearance from the body.

Acknowledgements

This work was supported by the National Center for Complementary and Alternative Medicine-sponsored Purdue-UAB Botanicals Center for Dietary Supplement Research (P50 AT-00477) and the USDA/IFAFS grant #00-52101-9695.

References

Allred, C.D., Ju, Y.H., Allred, K.E., Change, J. and Helferich, W.G. (2001) Dietary genistin stimulates growth of estrogen-dependent breast cancer tumors similar to that observed with genistein. Carcinogenesis 22, 1667–1673

Allred, C.D., Allred, K.E., Ju, Y.H., Virant, S.M. and Helferich, W.G. (2004) Dietary genistein results in larger MNU-induced, estrogen-dependent mammary tumors following ovarioectomy of Sprague-Dawley rats. Carcinogenesis 25, 211–218.

Aparico-Fernandez, X., Yousef, G., Loarca-Pina, G., de Mejia, E. and Lila, M. (2005) Characterization of polyphenolics in the seed coat of black Jamapa bean (*Phaseolus vulgaris. L.*). J. Ag. Food Chem. 53, 4615–4622.

Aparico-Fernandez, X., Garcia-Gasca, X., Yousef, G., Lila, M., de Mejia, E. and Loarca-Pina, G. (2006) Chemopreventive activity of polyphenolics from black Jamapa bean (*Phaseolus vulgaris* L.) on HeLa and HaCaT cells. J. Ag. Food Chem. 54, 2118–2122.

Aviram, M., Dornfeld, L., Rosenblat, M., Volkova, N., Kaplan, M., Coleman, R., Hayek, T., Presser, D. and Fuhrman, B. (2000) Pomegranate juice consumption reduces oxidative stress, atherogenic modifications to LDL, and platelet aggregation: studies in humans and in atherosclerotic apolipoprotein E-deficient mice. Am. J. Clin. Nutr. 71, 1062–1076.

Aviram, M., Dornfeld, L., Kaplan, M., Coleman, R., Gaitini, D., Nitecki, S., Hofman, A., Rosenblat, M., Volkova, N., Presser, D., Attias, J., Hayek, T. and Fuhrman B. (2002) Pomegranate juice flavonoids inhibit low-density lipoprotein oxidation and cardiovascular diseases: Studies in atherosclerotic mice and in humans. Drugs Exp. Clin. Res. 28, 40–62.

Boileau, T.W., Liao, Z., Kim, S., Lemeshow, S., Erdman, J.W. Jr., Clinton, D.K. (2003) Prostate carcinogenesis in N-methyl-N-nitrosourea (NMU)-testosterone-treated rats fed tomato powder, lycopene, or energy-restricted diets. J. Nat. Cancer Inst. 95, 1578–1586.

Brinker, A., Ma, J., Lipsky, P.E. and Raskin, I. (2007) Medicinal chemistry and pharmacology of genus *Tripterygium* (Celastraceae). Phytochemistry 68(6), 732–766.

Campbell, J., King, J., Harmston, M., Lila, M.A. and Erdman, J.W. Jr. (2006a) Synergistic effects of flavonoids on cell proliferation in Hepa-1c1c7 and LNCap cancer cell lines. J. Food Sci. 71, S358-S363.

Campbell, J., Lila, M.A., Nakamura, M. and Erdman, J. Jr. (2006b). Serum testosterone reduction following short-term phytofluene, lycopene, or tomato powder consumption in F344 rats. J. Nutr. 136, 2813–2819.

Canen-Adams, K., Lindshield, B., Wang, S., Jeffery, E., Clinton, S.K. and Erdman, J.W. Jr. (2007) Combinations of tomato and broccoli enhance antitumor activity in Dunning R3327-H prostate adenocarcinomas. Cancer Res. 67, 836–843.

Chan, M., Mattiacci, J., Hwang, H., Shah, A. and Fong, D. (2000) Synergy between ethanol and grape polyphenols, quercetin, and resveratrol, in the inhibition of the inducible nitric oxide synthase pathway. Biochem. Pharmacol. 60, 1539–1548.

Chen, L., Yang, X., Jiao, H. and Zhao, B. (2002) Tea catechins protect against lead-induce cytotoxicity, lipid peroxidation, and membrane fluidity in HepG2 cells. Toxicol. Sci. 69, 149–156.

Eder, M. and Mehnert, W. (1998) Bedeutung planzlicher begleitstoffe in extrackten. Pharmazie 53, 285–293.

Elattar, T. and Virji, A. (1999) The effect of red wine and its components on growth and proliferation of human oral squamous carcinoma cells. Anticancer Res. 19, 5407–5414.

Fang, Y., Smith, M.A., Rogers, R. and Pépin, M.-F. (1999) The effects of exogenous methyl jasmonate in elicited anthocyanin-producing cell cultures of ohelo (*Vaccinium pahalae*). In Vitro Cell. Devel. Biol. Plant 35, 106–113.

Franke, A., Cooney R., Custer L., Mordan, L. and Tanaka, Y. (1998) Inhibition of neoplastic transformation and bioavailability of dietary flavonoid agents Adv. Exp. Med. Biol. 439, 237–248.

Gartner, C., Stahl W. and Sies, H. (1997) Lycopene is more bioavailable from tomato paste than from fresh tomatoes. Am J. Clin. Nutri. 66, 116–122.

Gould, K. (2004) Nature's Swiss army knife. The diverse protective roles of anthocyanins in leaves. J. Biomed. Biotech. 5, 314–320.

Grusak, M., Rogers, R., Yousef, G., Erdman, J. Jr. and Lila, M. (2004) An enclosed-chamber labeling system for the safe ^{14}C-enrichment of phytochemicals in plant cell suspension cultures. In Vitro Cell. Devel. Biol. – Plant 40, 80–85.

Honda, K. (1990) Identification of host-plant chemicals stimulating oviposition by swallowtail butterfly, *Papilio protenor*. J. Chem. Ecol. 16, 325–337.

Honda, K., Nishii, W. and Hayashi, N. (1997) Oviposition stimulants for sulfur butterfly, *Colias erate poliographys*: cyanoglucosides as synergists involved in host preference. J. Chem. Ecol. 23, 323–331.

Hora, J., Maydew, E., Lansky, E. and Dwivedi C. (2003) Chemopreventive effects of pomegranate seed oil on skin tumor development in CD1 mice. J Med Food 6, 157–161.

Janle, E., Lila, M., Wood, L., Higgins, A., Yousef, G, Rogers, R., Kim, H., Jackson, G. and Weaver, C. (2007) Kinetics and tissue distribution of ^{14}C labeled grape polyphenol fractions. FASEB J. 21, 1070.

Jo, J.Y. (2005) Characterization, interaction, and chemoprevention of potent constituents from grape cell culture. PhD thesis dissertation, University of Illinois.

Jo, J.Y., de Mejia, E. and Lila, M.A. (2005) Effects of grape cell culture extracts on human topoisomerase II catalytic activity and characterization of active fractions. J. Agric. Food Chem. 53, 2489–2498.

Jo, J.Y., de Mejia, E. and Lila, M. (2006a). Catalytic inhibition of human DNA topoisomerase II by interactions of grape cell culture polyphenols. J. Agric. Food Chem. 54, 2083–2087.

Jo, J.Y., de Mejia, E. and Lila, M. (2006b) Cytotoxicity of bioactive polymeric fractions from grape cell culture on human hepatocellular carcinoma, murine leukemia, and non-cancerous PK15 kidney cells. Food Chem. Toxicol. 44, 1758–1767.

Joseph, J., Shukitt-Hale, B., Denisova, N., Bielinski, D., Martin, A., McEwen, J. and Bickford, P. (1999) Reversals of age-related declines in neuronal signal transduction, cognitive, and motor behavioral deficits with blueberry, spinach, or strawberry dietary supplementation. J. Neurosci. 19, 8114–8121.

Kraft, T.F.B., Schmidt, B., Knight, C., Cuendet, M., Kang, Y-H., Pezzuto, J., Seigler, D. and Lila, M. (2005) Chemopreventive potential of wild blueberry fruits in multiple stages of carcinogenesis. J. Food Sci. 70, S159–S166.

Krisa, S., Waffo Téguo, P., Decendit, A., Deffleux, G., Vercauteren, J. and Mérillon, J. (1999) Production of ^{13}C-labeled anthocyanins by *Vitis vinifera* cell suspension cultures. Phytochemistry 51, 651–656.

Kroon, P., Clifford, M., Crozier, A., Day, A., Donovan, J., Manach, C. and Williamson, G. (2004) How should we assess the effects of exposure to dietary polyphenols in vitro? Am. J. Clin. Nutr. 80, 15–21.

Lansky E., Jiang, W., Mo, H., Bravo, L., Froom, P., Yu, W., Harris, N., Neeman, I. and Campbell, M. (2005) Possible synergistic prostate cancer suppression by anatomically discrete pomegranate fractions. Investigational New Drugs 23, 11–20.

Lila, M.A. (2004) Anthocyanins and human health: an *in vitro* investigative approach. J. Biomed. Biotech. 5, 306–313.

Lila, M., Yousef, G., Jiang, Y. and Weaver, C. (2005) Sorting out bioactivity in flavonoid mixtures. J. Nutr. 135, 1231–1235.

Lila, M.A. and Raskin, I. (2005) Health-related interactions of phytochemicals. J. Food Sci. 7, R20–R27.

Liu, R. (2003) Health benefits of fruit and vegetables are from additive and synergistic combinations of phytochemicals. Am. J. Clin. Nutr. 78, 517S–520S.

Matito, C., Mastorakou, F., Centelles, J., Torres, J. and Cascante, M. (2003) Antiproliferative effect of antioxidant polyphenols from grape in murine Hepa-1c1c7. Eur. J. Nutr. 42, 43–49.

Mertens-Talcott S., Bomser, J., Romero, C., Talcott, S. and Percival S. (2005) Ellagic acid potentiates the effect of quercetin on p21$^{waf1/cip1}$, p53, and MAP-kinanses without affecting intracellular generation of reactive oxygen species in vitro. J Nutr. 135, 609–614.

Mertens-Talcott S., Talcott, S. and Percival S. (2003) Low concentrations of quercetin and ellagic acid synergistically influence proliferation, cytotoxicity and apoptosis in MOLT-4 human leukemia cells. J. Nutr. 133, 2669–2674.

Meyer, J., Pépin, M.-F. and Smith, M. (2002) Anthocyanin production from *Vaccinium pahalae*: limitations of the physical microenvironment. J. Biotech. 93, 45–57.

Miller, E., Biovannucci, E., Erdman Jr., J., Bahnson, R., Schwartz, S. and Clinton, S. (2002) Tomato products, lycopene, and prostate cancer risk. Urol. Clin. North Am. 29, 83–93.

Mouria, M., Gukovskaya, A., Jung. Y., Buechler, P., Hines, O., Reber, H. and Pandol, S. (2002) Food-derived polyphenols inhibit pancreatic cancer growth through mitochondrial cytochrome C release and apoptosis. Int. J. Cancer 98, 761–769.

Mullen, W., Graf, B., Caldwell, S., Hartley, R., Duthie, G., Edwards, C., Lean, M. and Crozier, A. (2002) Determination of flavonol metabolites in plasma and tissue of rats by HPLC radiocounting and tandem mass spectroscopy following oral ingestion of [2-14C] quercetin-4'-glucoside. J. Ag Food Chem. 50, 6902–6909.

Nishida, R., Ohsugi, T. and Fukami, H. (1990) Oviposition stimulant activity of tryptamine analogs on a Rutaceae-feeding swallowtail butterfly, *Papilio xuthus*. Agric. Biol. Chem. 54, 1853–1855.

Nishida, S. and Satoh, H. (2004) Comparative vasodilating actions among terpenoids and flavonoids contained in *Ginkgo biloba*. Clinica Chimica Acta 339, 129–133.

Pignatelli, P., Pulcinelli, F., Celestini, A., Lenti, L., Ghiselli, A., Gazzaniga, P. and Viola, F. (2000) The flavonoids quercetin and catechin synergistically inhibit platelet function by antagonizing the intracellular production of hydrogen peroxide. Am J. Clin. Nutr. 72, 1150–1155.

Reyes-Carmona, J., Yousef, G., Martínez-Peniche, R. and Lila, M. (2005) Antioxidant capacity of fruit extracts of blackberry (*Rubus* sp.) produced in different climatic regions. J. Food Sci. 70, S497-S503.

Schmidt, B., Howell, A., McEniry, B., Knight, C., Seigler, D., Erdman Jr., J. and Lila, M. (2004) Effective isolation of potent antiproliferation and antiadhesion components from wild blueberry (*Vaccinium angustifolium* Ait.) fruits. J. Ag. Food Chem. 52, 6433–6442.

Schmidt, B., Erdman Jr., J. and Lila, M.A. (2005) Effects of food processing on blueberry antiproliferative and antioxidant activity. J. Food Sci. 70, S389–S394.

Schmidt, B.M, Erdman Jr., J. and Lila, M. (2006) Differential effects of blueberry proantho-cyanidins on androgen sensitive and insensitive human prostate cancer cell lines. Cancer Lett. 231, 240–246.

Seeram, N.P., Adams, L.S., Hardy, M.I. and Heber, D. (2004) Total cranberry extract versus its phytochemical constituents: antiproliferative and synergistic effects against human tumor cell lines. J. Agric. Food Chem. 52, 2512–2517.

Shanmuganayagam, D., Beahm, M., Osman, H., Krueger, C., Reed, J. and Folts, J. (2002) Grape seed and grape skin extracts elicit a greater antiplatelet effect when used in combination than when used individually in dogs and humans. J. Nutr. 132, 3592–3598.

Sueiro, L, Yousef, G., Seigler, D., de Mejia, E., Grace, M. and Lila, M.A. (2006) Chemopreventive potential of flavonoid extracts from plantation-bred and wild *Aronia melanocarpa* (Black chokeberry) fruits. J. Food Sci. 71, C480-C488.

Thole, J., Kraft, T., Sueiro, L., Kang, Y.-H., Gills, J., Cuendet, M., Pezzuto, J., Seigler, D. and Lila, M.A. (2006) A comparative evaluation of the anticancer properties of European and American elderberry fruits. J. Med. Food 9, 498–504.

Vitrac, X., Desmouliere, A., Brouillaud, B., Krisa, S., Deffieux, G., Barthe, N., Rosenbaum, J., Mérillon, J. M. (2003) Distribution of [C-14] - *trans*-resveratrol, a cancer chemopreventive polyphenol, in mouse tissues after oral administration. Life Sci. 72, 2219–2233.

Vitrac, X., Krisa, S., Decendit, A., Vercauteren, J., Nührich, A., Monti, J.-P., Deffieux, G. and Mérillon, J.-M. (2002) Carbon-14 biolabelling of wine polyphenols in *Vitis vinifera* cell suspension cultures. J. Biotech. 95, 49–56.

Ye, X., Krohn, R., Liu, W., Joshi, S., Kuszynski, C., McGinn, T., Bagchi, M., Preuss, H., Stohs, S. and Bagchi, D. (1999) The cytotoxic effects of a novel IH636 grape seed proanthocyanidin extract on cultured human cancer cells. Mol. Cell. Biochem. 196, 99–108.

Yousef, G., Seigler, D., Grusak, M., Rogers, R., Knight, C., Kraft, T., Erdman Jr., J.and Lila, M. (2004) Biosynthesis and characterization of ^{14}C-enriched flavonoid fractions from plant cell suspension cultures. J. Agric. Food Chem. 52, 1138–1145.

Index

Printed in the United States of America